CHEMICAL ENGINEERING OF POLYMERS

Production of Functional and Flexible Materials

CHEMICAL ENGINEERING OF POLYMERS

OF POLYMERS

Production of
Functional and Flexible Materials

Edited by
Omari V. Mukbaniani, DSc
Marc J. M. Abadie, DSc
Tamara N. Tatrishvili, PhD

Apple Academic Press Inc.	Apple Academic Press Inc.
3333 Mistwell Crescent	9 Spinnaker Way
Oakville, ON L6L 0A2	Waretown, NJ 08758
Canada	USA

© 2018 by Apple Academic Press, Inc.

First issued in paperback 2021

No claim to original U.S. Government works

ISBN-13: 978-1-77463-044-0 (pbk)

ISBN-13: 978-1-77188-445-7 (hbk)

Library of Congress Control Number: 2017952488

Trademark Notice: Registered trademark of products or corporate names are used only for explanation and identification without intent to infringe.

Library and Archives Canada Cataloguing in Publication

International Caucasian Symposium on Polymers and Advanced Materials (4th : 2015 : Batumi, Georgia) Chemical engineering of polymers : production of functional and flexible materials / edited by Omari V. Mukbaniani, DSc, Marc J. M. Abadie, DSc, Tamara N. Tatrishvili, PhD.
Papers presented at the 4th International Caucasian Symposium on Polymers and Advanced Materials held at Batumi Black sea beach, Georgia July 1-4, 2015.
Includes bibliographical references and index.
Issued in print and electronic formats.

ISBN 978-1-77188-445-7 (hardcover).--ISBN 978-1-315-36598-5 (PDF)
1. Polymers--Congresses. 2. Chemical engineering--Congresses.
I. Mukbaniani, Omari V., editor II. Abadie, Marc J. M., editor III. Tatrishvili, Tamara N, editor IV. Title.

TP1081.I58 2017	668.9	C2017-905795-2	C2017-905796-0

CIP data on file with US Library of Congress

Apple Academic Press also publishes its books in a variety of electronic formats. Some content that appears in print may not be available in electronic format. For information about Apple Academic Press products, visit our website at **www.appleacademicpress.com** and the CRC Press website at **www.crcpress.com**

CONTENTS

LIST OF CONTRIBUTORS

Marc J. M. Abadie
Institute Charles Gerhardt of Montpellier – Agreggates, Interfaces and Materials for Energy (ICGM – AIME, UMR CNRS 5253), University Montpellier, Place Bataillon, 34095 Montpellier Cedex 5, France, E-mail: abadie@univ-montp2.fr & marc@ntu.edu.sg

Zh. A. Abilov
Al-Farabi Kazakh National University, Faculty of Chemistry and Chemical Technology, Al-Farabi ave, 71, Almaty, Republic of Kazakhstan

K. A. Ablayeva
Center of Physical and Chemical Methods of Research and Analysis, Tole bi Str. 96a, 050012, Almaty, Kazakhstan

T. Agladze
Georgian Technical University, Kostava Str., 77, 0175, Tbilisi, Georgia, E-mail: tamazagladze@emd.ge

O. Alekseeva
G.A. Krestov Institute of Solution Chemistry, Russian Academy of Sciences, Akademicheskaya str., 1, Ivanovo, 153045, Russia, E-mail: avn@isc-ras.ru

E. Alvarenga
NIBIO, Norwegian Institute of Bioeconomy Research, Pb 115, N-1431, Ås, Norway, E-mail: Emilio.Alvarenga@nibio.no

N. Ananiashvili
Ivane Javakhishvili Tbilisi State University, Rafiel Agladze Institute of Inorganic Chemistry and Electrochemistry. Mindeli str., 11, 0186, Tbilisi, Georgia

I. Andries
Moldova State University, 60 A. Mateevici str., MD-2009, Chisinau, Moldova

R. H. Arakelyan
Yerevan State University, Alek Manoogian 1, Yerevan, 375025, Republic of Armenia

G. S. Askarova
Center of Physical and Chemical Methods of Research and Analysis, Tole bi Str. 96a, 050012, Almaty, Kazakhstan

N. Bagrovskaya
G. A. Krestov Institute of Solution Chemistry, Russian Academy of Sciences, Akademicheskaya str., 1, Ivanovo, 153045, Russia

B. Bendeliani
Ilia Vekua Sukhumi Institute of Physics and Technology, Laboratory of Cryogenic Technique and Technologies, Mindeli Str. 7, 0186 Tbilisi, Georgia

A. Berezhnytska
V. I. Vernadsky Institute of General and Inorganic Chemistry NASU, Kyiv, Ukraine

A. Chirita
Moldova State University, 60 A. Mateevici str., MD-2009, Chisinau, Moldova

I. Chitrekashvili
Petre Melikishvili Institute of Physical and Organic Chemistry of Ivane. Javakhishvili Tbilisi, State University, I. Chavchavadze Ave. 1, 0179 Tbilisi, Georgia

N. Chorna
Chuiko Institute of Surface Chemistry, National Academy of Science of Ukraine, General Naumov str, 17, 03164 Kyiv, Ukraine

K. Chubinidze
Tbilisi State University, 1 Ilia Chavchavadze Ave, Tbilisi 0179, Georgia, E-mail: chubinidzeketino@yahoo.com

Vasyl Chumachenko
Problem Research Laboratory, National University of Food Technology, 68, Volodymyrska Street 01601, Kyiv, Ukraine

O. Corsac
Moldova State University, 60 A. Mateevici str., MD-2009, Chisinau, Moldova

A. Covali
Faculty of Physics and Engineering, Moldova State University, 60 A. Mateevici str., MD 2009, Chisinau, Moldova

J. Datta
Polymer Technology Department, Chemical Faculty, Gdansk University of Technology, Gabriela Narutowicza Street 11/12, 80-233 Gdansk, Poland

S. P. Davtyan
Armenian National Polytechnic University 105, Teryana Str., 105 375009 Yerevan, Armenia

G. Dgebuadze
Ilia Vekua Sukhumi Institute of Physics and Technology, Laboratory of Cryogenic Technique and Technologies, Mindeli Str. 7, 0186 Tbilisi, Georgia

N. Dokhturishvili
Petre Melikishvili Institute of Physical and Organic Chemistry of Ivane. Javakhishvili Tbilisi, State University, I. Chavchavadze Ave. 1, 0179 Tbilisi, Georgia

M. Donadze
Georgian Technical University, Kostava Str., 77, 0175, Tbilisi, Georgia

G. Dragalina
Faculty of Chemistry and Chemical Technology

A. H. Durgaryan
Yerevan State University, Alek Manoogian 1, Yerevan, 375025, Republic of Armenia, E-mail: durnar63@yahoo.com

N. A. Durgaryan
Yerevan State University, Alek Manoogian 1, Yerevan, 375025, Republic of Armenia

Ya. Fedorov
V. I. Vernadsky Institute of General and Inorganic Chemistry NASU, Kyiv, Ukraine

S. Filipchenko
Taras Shevchenko National University, Faculty of Chemistry, 60 Volodymyrska Str., Kyiv 0160, Ukraine

K. Formela
Polymer Technology Department, Chemical Faculty, Gdansk University of Technology, Gabriela Narutowicza Street 11/12, 80-233 Gdansk, Poland

V. Gabunia
Ilia Vekua Sukhumi Institute of Physics and Technology, Laboratory of Cryogenic Technique and Technologies, Mindeli str, 7, 0186 Tbilisi, Georgia

M. Gachechiladze
Ivane Javakhishvili Tbilisi State University, Rafiel Agladze Institute of Inorganic Chemistry and Electrochemistry. Mindeli str, 11, 0186, Tbilisi, Georgia

E. Gavashelidze
Petre Melikishvili Institute of Physical and Organic Chemistry of Ivane, Javakhishvili Tbilisi, State University, I. Chavchavadze Ave. 1, 0179 Tbilisi, Georgia

N. Gelashvili
Petre Melikishvili Institute of Physical and Organic Chemistry of Ivane. Javakhishvili Tbilisi, State University, I. Chavchavadze Ave. 1, 0179 Tbilisi, Georgia

E. Govasmark
Energy Recovery Agency of Oslo County, PO-BOX 54 Mortensrud, NO-1215, Oslo, Norway, E-mail: Espen.Govasmark@ege.oslo.kommune.no

G. K. Grigoryan
The Scientific Technological Centre of Organic and Pharmaceutical Chemistry NAS RA, Institute of Organic Chemistry, 26, Azatutyan STR, 0014, Yerevan, Armenia

N. G. Grigoryan
The Scientific Technological Centre of Organic and Pharmaceutical Chemistry NAS RA, Institute of Organic Chemistry 26, Azatutyan STR, 0014, Yerevan, Armenia

M. Gurgenishvili
Petre Melikishvili Institute of Physical and Organic Chemistry of Ivane, Javakhishvili Tbilisi, State University, I. Chavchavadze Ave. 1, 0179 Tbilisi, Georgia, E-mail: marina.gurgenishvili@yahoo.com

B. Shirkavand Hadavand
Department of Resin and Additives, Institute for Color Science and Technology, Tehran, Iran, E-mail: zeinab.hesari@yahoo.com

J. T. Haponiuk
Polymer Technology Department, Chemical Faculty, Gdansk University of Technology, Gabriela Narutowicza Street 11/12, 80-233 Gdansk, Poland, E-mail: jozef.haponiuk@pg.gda.pl

Yuliia Harahuts
Taras Shevchenko National University, Faculty of Chemistry, 60 Volodymyrska Str., Kyiv 0160, Ukraine

L. Hayrapetyan
Yerevan State University, A. Manookyan St. 1, 0025, Armenia, E-mail: Scirec@mail.ru

S. Hayrapetyan
Yerevan State University, A. Manookyan St. 1, 0025, Armenia, E-mail: Scirec@mail.ru

A. Hejna
Polymer Technology Department, Chemical Faculty, Gdansk University of Technology, Gabriela Narutowicza Street 11/12, 80-233 Gdansk, Poland

Z. Hesari
Department of Chemistry, Faculty of Science, Science and Research of Tehran Branch Islamic Azad University, Tehran, Iran, E-mail: zeinab.hesari@yahoo.com

A. A. Hovhannisyan
The Scientific Technological Centre of Organic and Pharmaceutical Chemistry NAS RA, Institute of Organic Chemistry 26, Azatutyan STR, 0014, Yerevan, Armenia, E-mail: hovarnos@gmail.com

A. Ivancic
Moldova State University, 60 A. Mateevici str., MD-2009, Chisinau, Moldova

T. K. Jumadilov
JSC, "Institue of Chemical Sciences after A.B. Bekturov", Sh. Valikhanov st. 106, Almaty, Republic of Kazakhstan, E-mail: jumadilov@mail.ru

M. Khaddazh
Peoples' Friendship University of Russia, Scientific-Educational Centre of Nanotechnology, Miklukho-Maklaya 10/2, 117198, Moscow, Russia

A. Kolendo
Kyiv Taras Shevchenko National University, Volodymyrska Str. 60, 01033, Kyiv, Ukraine

A. Yu. Kolendo
Department of Macromolecular Chemistry, Taras Shevchenko National University of Kyiv, 60 Volodymyrska Street, 01033, Kyiv, Ukraine

R. G. Kondaurov
Al-Farabi Kazakh National University, Faculty of Chemistry and Chemical Technology, Al-Farabi ave. 71, Almaty, Republic of Kazakhstan

O. Krupka
Kyiv Taras Shevchenko National University, Volodymyrska Str. 60, 01033, Kyiv, Ukraine

A. P. Kurbatov
Center of Physical and Chemical Methods of Research and Analysis, Tole bi Str. 96a, 050012, Almaty, Kazakhstan

N. Kutsevol
Taras Shevchenko National University, Faculty of Chemistry, 60 Volodymyrska Str., Kyiv 0160, Ukraine

Nataliya Kutsevol
Taras Shevchenko National University, Faculty of Chemistry, 60 Volodymyrska Str., Kyiv 0160, Ukraine, Email: kutsevol@ukr.net

O. Kychkiruk
Department of Sciences, Ivan Franko Zhytomyr State University, 40 Velyka Berdychivska Str., 10008 Zhytomyr, Ukraine

O. Linnik
Chuiko Institute of Surface Chemistry, National Academy of Science of Ukraine, General Naumov Str. 17, 03164 Kyiv, Ukraine, E-mail: okslinnik@yahoo.co.uk

T. Lobzhanidze
Iv. Javakhishvili Tbilisi State University, Department of Chemistry, I. Chavchavadze Ave, 1, 0179 Tbilisi, Georgia

Matanat Ya. Magerramova
Nagiyev Institute of Catalysis & Inorganic Chemistry, Azerbaijan National Academy of Sciences, H. Javid Ave., 113, AZ1143 Baku, Azerbaijan

N. Maisuradze
Petre Melikishvili Institute of Physical and Organic Chemistry of Ivane, Javakhishvili Tbilisi, State University, I. Chavchavadze Ave. 1, 0179 Tbilisi, Georgia

G. Makharadze
Ivane Javakhishvili Tbilisi State University Chavchavadze Avenue 1, Tbilisi 0128, Georgia

T. Makharadze
Ivane Javakhishvili Tbilisi State University Chavchavadze Avenue 1, Tbilisi, 0128, Georgia

G. Mamniashvili
Ivane Javakhishvili Tbilisi State University, E. Andronikashvili Institute of Physics, Tamarashvili str, 6, 0162, Tbilisi, Georgia

Andriy Marinin
Problem Research Laboratory, National University of Food Technology, 68, Volodymyrska Street 01601 Kyiv, Ukraine

T. Marsagishvili
Ivane Javakhishvili Tbilisi State University, Rafiel Agladze Institute of Inorganic Chemistry and Electrochemistry. Mindeli str, 11, 0186, Tbilisi, Georgia, E-mail: tamazmarsagishvili@gmail.com

M. Melnyk
Chemistry Faculty, Kyiv National Taras Shevchenko University, 12 L. Tolstogo Str., 01033 Kyiv, Ukraine

J. Metreveli
Ivane Javakhishvili Tbilisi State University, Rafiel Agladze Institute of Inorganic Chemistry and Electrochemistry. Mindeli str., 11, 0186, Tbilisi, Georgia

I. Metskhvarishvili
Ilia Vekua Sukhumi Institute of Physics and Technology, Laboratory of Cryogenic Technique and Technologies, Mindeli Str. 7, 0186 Tbilisi, Georgia

M. Metskhvarishvili
Georgian Technical University, Department of Engineering Physics, Kostava Str. 77, 0175 Tbilisi, Georgia

N. A. Miraqyan
Yerevan State University, Alek Manoogian 1, Yerevan, 375025, Republic of Armenia

L. Nadareishvili
Georgian Technical University, V. Chavchanidze Institute of Cybernetics, 5 S. Euli, St, 0186 Tbilisi, Georgia

O. Nadtoka
Taras Shevchenko National University of Kyiv, Department of Chemistry, Volodymyrska Str., 60, 01033 Kyiv, Ukraine

N. Nasedkina
Faculty of Physics and Engineering, Moldova State University, 60 A. Mateevici str., MD 2009, Chisinau, Moldova

O. Nikolaeva
Chemistry Faculty, Kyiv National Taras Shevchenko University, 12 L. Tolstogo Str., 01033 Kyiv, Ukraine

A. Noskov
G.A. Krestov Institute of Solution Chemistry, Russian Academy of Sciences, Akademicheskaya Str., 1, Ivanovo, 153045, Russia

G. Papava
Petre Melikishvili Institute of Physical and Organic Chemistry of Ivane, Javakhishvili Tbilisi, State University, I. Chavchavadze Ave. 1, 0179 Tbilisi, Georgia

K. Papava
Petre Melikishvili Institute of Physical and Organic Chemistry of Ivane. Javakhishvili Tbilisi, State University, I. Chavchavadze Ave. 1, 0179 Tbilisi, Georgia

Sh. Papava
Petre Melikishvili Institute of Physical and Organic Chemistry of Ivane, Javakhishvili Tbilisi, State University, I. Chavchavadze Ave. 1, 0179 Tbilisi, Georgia

I. Pavlenishvili
Georgian Technical University, V. Chavchanidze Institute of Cybernetics, 5 S. Euli, St, 0186 Tbilisi, Georgia

G. Petriashvili
Georgian Technical University, 68 Merab Kostava Street, Tbilisi, Georgia Tbilisi, 0175, Georgia

G. Pirumyan
Yerevan State University, A. Manookyan St. 1, 0025, Armenia, E-mail: Scirec@mail.ru

Ł. Piszczyk
Polymer Technology Department, Chemical Faculty, Gdansk University of Technology, Gabriela Narutowicza Street 11/12, 80-233 Gdansk, Poland

N. Ponjavidze
Tbilisi State University, 1 Ilia Chavchavadze Ave, Tbilisi 0179, Georgia

T. Potlog
Faculty of Physics and Engineering, Moldova State University, 60 A. Mateevici str., MD 2009, Chisinau, Moldova

S. Robu
Moldova State University, 60 A. Mateevici str, MD-2009, Chisinau, Moldova, E-mail: s.v.robu@mail.ru

B. Salbu
Norwegian University of Life Sciences (NMBU), Department of Env. Sci., PO-Box 5003, No. 1432, Ås, Norway, E-mail: Brit.Salbu@nmbu.no

I. Savchenko
Kyiv National Taras Shevchenko University, Department of Chemistry, Kyiv, Ukraine, E-mail: iras@univ.kiev.ua

L. Sharashidze
Georgian Technical University, V. Chavchanidze Institute of Cybernetics, 5 S. Euli St, 0186 Tbilisi, Georgia

V. Sherozia
Petre Melikishvili Institute of Physical and Organic Chemistry of Ivane, Javakhishvili Tbilisi, State University, I. Chavchavadze Ave. 1, 0179 Tbilisi, Georgia

Ye. S. Sivokhina
Center of Physical and Chemical Methods of Research and Analysis, Tole bi Str. 96a, 050012, Almaty, Kazakhstan

N. Smirnova
Chuiko Institute of Surface Chemistry, National Academy of Science of Ukraine, General Naumov str, 17, 03164 Kyiv, Ukraine

V. Smokal
Kyiv Taras Shevchenko National University, Volodymyrska Str., 60, 01033, Kyiv, Ukraine, E-mail: vitaliismokal@gmail.com

D. Sternik
Maria Curie-Skłodowska University, pl. Maria Curie-Skłodowskiej 3, 20-031 Lublin, Poland

G. Supatashvili
Ivane Javakhishvili Tbilisi State University Chavchavadze Avenue 1, Tbilisi, 0128, Georgia

V. G. Syromyatnikov
Department of Macromolecular Chemistry, Taras Shevchenko National University of Kyiv, 60 Volodymyrska Street, 01033, Kyiv, Ukraine, E-mail: svg@univ.kiev.ua

Z. Tabukashvili
Petre Melikishvili Institute of Physical and Organic Chemistry of Ivane. Javakhishvili Tbilisi, State University, I. Chavchavadze Ave. 1, 0179 Tbilisi, Georgia

G. Tatishvili
Ivane Javakhishvili Tbilisi State University, Rafiel Agladze Institute of Inorganic Chemistry and Electrochemistry. Mindeli str, 11, 0186, Tbilisi, Georgia

G. Telegeev
Institute of Molecular Biology and Genetics of Natl. Acad. of Sci. of Ukraine, 150 Zabolotnogo str.150, Kyiv 03680, Ukraine, E-mail: gtelegeev@ukr.net

P. Telegeeva
Taras Shevchenko National University, Faculty of Chemistry, 60 Volodymyrska Str., Kyiv 0160, Ukraine

A. Przybytek
Polymer Technology Department, Chemical Faculty, Gdansk University of Technology, Gabriela Narutowicza Street 11/12, 80–233 Gdansk, Poland

A. O. Tonoyan
Armenian National Polytechnic University, 105 Teryana Str., 375009 Yerevan, Armenia, E-mail: atonoyan@mail.ru

N. Topuridze
Georgian Technical University, V. Chavchanidze Institute of Cybernetics, 5 S. Euli, St, 0186 Tbilisi, Georgia

L. Vretik
Chemistry Faculty, Kyiv National Taras Shevchenko University, 12 L. Tolstogo Str., 01033 Kyiv, Ukraine, E-mail: lvretik@gmail.com

D. G. Vyshnevsky
Department of Macromolecular Chemistry, Taras Shevchenko National University of Kyiv, 60 Volodymyrska Street, 01033, Kyiv, Ukraine

E. Yanovska
Chemistry Faculty, Kyiv National Taras Shevchenko University, 12 L. Tolstogo Str., 01033 Kyiv, Ukraine

Eldar B. Zeynalov
Nagiyev Institute of Catalysis & Inorganic Chemistry, Azerbaijan National Academy of Sciences, H. Javid Ave., 113, AZ1143 Baku, Azerbaijan, E-mail: zeynalov 2000@yahoo.com

O. A. Zhuravleva
Peoples' Friendship University of Russia, Scientific-Educational Centre of Nanotechnology, Miklukho-Maklaya 10/2, 117198, Moscow, Russia

LIST OF ABBREVIATIONS

AAM	acrylamide
AIBN	azoisobutyronitrile
AY	acridine yellow
BHT	3,5-di-tert-butyl-4-hydroxyphenyl
BPO	benzoyl peroxide
CB	conduction band
CD	charge density
CPE	composite polymer electrolyte
CPO	cumyl peroxide
DCPC	dicyclohexylperoxide carbonate
DLS	dynamic light scattering
DMF	dimethylformamide
DSA	differential scanning analyzer
DSC	differential scanning calorimetry
DSC	dye-sensitised solar cells
DTA	differential thermal analysis
EAS	electronic absorption spectra
EOS	epoxidized soybean oil
EP	emulsion polymerization
EPC	epoxypropyl carbazole
ES	emeraldine salt
FAME	fatty acids methyl esters
FFA	free fatty acids
FTIR	Fourier transform infrared
FTU	turbidity
GA	arabic gum
GZSD	graded zone stretching device
$HAuCl_4$	thetrachloroauric acid
HCl	hydrochloric acid
HCOOH	formic acid
HDODA	hexandiol diacrylate
HNO_3	nitric acid
HNPs	hybrid nanoparticles

HOMO	highest occupied molecular orbital
ITO	indium-tin oxide
KOH	potassium hydroxide
LAAP	liquid acetylacetone peroxide
LCHP	liquid cyclo hexanone peroxide
LDPE	low density polyethylene
LKP	liquid ketone peroxide
LUMO	lowest unoccupied molecular orbital
MEKP	methy ethyl kentone peroxide
MHM	multifunctional hybrid materials
MRO	maintenace, repairing and overhall
MW	molecular weight
NH_4	ammonium per-sulphate
NMP	N-methylpyrrolidinone
NPs	nanoparticles
OA	surfactant
OC	octene
OEDMA	γ,β-dimethacrylate
OH	hydroxyl
OLEDs	organic light emitting diodes
OMA	octylmethacrylate
PAA	polyacrylamide
PANi	polyaniline
PDS	peroxydisulphate
PDT	photodynamic therapy
PE	polyethylene
PEPC	poliepoxipropil carbazole
PL	photoluminescence
PMMA	polymethylmethacrylate
PMP	polymer-monomer particles
PPDA	phenylenediamine
PS	polystyrene
PSD	particle size distribution
PTP	photothermoplastic
PU	polyurethane
PUF	polyurethane foam
PVDF	polyvinylidene fluoride
PVP	polyvinylpyrrolidone

QELS	quasy elastic light scattering
RAF	rigid amorphous fraction
SAPF	solid amorphous polymer fraction
SC	superconducting
SEM	scanning electron microscope
SPE	solid polymer electrolyte
TAP	technical advisory panel
TBO-Pc-Zn	tetrabenzoxyphthalocyanine zinc
TEM	transmission electron microscopy
TG	thermogravimetric
TGA	thermal gravimetric analysis
TMS	tetramethylsilane
TNF	trinitrofluorenone
TPS	thermoplastic starch
UP	unsaturated polyester
UPR	unsaturated polyeter
USA	sigma aldrich
VB	valence band
VC	vinylcarbazole
VER	vinyl ester resins
WPC	wood polymer composites
XRD	x-ray diffraction

ABOUT THE EDITORS

Omari V. Mukbaniani, DSc
Full Professor, Iv. Javakhishvili Tbilisi State University,
Faculty of Exact and Natural Sciences, Department of Chemistry;
Chair of Macromolecular Chemistry; Director, Institute of
Macromolecular Chemistry and Polymeric Materials, Tbilisi, Georgia

Omari Vasilii Mukbaniani, DSc, is Professor and Head of the Macromolecular Chemistry Department of Iv. Javakhishvili Tbilisi State University, Tbilisi, Georgia. He is also the Director of the Institute of Macromolecular Chemistry and Polymeric Materials. He is a member of the Academy of Natural Sciences of the Georgian Republic. For several years he was a member of the advisory board of the Journal Proceedings of Iv. Javakhishvili Tbilisi State University (Chemical Series) and contributing editor of the journals *Polymer News* and the *Polymers Research Journal*. He is member of editorial board of the *Journal of Chemistry and Chemical Technology*. His research interests include polymer chemistry, polymeric materials, and chemistry of organosilicon compounds. He is an author more than 420 publication, 13 books, four monographs, and 10 inventions. He created in 2007 the "International Caucasian Symposium on Polymers and Advanced Materials," ICSP, which takes place every other two years in Georgia. The last symposium, ICSP 5, was held in July 2017 in Tbilisi, Georgia.

Marc J. M. Abadie, DSc
Emeritus Professor, Institute Charles Gerhardt of Montpelier—Aggregates,
Interfaces & Materials for Energy (ICGM-AIME, UMR CNRS 5253),
University Montpelier, France

Professor Marc J. M. Abadie is Emeritus Professor at the University Montpellier, Institute Charles Gerhardt of Montpelier, Aggregates, Interfaces and Materials Energy (ICGM-AIME, UMR CNRS 5253), France. He was head of the Laboratory of Polymer Science and Advanced Organic Materials

– LEMP/MAO. He is currently "Michael Fam" Visiting Professor at the School of Materials Sciences and Engineering, Nanyang Technological University NTU, Singapore. His present activity concerns high performance polymers for PEMFCs, composites and nanocomposites, UV/EB coatings, and biomaterials. He has published 11 books and 11 patents. He has advised nearly 95 MS and 52 PhD students with whom he has published over 402 papers. He has more than 40 years of experience in polymer science with 10 years in the industry (IBM, USA – MOD, UK & SNPA/Total, France). He created in the 1980s the "International Symposium on Polyimides and High Temperature Polymers," a.k.a. STEPI, which takes place every other three years in Montpellier, France. A recent symposium, STEPI 10, look place in June 5–8, 2016.

Tamara N. Tatrishvili, PhD

Senior Specialist, Unite of Academic Process Management
(Faculty of Exact and Natural Sciences), Ivane Javakhishvili Tbilisi State
University; Senior Researcher, Institute of Macromolecular Chemistry
and Polymeric Materials, Georgia

Tamara Tatrishvili, PhD, is Senior Specialist at the Unite of Academic Process Management (Faculty of Exact and Natural Sciences) at Ivane Javakhishvili Tbilisi State University as well as Senior Researcher of the Institute of Macromolecular Chemistry and Polymeric Materials in Tbilisi, Georgia.

PREFACE

Increasing interest in lightweight and high performance materials is leading to significant research activity in the area of polymers and composites. One recent focus is to develop multifunctional materials that have more than one property tailored as per the design requirements in addition to achieving low density.

The possibility of simultaneously tailoring several desired properties is attractive but very challenging, and it requires significant advancements in the science and technology of high performance functional polymers and composites.

The structures and functions of these advanced polymer and composite systems are evaluated with respect to improved or novel performance, and the potential implications of those developments for the future of polymer-based composites and multifunctional materials are discussed.

The ensemble selected papers presented at the 4th International Caucasian Symposium on Polymers and Advanced Materials are collected, as original unpublished research, in this book. ICSP1AM-4 took place in Batumi Black sea beach, Georgia July 1–4, 2015. This book focuses exclusively on the latest research related to polymer and composite materials, especially new trends in frontal polymerization and copolymerization synthesis, functionalization of polymers, physical properties and hybrid systems. A large part is devoted to composites and nanocomposites; their mechanical impact on their properties are evaluated and discussed. Applications in medicine as drug-release for cancer treatment or as development of green materials and as renewable resources with a particular emphasis on the research of the authors' group(s) are presented. Coating and curing processes are also covered in this book in order to have a better control and understanding of the crosslinking reactions.

The progress captured in the current set of articles shows promise for developing materials that seem capable of moving this field from laboratory-scale prototypes to actual industrial applications.

This book serves as a valuable and informative reference for scientists, engineers, medical technologists and practioners engaged in the teaching, research, development and use of functional polymers and composites.

Professor, Dr. of Science, Omari Mukbaniani
Iv. Javakhishvili Tbilisi State University,
Faculty of Exact and Natural Sciences,
Department of Chemistry, Chair of Macromolecular Chemistry,
Director of the Institute of Macromolecular Chemistry
and Polymeric Materials

Dr. Marc J. M. Abadie, P. Eng.
Professor Emeritus, Doctor Honoris Causa
Institut Charles Gerhardt de Montpellier – Agrégats, Interfaces et
Matériaux pour l'Energie (IGCM AIME UMR CNRS 5253)
STEPI General Chairman, Expert près la Cour d'Appel
"Michael Fam" Visiting Professor @ NTU/MSE, Singapore

Dr. Tamar Tatrishvili
Ivane Javakhishvili Tbilisi State University,
Senior Specialist of the Unite of Academic Process Management
(Faculty of Exact and Natural Sciences),
Senior Researcher of the Institute of Macromolecular Chemistry and
Polymeric Materials

PART I

SYNTHESIS AND APPLICATION

CHAPTER 1

COPOLYMERIZATION OF ANILINE WITH P-PHENYLENEDIAMINE IN AN ACETIC ACID MEDIUM

A. H. DURGARYAN, N. A. DURGARYAN, R. H. ARAKELYAN, and N. A. MIRAQYAN

Yerevan State University, Alek Manoogian 1, Yerevan, 375025, Republic of Armenia, E-mail: durnar63@yahoo.com

CONTENTS

ABSTRACT

The oxidative copolymerization of aniline with *p*-phenylenediamine in an acetic acid medium has been investigated for the first time. It has been determined that as a result of copolymerization; a polymer having a structure analogous to the polyaniline called emeraldine is formed. The obtained copolymer doped with 3N hydrochloric acid has an electrical conductivity five times higher than that of polyaniline prepared by the usual method. Aniline polymerization proceeds more slowly at 273 K under the same copolymer-ization conditions and the obtained polymer has low conductivity. However, in a mixture of acetic acid-methanol, the reaction proceeds faster and the obtained polymer has conductivity almost equal to that of emeraldine. It has been observed that the conductivity of polymers doped with formic acid is lower than that of polymers obtained by doping with hydrochloric acid by 2 orders of magnitude.

1.1 INTRODUCTION

Among the electro active polymers, the emeraldine form of polyaniline (PANI) attracts intense interest [1, 2]. PANI occupies a particular place among the electro active conjugated polymers owing to its environmental stability, easy and cheap method of preparation and its unique properties. It is extensively used in various fields of technology [3–14]. To improve the properties of PANI, studies of aniline copolymerization with different monomers have been conducted [12–19]; however, the conductivity of the obtained copolymers is inferior to that of PANI in spite of high solubility.

According to the literature data, copolymer I[20] is formed during the copolymerization of *p*-phenylenediamine (PPDA) with aniline.

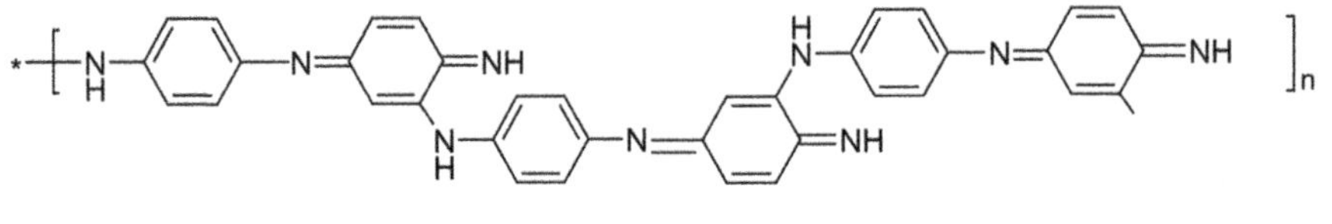

COPOLYMER I

In the formation of copolymer I, for 1 mole each of aniline and PPDA, 3 moles of peroxydisulfate (PDS) are required [20] because when using 1 mole of PDS for 1 mole of aniline, the calculated yield for aniline cannot be higher than 33%, while a yield of 62% to 75% was actually obtained [20]. It may be concluded from these data that the reaction does not proceed accord-ing to Figure 4 given in the Ref. [20].

According to other literature data, in the case of a molar ratio of aniline/PPDA of 50:1, the presence of PPDA greatly increases the rate of polymer-ization and does not affect the structure and crystalline of PANI [21].

These data strengthen our opinion that the reaction mechanism differs from that described in the literature [20], and considering the scheme for the oxidative polymerization of PPDA [22]; we have proposed that it proceeds with a mechanism similar to that depicted in Scheme 1[23].

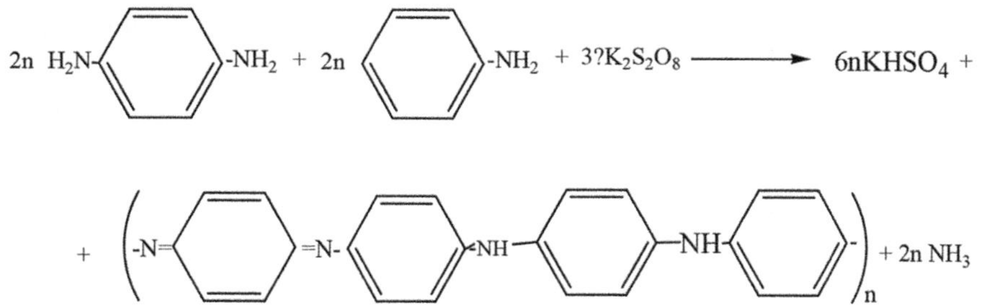

$2n \ H_2N-\langle \bigcirc \rangle-NH_2 \ + \ 2n \ \langle \bigcirc \rangle-NH_2 \ + \ 3?K_2S_2O_8 \longrightarrow \ 6nKHSO_4 \ +$

$+ \ \left(-N=\langle \bigcirc \rangle=N-\langle \bigcirc \rangle-NH-\langle \bigcirc \rangle-NH-\langle \bigcirc \rangle- \right)_n \ + \ 2n \ NH_3$

SCHEME 1

If the reaction occurs according to Scheme 1, the maximum yield calculated for aniline will be 67%, while in the case of a ratio of PPDA-aniline in the copolymer of more than 1; the yield will be above 67%.

The purpose of the present work is to verify the possibility that the reaction proceeds according to Scheme 1.

1.2 EXPERIMENTAL PART

1.2.1 MEASUREMENTS

The inherent viscosity of DMSO solution was determined at 25°C, using Ubbelohde viscometer.

The UV/vis spectra of the polymer samples were recorded in 1 cm quartz curettes with Secord 65 spectrometer. FT IR Nicolet Nexus spectrometer was served for obtaining FT-IR spectra in the range of 5000–600 cm^{-1}(KBr pellets). ^{1}H-NMR spectra were obtained in deutera teddi methyl sulfoxide using Mercury 300 Varian NMR spectrometer.

Electrical conductivity was measured on the preliminarily prepared pellets by two-probe method using Teraohmmeter E6–137.

1.2.2 MATERIALS

PPDA was purified by sublimation (mp 416–418 K) and PPDA sulfate was recrystallized from water. Aniline was used after double-distillation. All other chemicals were of analytical grade and were used as received without any further purification

1.2.3 OXIDATIVE COPOLYMERIZATION OF PPHDA WITH ANILINE BY POTASSIUM PEROXYDISULFATE

• *Potassium Peroxydisulfate in the hydrochloric acid medium [20]*

To 1.08 g (10 mmol) of PPhDA and 0.91 ml (10 mmol) of aniline in 100 ml (0.1N) of hydrochloric acid, solution of 2.7 g (10 mmol) potassium peroxydisulfate in 190 ml (0.1N) of hydrochloric acid has been added within 3 hours under continuous magnetic stirring at (273–275 K). The solution was kept in refrigerator (270 K) at night. Then, the solution was filtered and the precipitate was washed with distilled water until neutral pH. 190 ml of 0.1 N solution of ammonium hydroxide was added on the precipitate and the mixture was filtered after 13 hour stirring. The precipitate was washed with distilled water until neutral pH and absence of SO_4^{2-}ions, dried in vacuum(323 K/2 kPa) and stored in desiccators over phosphorus pentoxide.The yield is 1.02g (60%).After getting rid of water from the filtrate the ammonia was determined.

• *Potassium Peroxydisulfate in the glacial acetic acid medium*

  a. **Copolymer (series C-I)**. To a solution of 0.43 g (4.6 mmol) aniline, 0.93 g (4.3 mmol) PPDA sulfate in 8 ml acetic acid. 1.44 g (6.45 mmol) potassium per sulfate was added under contentious magnetic stirring at 27 K, the mixture was stirred for 3.3 h. After that, the reaction mixture was kept in refrigerator at 270 K. Then the cooled water was added into the mixture. Subsequent procedure was similar to the mentioned above. The yield is 0.42 g (51%).

  b. **Copolymer (series C-II)**. The oxidative copolymerization was carried out by the above-mentioned procedure with the difference in the duration of the reaction (*see* Table 1.1).

  c. **Copolymer (series C-III)**. The oxidative copolymerization was carried out by the above-mentioned procedure using acetic acid (8 ml)-methanol (0.8 ml) medium (Table 1.1).

1.2.4 SEMI-QUANTITATIVE DETERMINATION OF AMMONIA

Ammonia was determined using distillation apparatus equipped with liquid trap, receiver filled with 120 ml 0.1N hydrochloric acid for ammonia absorption and Tishchenko bottle filled with water.

TABLE 1.1 Data for Yields, Specific Viscosities and Electrical Conductivities of Aniline Homopolymers (P) and Copolymers (C) with PPDA

Homopolymer, (P); copolymer(C)	Reaction time		Yield %	$[\eta]$ dl/g	σ; S/cm	
	Hours	Days			Dopant	
					3 N hydrochloric acid	Formic acid
C-I[a]	3,3	1	51		0,1	
C-I[b]	3,3	1	51		0.15	
P-I	15	4	14.3	0.074	3.4×10^{-3}	
C-II[a]	2.3	7	61		0.13	
C-II[b]	2.3	7	61		0.17	
C-III[b]	9	2	54	0.20	0.56; 0.46	1×10^{-3}
P-III	36	8	59	0.16	0.06; 0.052; 0.0459	1.8×10^{-3}

[a, b] *see* experimental section.

On a dry precipitate obtained after evaporation of the liquid from the filtrate, solution of 2.4 g of KOH in 6 ml of water was added through the dropping funnel. Heating was carried out so, that liquid distillation didn't proceed. The ammonia content was determined by titration of hydrochloric acid with alkali and was found to be 8.96 mmol.

1.2.5 OXIDATIVE POLYMERIZATION OF ANILINE

- **Aniline in acetic acid medium.** About 4.01 g (15 mmol) of potassium peroxidisulfate was added to the mixture of 1.04 g (11.16 mmol) of aniline, 0.483 g (3.72 mmol) of aniline hydrochloride in 5 ml of acetic acid under contentious magnetic stirring 15 hour within 5 days, at 273–275 K. Every time when stirring was interrupted, the reaction mixture was kept in refrigerator at 270 K. Subsequent procedure was similar to the mentioned above. Yield is 0.11 g (8.6%) in the mixture of acetic acid-methanol.

  About 3.64 g (0.012 mol) of ammonium peroxidisulfate was added to a solution of 1.11 g(0.012 mol) aniline, 0.523 g(0.004 mol) of aniline hydrochloride in 17 ml of acetic acid and 1.7 ml of methanol under continuous magnetic stirring 36 hour within 8 days, at

273–275 K. Every time when stirring was interrupted, the reaction mixture was kept in freezer at 1–2°C. Further procedure was carried out as described above. Yield is 0.82 g (57%).

1.2.6 PROCEDURE FOR DOPING

- **Doping with hydrochloric acid.** About 1.9 ml of 3N hydrochloric acid was added on the fine powder of polymer and left 7 days at room temperature. Then the precipitate was filtered and washed twice with a small amount of ethanol. The precipitate was dried until constant weight under vacuum in desiccators over phosphorus pentoxide (0.2kPa).
- **Doping with formic acid.** About 0.05 g of fine powder of polymer was dissolved in 5 ml of formic acid at room temperature. The solvent was removed at room temperature and the remained powder was dried until constant weight under vacuum in a desiccators over phosphorus pentoxide (0.2 kPa).

1.3 RESULTS AND DISCUSSION

During the course of the reaction according to Scheme, ammonia is formed as a result. For the purpose of checking these data, we have repeated the experiment described in the literature [20] and we have semi quantitatively determined the resulting ammonia content and found that 1.43 mmol of ammonia were formed from 10 mmol of PPDA.

If the reaction proceeds according to Scheme 1, a polymer analogous to that of emeraldine would be formed. However, the conductivity of the obtained chloride salt form of the polymer is equal to 2.719×10^{-5} S/cm [20], which is much lower than that of emeraldine. Because the mentioned value is related to the conductivity of the as-synthesized salt form of the polymer, and consequently, there are no data about doping levels, the as-synthesized polymer was converted to its base form and doped with 3 N hydrochloric acid to reveal conductivity of 2.9×10^{-7} S/cm, which is, as mentioned above, far inferior to that of emeraldine.

Moreover, PPDA-aniline copolymer solutions in methanol and DMSO exhibit absorption bands at 308 and 549 nm [20], and 320 and 585 nm (Figure 1.1), respectively. The emeraldine base in N-methylpyrrolidinone

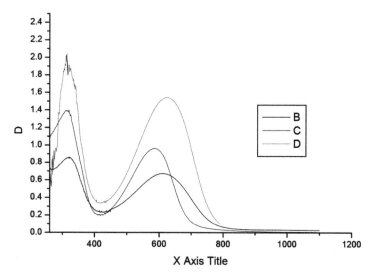

FIGURE 1.1 (A) UV spectra (l, nm) of (B) PPDA-aniline copolymer, obtained in a mixture of acetic acid-methanol, in NMP, λ_{max} = 320,611; (C) PPDA-aniline copolymer, obtained in hydrochloric acid, in DMSO, λ_{max} = 320,585; (D) Polyaniline, obtained in a mixture of acetic acid-methanol, in NMP, λ_{max} = 320,629.

(NMP) is characterized by two absorption bands at 320 and 634 nm [24], and the film shows absorption bands at 326 and 630 nm [25].

It follows from the presented data that the polymer obtained by chemical oxidative copolymerization of PPDA and aniline does not have a structure similar to that of emeraldine. These data could be explained by the occurrence of side reactions that occur during the oxidative polymerization of PPDA [26].

To reduce the proportion of side reactions, we performed the copolymerization reaction of PPDA and aniline in glacial acetic acid and acetic acid-methanol media. The reaction in glacial acetic acid was conducted in three variations designed to investigate the effect of reaction time on the yields and electrical conductivities of the obtained copolymers (Table 1.1). An acetic acid-methanol medium was applied to decrease the melting point of acetic acid.

The semi quantitative determination of ammonia content shows that 2.3 mmol of PPhDA and 0.19 mmol of ammonia were formed.

Most likely, the reaction of PPDA with PDS takes place at first, and 1,4-benzoquinonediimine is formed, which can participate in the subsequent reaction (Scheme 2).Finally, as a result of copolymerization, poly(1,4-benzoquinonediimine-*N,N'*-diyl-1,4-phenyleneimino-1,4-phenyl-eneimino-1,4-phenylene), which has a structure analogous to emeraldine PANI, is formed (Scheme 1). The copolymers are soluble in formic acid, NMP and DMF, and partially soluble in DMSO.

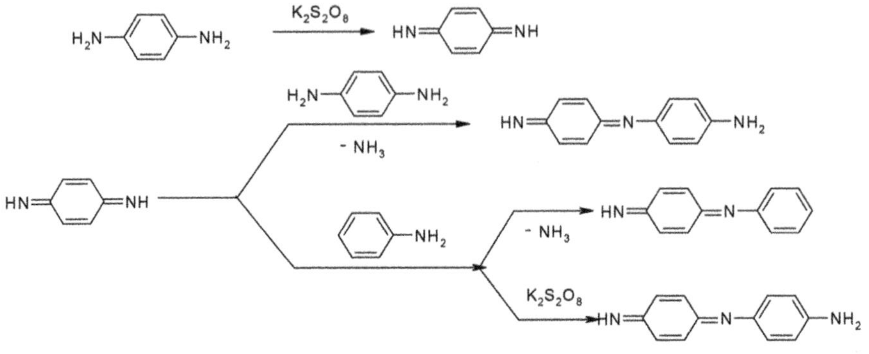

SCHEME 2

The PMR, IR and UV spectra of the obtained copolymers are in close agreement with those published in the literature for emeraldine.

The UV spectrum of the copolymer obtained in a medium of acetic acid-methanol shows absorption bands at 320 and 611 nm in NMP (Figure 1.1). Accordingly, as cited above, a solution of emeraldine in NMP exhibits absorption bands at 320 and 634 nm [23], while the film shows absorption bands at 326 and 630 nm [24]. Polyaniline obtained in a mixture of acetic acid-methanol, in NMP shows absorption bands at 320 and 629 nm.

The ^{1}H NMR spectrum of the copolymer (Figure 1.2) is very similar to that reported in the literature for emeraldine [27], with the difference being that there is an additional very weak chemical shift at 6.6 ppm, which may most likely be assigned to the *ortho* protons of the terminal primary amino groups.

IR spectrum: 506,696,717,831,956, 1008, 1162, 1310, 1331, 1498, 1590, 3033, 3331 cm^{-1}. These data almost completely coincide with those reported in the literature for emeraldine [25].

To make a comparison with aniline-PPDA copolymerization, the polymerization of aniline in acetic acid and in a mixture of acetic acid-methanol has been investigated. The oxidative polymerization reaction of aniline proceeds at a much slower rate compared to copolymerization, and it depends on the medium in which the reaction occurs. The specific electro conductivity of

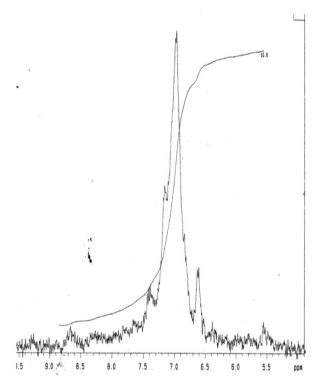

FIGURE 1.2 ^{1}H NMR spectrum (DMSO-d_6, δ, ppm) of the PPDA-aniline copolymer.

the copolymers doped with 3N hydrochloric acid changed, depending on the copolymerization conditions, from 0.1 to 0.51 S/cm. The PANI obtained in acetic acid (P-I) has an electrical conductivity of 3.4×10^{-4} S/cm, and that obtained in a mixture of acetic acid-methanol was 0.06 S/cm. The latter value is comparable to that obtained for the 3N hydrochloric acid-doped emeraldine PANI (0.09 S/cm) having been synthesized by the standard method (in a solution of hydrochloric acid at 0°C) and to that according to the literature data (0.1 S/cm)[20].

The great difference in the conductivities of P-I and P-II could mainly be caused by the low molecular mass of the former (see Table 1.1). It should be noted that the characteristic viscosities of the copolymer(C-III) and the polymer (P-III) differ only slightly. The obtained data hereby indicate that PPDA-aniline copolymers are, in fact, analogous to emeraldine PANI structures. The fact that the conductivity of the copolymer is slightly higher than that of PANI can be caused by fewer defects in the former compound's structure.

According to the literature, the conductivity of PANI salts obtained in formic acid is higher than that obtained in hydrochloric acid [28]. It should be noted that, during the oxidative polymerization of aniline, ammonium or potassium bisulfate is formed from peroxydisulfate, and it also forms salts with the polymer, along with hydrochloric acid and formic acid [29].

The conductivities of both polymers and copolymers are two orders of magnitude lower in the case of doping with pure formic acid than in a 3N solution of hydrochloric acid (Table 1.1).

ACKNOWLEDGMENTS

The financial support of The State Committee of Science of MES RA, research project № SCS_13-1D298 is gratefully acknowledged.

KEYWORDS

- aniline
- conductivity
- conjugated polymer
- copolymerization
- emeraldine
- polyaniline
- p-phenylene-diamine

REFERENCES

1. Skorheim, T. A., Elsenbaumer, R. L., & Reynolds, J. R. (Eds.), (1998). Handbook of Conducting Polymers, 2nd ed., Marcel Dekker Inc.: New York, Basel, Hong Kong.
2. Pron, A., & Rannou, P. (2002). Processable conjugated polymers: from organic semi-conductors to organic metals and superconductors, *Prog. Polym. Sci., 27*, 135.
3. MacDiarmid, A. G. (1985). Short Course on Conductive Polymers, Suny: New Paltz, NY, USA.
4. Tallman, D. E., Spinks, G., Dominis, A., & Wallace, G. G. (2002). Electro active conducting polymers for corrosion control. Part 1: General introduction and a review of non-ferrous metals. *J. Solid State Electrochem., 6*, 73–84.

5. MacDiarmid, A. G., Mu, S. L., Somasiri, M. L. D., & Wu, W. (1985). Electrochemical characteristics of polyaniline cathodes and anodes in aqueous electrolytes. *Cryst. Liq. Cryst., 121*, 187–190.

6. Rahmanifar, M. S., Mousavi, M. F., & Shamsipur, M. (2005). Heli: A study on circuit voltage reduction as a main drawback of Zn-polyaniline rechargeable batteries. *Synth. Met., 155*, 480.

7. Abdolreza Mirmohseni, &Reza Solhjo. (2003). Preparation and characterization of aqueous polyaniline battery using a modified polyaniline electrode.Eur. Polym. J., 39, 219.

8. Kanatzidis, M. G. (1990). Conductive polymers. Chem. Eng. News, 68, 36.

9. Anderson, M. R., Mattes, B. R., Reiss, H., & Kaner, R. B. (1991). Gas separation membranes: A novel application for conducting polymers. *Synth. Met., 41*, 1151.

10. Illing, G., Hellgardt, K., Schonert, M., Wakeman, R. J., & Jungbauer, A.(2005). Towards ultrathin polyaniline films for gas separation. *Membr. Sci., 253*, 199.

11. Gerard, M., Chaubey, A., & Malhotra, B. D. (2002). Application of conducting polymers to biosensors. *Biosens. Bioelectron., 17*, 345. http://dx.doi.org/10.1016/S0956-5663(01)00312-8.

12. Karyakin, A. A., Vuki, M., Lukachova, L. V., Karyakina, E. E., Orlov, A. V., Karpachova G. P., et al. (1999). Processible polyaniline as an advanced potentiometric pH transducer: Application to biosensors. *Anal. Chem., 71*, 2534.

13. Tzung-Hua Lin, & Kuo-Chuan Ho. (2006). A complementary electro chromic device based on polyaniline and poly (3, 4-ethylenedioxythiophene). *Solar Energy Materials and Solar Cells, 90*(4–6), 506.

14. Angelopoulos, M., Shaw, J. M., Kaplan, R. D., & Perreault, S. (1989). Conducting polyanilines: Discharge layers for electron-beam lithography, *JVST, B7*, 1519.

15. Jyongsik Jang, Joonwon Bae, Moonjung Choi, Seong-Ho Yoon. (2005). Carbon, fabrication and characterization of polyaniline coated carbon nanofiber for super capacitor, *Carbon, 43*, 2730.

16. Arias-Pardilla, J., Salavagione, H. J., Barbero, C., Morallorn, E., & Varzquez J. L. (2006). Study of the chemical copolymerization of 2-aminoterephthalic acid and aniline: Synthesis and copolymer properties. *Eur. Polym. J., 42*, 1521.

17. Ramanujam, P. S., Hvilsted, S., Ujhelyi, F., Koppa, P., Lörincz, E., Erdei, G., et al. (2001). Physics and technology of optical storage in polymer thin. *Synth. Met., 124*(1), 145.

18. Junqing, Zh., Shijie, X., Shenghao, H., Zhiwei, Y., Lina, Y., & Tianlin Y. (2000). Organic light-emitting diodes with AZO films as electrodes. *Synth Met., 114*(3), 251.

19. Xin-Gui Li, Mei-Rong Huang, & Yi-Min Hua. (2005). Facile synthesis of processible aminoquinoline/phenetidine copolymers and their pure semiconducting. *Macromolecules, 38*, 4211–4219.

20. Rani, M., Ramachandran, R., & Kabilan, S. (2010). A facile synthesis and characterization on semiconducting *p*-phenylenediamine-aniline copolymer. *Synth. Met., 160*, 678–684.

21. Shenashen, M. A., Ayad, M. M., Salahuddin, N., & Youssif, M. A. (2010). Usage of quartz crystal microbalance technique to study the polyaniline films formation in the presence of *p*-phenylenediamine. *React. & Funct. Polym. 70*, 843–848.

22. Durgaryan, A. A., Arakelyan, R. H., & Durgaryan, N. A.(2010). On the structure of poly(p-phenylenediamine), Second International Caucasian Symposium on Polymers and Advanced Materials, Tbilisi, Georgia, 31.

23. Durgaryan, N. A., Arakelyan, R. H., & Durgaryan, A. H. (2011). Concerning the mechanism of oxidative polymerization of *p*-phenylenediamine Prague, July 10–14 Czech Republic, SL02, 46.
24. Albuquerque, J. E., Matioso, L. H. C., Baloqh, D. T., Faria, R. M., Masters, J. G., & MacDiarmid, A. G. (2000). A simple method to estimate the oxidation state of polyanilines. *Synth. Met., 113*, 19.
25. Gao, Y. (1990). Spectroscopic studies of acceptor and donor doping of polyaniline in the emeraldine base and pernigraniline forms. *Synth. Metals, 35*, 319–332.
26. Durgaryan, A. A., Arakelyan, R. A., & Durgaryan, N. A. (2014). Oxidative polymerization of *p*-phenylenediamine. *Russ. J. Org. Chem., 84*, 912.
27. Focke, W. (1987). Conduction Mechanism in Polyaniline, p. 58.
28. Gomes, E. C., & Oliveira, M. A. S. (2012). Chemical polymerization of aniline in hydrochloric acid (HCl) and formic acid (HCOOH) media: Differences between the two synthesized polyanilines. *Am. J. Polym. Sci., 2*, 5.
29. Durgaryan, A. H., Durgaryan, N. A., Arakelyan, R. H., Vanyan, S. G., Asaturian, R. A., & Avetissyan, A. A. (2006). Comparative investigation of electric conductivity, spine properties and concentration conjugated and nonconjugated polymers. *Chem. J. Arm., 59*, 84 (in Russian).

CHAPTER 2

FREE-RADICAL FRONTAL POLYMERIZATION OF ACRYLAMIDE IN THE PRESENCE OF NANO AND MICRO ADDITIVES

A. O. TONOYAN and S. P. DAVTYAN

Armenian National Polytechnic University, Teryana Str., 105, 375009, Yerevan, Armenia, E-mail: atonoyan@mail.ru

CONTENTS

ABSTRACT

The effect of different nanoparticles (bentonite, SiO_2) and finely disperse fillers (chalk, diatomite) on peculiarities of acrylamide frontal free-radical polymerization initiated by azo-iso-butyronitrile and benzoyl peroxide was studied. It was determined that the increase in an order with respect to initiator is conditioned by influence of nano and micro additives not only on the mechanism of initiator decomposition but also on the reaction of chains bimolecular termination.

2.1 INTRODUCTION

It has been known that according to the theory of heat waves propagation [1, 2] the velocity of vinyl monomers free-radical frontal polymerization depends on concentration of initiator as shown in the equation: u = [Ionf (xi)] 1/2 = Ion/2[f (xi)] 1/2. Where u is front velocity, Io and n are initial concentration of initiator and power of initiator, f(xi) is function depending on monomer concentration, temperature and polymerization thermal mode, heat effect, ultimate conversion, temperature of adiabatic heating-up and other parameters [2].

The numerical calculations [1] give 0.40 for the n value versus 0.48 obtained from analytical evaluations [2]. However, earlier experimental studies of free-radical frontal polymerization of 3-(oxyethylen)—γ, β-dimethacrylate (OEDMA) and methyl methacrylate under high pressure (up to 5 kbar) [3, 4] have shown that value of n depends both on initiator and monomer type. Thus in the case of frontal polymerization of OEDMA the orders with respect to initiator were 0.22, 0.32, 0.34 for t-butyl peroxide (t-BP), benzyl peroxide (BPO) and dicyclohexylperoxide carbonate (DCPC) respectively while the value of n for methyl methacrylate polymerization initiated by BPO was 0.36. Based on these results it could be assumed that such change of n value is conditioned by specific impact of high pressure on initiation efficiency and chains termination and etc. Later the frontal polymerization of methacrylic acid and triethylene glycol dimethacrylate without applying of high pressure has been studied [5]. The study of methacrylic acid polymerization showed that power dependences of front velocity on concentration of initiators were 0.24, 0.25, 0.27, 0.26 for azoisobutyronitrile (AIBN), cumyl peroxide (CPO), lauroyl peroxide (LPO), t-butyl peroxides respectively, while in the case of triethylene glycol dimethacrylate the exponents were 0.2, 0.23, 0.31 for AIBN, BPO and LPO. Interestingly that in the case of acrylamide free-radical frontal polymerization the value of n for BPO was 0.43±0.02. The observed data indicate that the power dependence of front velocity on initiator concentration substantially depends on the nature of the initiator. As a matter of course it is very important to ascertain the effect of different kind of finely disperse and nanosized fillers on the order with respect to initiator and accordingly on the elementary reactions of chains initiation, propagation and termination. This problem becomes especially crucial if to consider the formation of the solid amorphous polymer fraction (SAPF) at interface of nanocomposites due to strong interaction of macromolecules with nanoparticles surface taking place in the course of synthesis of polymer nanocomposites by free-radical polymerization of vinyl monomers [6–9].

Thus in this study we focus on acrylamide (AAM) frontal polymerization initiated by BPO and AIBN, in particular, on the effect of finely disperse fillers (diatomite, chalk) and nanoparticles (SiO_2, bentonite) on power dependence of front velocity on initiator concentration and mechanism of elementary reactions of chain initiation and termination.

2.2 EXPERIMENTAL PART

AAM (Aldrich) was purified following a protocol described elsewhere [5]. AIBN and BPO were used after double recrystallization from ethanol and drying in vacuum oven. Powdery SiO_2 (Sigma Aldrich, 10 nm), bentonite (~15 μm), diatomite (~30 μm) and chalk (~50 μm) were used as fillers. The necessary amounts of acryl amide, polymerization initiators (to provide steady-state regime of free-radical frontal polymerization) and specified additives were thoroughly mixed [8] and placed by small portions into glass tubes with further compacting. Frontal polymerization of AAM filled mixtures was carried out as we described previously. Reaction was carried out in 10×100-mm vertically set tubes. Frontal polymerization was initiated by application of hot metal surface to the top (or bottom) of reaction mixture [8, 9].

2.3 RESULTS AND DISCUSSION

The frontal polymerization linear velocity dependence on AIBN concentration for acrylamide-bentonite mixture containing 30.0 wt% of filler is shown in Figure 2.1.

From the data presented in Figure 2.1 the ~0.6-power dependence of the front velocity on the AIBN concentration was determined for AAM mixture containing 30.0 wt% of bentonite. By the same way the power dependences on initiators (BPO, AIBN) can be calculated for the filled mixtures of AAM containing different amounts of bentonite. These results are shown in Figure 2.2a,b.

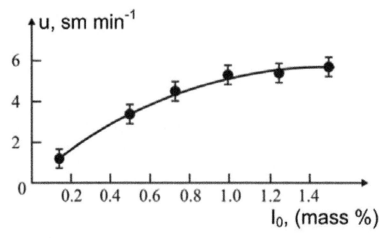

FIGURE 2.1 Linear velocity of AAM-bentonite mixture polymerization front as a function of initiator (AIBN) concentration. Filler content is 30 wt%.

The order with respect to initiator has complex dependence on composition of filled mixtures and nature of initiator as it follows from the data obtained. As it is shown in Figure 2.2, the order does not depends on direction of front propagating – vertically from top to bottom or vice versa. For both types of initiators the value of n increases with increasing the content

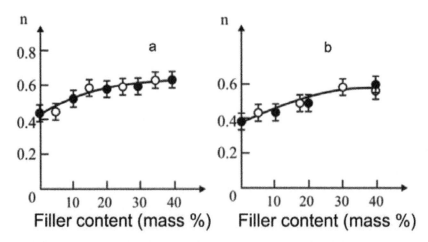

FIGURE 2.2 Power of initiator dependence on filler loading for AAM-bentonite mixtures frontal polymerization initiated by BPO (a) and AIBN (b). (●) and (◌) – ascending and descending heat waves, respectively.

of filler in mixtures. In the case of BPO the filler content increasing leads to increase in power of initiator up to 0.61±0.02 while for AIBN the order increases from 0.38 to 0.58±0.02. Observed increase in order with respect to initiator above 0.5 with filler loading increasing can be explained by interaction of initiator molecules with surface of bentonite separate layers. Further decomposition of initiators as well as the initiating and propagation of polymer chains take place on the surface of bentonite separate layers. It is assumed that the occlusion of polymerization reactive species can be promoted by interaction between binder macromolecules and separate layers of bentonite resulting in formation of the solid amorphous polymer fraction [9].

In order to clarify the influence of filler structure (layered or porous) on frontal polymerization behavior the similar studies were carried out for AAM mixtures with chalk fine powder using the same polymerization initiators. The obtained data are presented in Figure 2.3 and show the effect of filler loading on power of initiator.

Regardless of the filler absence or presence the orders of dependence of AAM frontal polymerization velocity on BPO and AIBN concentration coincide both for rising and descending heat waves in the all investigated range of filler loading. However, the absolute values of exponents are

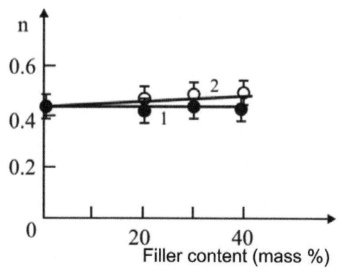

FIGURE 2.3 Power of initiator dependence on filler loading for AAM-chalk mixtures frontal polymerization initiated by BPO (curve 1) and AIBN (curve 2).

slightly different for each type of initiator. The front velocity dependence on BPO initial concentration for chalk filled AAM (filler content is from 20 to 40 wt%) is described by following equation u ~I00.43±0.02. In the case of using AIBN the order of dependence on initiator increases up to 0.48 (filler content is 50 wt.%).

Interestingly that for frontal polymerization of AAM with diatomite additives the following results were obtained: n = 0.65 ± 0.02 for BPO and n = 0.6 ± 0.02 for AIBN. The observed increase in n value can be explained by intercalation of polyacrylamide macromolecules into micro- and nanosized pores of diatomite, their interaction with surface of pores and occlusion of reactive species in the pores. The difference in absolute values of powers of initiators for mixtures filled with bentonite, diatomite and chalk is probably conditioned by both the filler structure and the nature of chemical compounds of the surface layers of specified fillers.

The results of impact of SiO_2 spherical nanoparticles on the order with respect to initiator are not presented because they are similar to the data shown in Figure 2.2. It should be only noted that for BPO initiated AAM frontal polymerization the power of initiator concentration increases to 0.68±0.03 with increasing SiO_2 loading while in the case of AIBN the exponent value reaches up to 0.6±0.02.

Thus the data obtained for considered fillers suggest that decomposition of initiators takes place not only in the bulk of reaction medium but also on the surface of nanoparticles and macro radicals termination occurs by bimolecular and monomolecular (occlusion of radicals) reaction mechanisms. These facts are the main reason of marked increase in the power of initiator concentration observed for frontal polymerization of AAM mixtures with SiO_2 and fine powders of chalk and diatomite. Thus, these experimental and theoretical studies have both scientific and methodical interest and are aimed at ascertaining the influence of initiator nature, micro- and nanosized fillers and their geometrical shape on macro kinetic peculiarities of vinyl monomers frontal polymerization, particularly, on n value and clarification of mechanism of such a typical phenomenon. The development of theoretical and experimental methods of frontal polymerization gives the possibility to get additional information concerning polymer chains initiation and termination at high temperatures typical for frontal polymerization processes.

KEYWORDS

- acryl amide
- adiabatic polymerization
- frontal free-radical polymerization
- heat waves propagation
- intercalated nanocomposites
- nano and micro particles
- order with respect to initiator

REFERENCES

1. Davtyan, S. P., Zhirkov, V. P., Wolfson, S. A. (1984). Problems of Non-isothermal Character in Polymerization Processes, *Rus Chem. Rev., 53*(2), 150–163.
2. Sevan Davtyan, Anahit Tonoyan, (2014). Theory and Practice of Adiabatic and Frontal Polymerization, Palmarium Academic Publishing. 660 p
3. Chechilo, N. M., Enikolopyan, N. S. (1975). Effect of the Concentration and Nature of Initiators on the Propagation Process in Polymerization, *Dokl. Phys. Chem. 221*, 392–394.
4. Chechilo, N. M., Enikolopyan, N. S. (1976). Effect of Pressure and Initial Temperature of the Reaction Mixture during Propagation of a Polymerization Reaction, *Dokl. Phys. Chem., 230*, 840–843.
5. Pojman, J. A., Willis, J., Fortenberry, D., Ilyashenko, V., Khan, A. (1995). Factors Affecting Propagating Fronts of Addition Polymerization: Velocity, Front Curvature, Temperature Profile, Conversion and Molecular Weight Distribution, *J. Polym. Sci. Part A: Polym. Chem. 33*, 643–652.
6. Sargsyan, A. G., Tonoyan, A. O., Davtyan, S. P., Schick, C., (2007). The amount of immobilized polymer in PMMA SiO$_2$ nanocomposites determined from calorimetric data. *European Polymer Journal, 8*, 3113–3125.
7. Davtyan, S. P., Berlin, Al. Al, Schick, C., Tonoyan, A. O., Rogovina, S. Z. (2009). Polymer Nanocomposites with a Uniform Distribution of Nanoparticles in a Polymer Matrix Synthesized by the Frontal Polymerization Technique *Russian Nanotechnologies. 4*(7), 489–498.
8. Davtyan, S. P., Berlin, A. A., Tonoyan, A. O. (2010). On Principal Approximations in the Theory of Frontal Radical Polymerization of Vinyl Monomers, *Russ. Chem. Rev. 79*, 234–248.
9. Davtyan, S. P., Hambardzumyan, A. F., Davtyan, D. S., Tonoyan, A. O., Hayrapetyan, S. M., Bagyan, S. E., Manukyan, L. S. (2002). The Structure, Stability of Auto waves during Polymerization of Co Metal-Complexes with Acryl Amide. *European Polymer Journal, 38*(12), 2423–2434.

CHAPTER 3

INFLUENCE OF pH ON THE STRUCTURING OF ZETAG 9014® TYPE OF CATIONIC POLYMERS

E. ALVARENGA,[1] L. HAYRAPETYAN,[2] S. HAYRAPETYAN,[2] E. GOVASMARK,[3] G.PIRUMYAN,[2] and B. SALBU[4]

[1]NIBIO, Norwegian Institute of Bioeconomy Research, Pb 115, N-1431, Ås, Norway, E-mail: Emilio.Alvarenga@nibio.no

[2]Yerevan State University, A. Manookyan St. 1, 0025, Armenia, E-mail: Scirec@mail.ru

[3]Energy Recovery Agency of Oslo County, PO-BOX 54 Mortensrud, NO-1215, Oslo, Norway, E-mail: Espen.Govasmark@ege.oslo.kommune.no

[4]Norwegian University of Life Sciences (NMBU), Dept. of Env. Sci., PO-BOX 5003, NO-1432, Ås, Norway, E-mail: Brit.Salbu@nmbu.no

CONTENTS

ABSTRACT

The structure of the cationic polymer ZETAG 9014® under the influence of changes in pH of the medium was studied in an aqueous solution. Cationic polymer type ZETAG 9014® is structured under the influence of pH changes. When reducing the pH of the medium, there is a clear trend, in which the average particle size reduces and the size distribution narrows. Conversely, at higher pH values (alkaline region), the average particle diameter broadens the PSD. It is important to note that the type of pH modifying agent (in particular, acid) can also affect the polymer structuring process. The chemical agents used to modify the pH of the suspension (medium and polymer) were sodium hydroxide (NaOH), potassium hydroxide (KOH), hydrated lime ($Ca(OH)_2$) and ammonium hydroxide NH_4OH.

3.1   INTRODUCTION

Dewatering of wastewater treatment sludges is commonly used after the coagulation and flocculation steps. Organic polyelectrolytes are applied for the enhancement of the flocculation of organic matter and thus improving the dewatering performance of the solid–liquid separation units. The latter are usually found in the wastewater industry as either sedimentation basins, filter or centrifugation steps of the process. A high water content prevails among sludge, which are obtained from different separation processes. The dry matter content must hence be increased in order to minimize transportation costs. Polymers have hence a relevant role in this sludge conditioning [1].

When compared with inorganic coagulants, there are advantages when using polymers in water treatment. Among these, there are for instance, lower coagulant dose requirements, a smaller volume of produced sludge, and a slight increase in the ionic load of the treated water, a reduced level of coagulants in the treated water and cost savings of up to 25–30% [1–5].

Water processing either of raw or wastewater, normally involves physicochemical processes, in which suspended solids and colloids agglomerate due to destabilization of the suspension by means of coagulation and flocculation. The use of polymers for the aggregation of particles in water and wastewater treatment could have different mechanisms: (i) polymer bridging; (ii) charge neutralization, including 'electrostatic patch' effects; (iii) depletion flocculation [15]. The water treatment chemicals including flocculants and coagulants are particularly designed for the solid–liquid separation processes in the industrial and municipal wastewater treatment sector [3–5]. Cationic polyacrylamide among other polymeric flocculants has been extensively used in the sludge dewatering process.

Molecular weight (MW) is known as the main feature of polymeric flocculants and values vary broadly from a few thousand up to tens of millions of grams per mol. On the other hand, polyelectrolytes have the charge density (CD) as their main characteristic, which can be experimentally determined by the technique of colloid titration [8].A convenient measure of the "diameter" of a polymer molecule is the root mean square (rms) value of the end-to-end distance, r. The rms is roughly estimated (in nm) by $0.06M^{1/2}$ for many common non-ionic polymers, where M is the MW [3,6]. For M = 1 million, the rms end to-end distance value would be about 60 nm approximately. The extent of the random coil or the most likely configuration depends on the interaction between polymer segments. When the coil expands,itis limited by the repulsion between segments. This fact becomes even more remarkable for polyelectrolytes in which the segments are charged. Important ionic strength effects could expand considerably or not the polymer coil [3]. However, it is not well understood how the ionic strength affects the structuring of the polymer when different cations are present in the water. There is a knowledge gap as well on the effect of changes of chemical properties (i.e., pH) on the polymer structure in a suspension. Moreover, a polymer chain in a random coil configuration effectively occupies a much larger volume than the equivalent number of isolated monomer units. This is due to the aqueous solution which remains within the coil configuration. For this reason, polymer solutions can scatter light and show a significantly higher viscosity than water [7].

Among the varieties of cationic polymers available, the ones that contain quaternary ammonium groups have a formal positive charge independently of pH. These are called strong electrolyte polymers as reviewed in detail by Bolto[9]. One example of such cationic polymers are copolymers of acrylamide and the cationic ester acryloxyethyltrimethyl-ammonium chloride, which are extensively used in the water industry [10].

In alkaline conditions, hydrolysis of the ester groups is enhanced and consequent loss of cationic charge has been found to be CD and pH dependent [3]. Degradation can occur to certain extent even at pH 6 for polymers having a CD of 24%, with a half-life of 24 h at pH 7 and 0.25 h at pH 8.5

[11]. However, stability of the polymer is reached at pH 4. There is no hydrolysis of acrylamide units up to pH 8.5. Smith-Palmer et al. [12] have investigated without specifying the pH of the suspension, ester hydrolysis on less ionic polymers having a CD of 6% with a half-life of 22 months. Charge densities of 30% have shown a significant decrease in hydrolysis rate, particularly for pure solutions. Whereas for pH levels above 8, a substantial degradation is caused that is not observed for a polymer of 100% charge [13]. Anionic carboxylate groups are formed in addition to the loss of cationic sites. There is hence a change in the chain conformation as a result of hydrolysis, which reduces the chain extension and makes the polymer less efficient as a flocculent. However, for the homopolymers, the corresponding methacrylate formed is resistant towards hydrolysis degradation [14].

The goal of this study was to prepare to investigate the changes in the structure of the polyelectrolyte ZETAG® 9014 in an aqueous solution with a polymer concentration of 0.3 g/l in order to see the behavior of the particle size, particle size distribution (PSD) and turbidity in a broad alkaline range of pH. Such pH was modified with different bases (weak and strong), which increased the aggregation of particles in the polymeric suspension in order reach the iso-electric point at a certain pH value above 7.

3.2   MATERIALS AND METHODS

3.2.1   CONDITIONING OF THE SUBSTRATE

The substrate that was taken into account was deionized water. The polymer (ZETAG® 9014) content added to the substrate was 300 mg and dispersed in 1 liter of water. This polymer was produced by BASF SE(Germany). The substrate became a turbid suspension by addition of the polymer. The pH was modified by means NaOH, KOH, $Ca(OH)_2$ and NH_4OH addition by volume. The $Ca(OH)_2$ was added to the suspension as a slurry due to its low solubility in water. The rest of the bases were completely solubilized in water prior to the addition to the suspension. The concentration of the reagents solutions was 0.1 N. The pH was measured at room temperature (25°C) with a pH meter Thermo Orion model Dual Star after mixing the polymer with the substrate with an exposition time of 1 min in a magnetic stirrer.

3.2.2   TURBIDITY, ZETA POTENTIAL, AND SIZE DETERMINATION

The turbidity (FTU) was measured after exposition with acids or the bases in a Hanna turbid meter model HI 93703. In addition, an aliquot of 1 mL was taken and was diluted 10 times with the substrate. The purpose was to determine the size distributions in the diluted suspension as well as the zeta potential. For that purpose, a Mavelrn Zeta SizerNano Series instrument was utilized. A 12 mm polystyrene cell was used to expose the samples to the light beam for the further dynamic light scattering. A characterization of the particle sizes was intended due to the turbid suspension obtained by the addition of the polymer to the substrate.

3.3   RESULTS AND DISCUSSION

3.3.1   pH ADJUSTMENT WITH 0.1 N KOH

Figure 3.1 shows the PSD of an aqueous solution of a cationic polyelectrolyte ZETAG 9014®. PSD measurements were carried out immediately after dilution of the polyelectrolyte by means of deionized water to 0.3 g/L. Apparently, the system is not fully stabilized as shown in Figure 3.1–3.1a; thus there is not precise reproducibility of PSD. In this case, there are bi (tri) modal structure distributions ranging from 80–105 nm and structured particle – 600–900 nm. It is suggested that the initial size of the polymer particles is around 100 nm. The second size range from 600–900 nm suggest shows then structured particles 6–9 times higher than the initial size.

In the initial state, cationic polyelectrolyte ZETAG 9014® diluted in water has bi (tri) modal PSD as seen from see Figure 3.1a. Moreover, it has 491 FTU of turbidity at pH 4.06 (25°C). The third peak of PSD has the range 23.5 ± 6.5 nm most likely due to the destruction of the polymer chains as a result of dilution, which after a time the newly form aggregates of bigger size.

It is known that the most important characteristic of a polymer floccu-lants is their MW and in the case of polyelectrolytes the CD as well. Usually the MW of these systems ranges from a few thousand to tens of millions Dalton (low molecular weight – <105, middle – 105–106 and high —> 106). In aqueous solution, the polymer solutions often take a random coil configu-ration. For high MW polymers, the size of these coils is about 100 nm [3, 6].

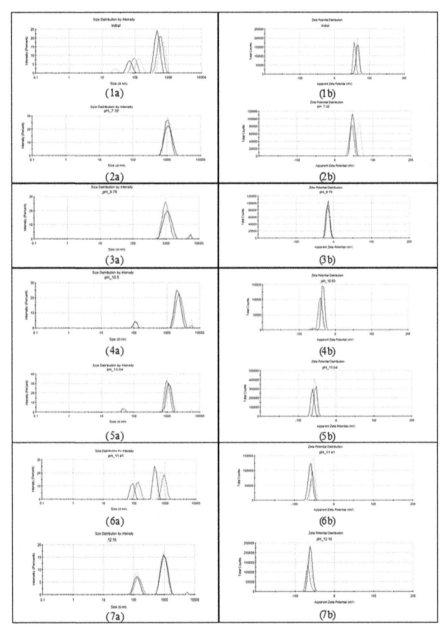

FIGURE 3.1 Particle size distribution of the aqueous solution (0.3 g/L) and charge distribution on the surface of the particles of cationic polyelectrolyte ZETAG9014® at different values of pH. The pH was adjusted by means of KOH 0.1 N.

In order to increase the pH, KOH is added and the resulting pH is raised from 4.06 to 7.32. From Figures 3.1–3.2a, it is shown that the PSD becomes unimodal. In this case, as a result the turbidity values are reduced from 491 to 337 FTU as seen in Table 3.1. This can be explained by the average particle size which is also increased from 580.7 nm (Figure 3.1) to 1057 nm (Figure 3.2). There is a sharp decrease in the number of particles in the resulting structured system.

On the Figure 3.2b, it is presented the distribution of charges on the surface of the particles of the cationic polyelectrolyte ZETAG 9014® at pH 7.32. From Figure 3.2b, it follows that the z-potential on the surface of the polymer particles is 43.6±5.18 mV.

Figure 3.3a shows the PSD of the cationic polyelectrolyte ZETAG 9014® at pH 9.76. At pH 7.32, a monomodalPSD of 1057±200.8 nm does not change considerably in comparison with 1003±202.7 nm at pH 9.7. Minor bimodality was observed for larger particles (5–6 microns). This is likely the result of a fluctuating cluster. The turbidity of the solution is increased from 337 to 442 FTU as shown in Table 3.1, which most likely occurs due firstly to some reduction in the average particle size(from 1003 nm to 1057 nm) and secondly to a slight increase in the standard deviation from the average size of particle sizes (from 200.8 to 202.7 nm). It is important to note that in the pH range 7.32–9.76 the structuring of the polyelectrolyte does not undergo drastic changes and for instance this system in this pH range is quite stable. Moreover, in this pH range there is a change of sign of the surface charge (up from 43.6 to –14.8 mV) that explains such stability as shown in Figure 3.3b and Table 3.2.

A further increase in pH from 9.76 to 10.50 leads to the formation tri-modal structuring in 111.2 ± 11.78 nm (peak intensity 9.2%), 2425±507.8

TABLE 3.1 Turbidity of the Polymer Suspension when the pH was Adjusted with KOH 0.1 N

pH (20°C)	Turbidity (FTU)
7.32	337
9.76	442
10.50	385
11.04	320
11.41	326
12.16	303

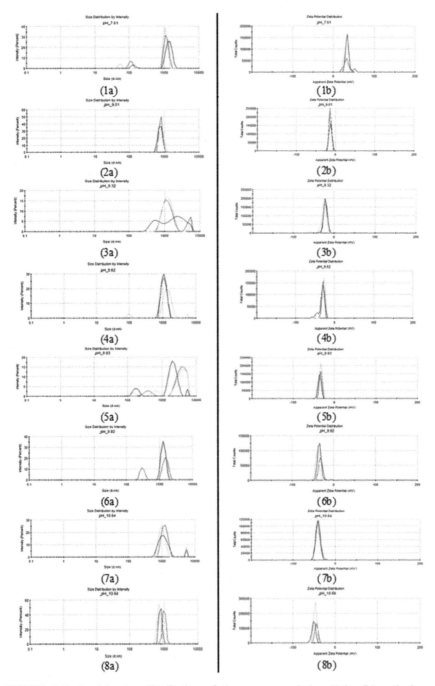

FIGURE 3.2 Particle size distribution of the aqueous solution (0.3 g/L) and charge distribution on the surface of the particles of cationic polyelectrolyte ZETAG 9014® at different values of pH. The pH was adjusted by means of NH₄OH.

nm (peak intensity 89.0%) and 5560 nm (peak intensity 1.8%). Increasing the average particle diameter of up to 2425 nm (see Figure 3.4a).

The appearance of a peak in the PSD at 112 nm in Figure 3.4a indicates that in this range of pH, there is some de-structuration of agglomerates to previous initial particle size(see Figure 3.3a) of the polyelectrolyte followed by the formation of larger aggregates. There is a decrease of turbidity values

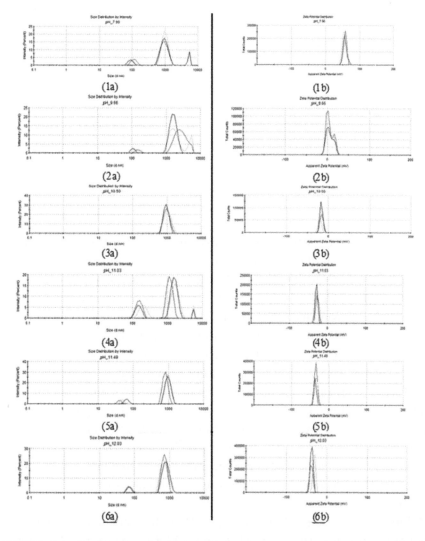

FIGURE 3.3 Particle size distribution of the aqueous solution (0.3 g/L) and charge distribution on the surface of the particles of cationic polyelectrolyte ZETAG 9014® at different values of pH.pH was adjusted by means of Ca(OH)$_2$.

TABLE 3.2 Distribution of Particle Intensity and Z-Potential for the pH Adjustment of the Polymer Suspension with KOH 0.1 N

KOH							
Size Distribution by Intensity				**Zeta Potential Distribution**			
Size (d,nm)	**% Intensity**	**St. Dev (d,nm)**	**St. Dev/ Size, K**	**pH**	**Mean (mV)**	**Area (%)**	**St. Dev (mV)**
580.7	72.8	103.3	0.178	4.34	56.8	100	3.35
94.56	27.2	15.39	0.163				
1057.0	100	200.8	0.190	7.32	43.6	100	5.18
1003.0	100	202.7	0.202	9.76	−14.8	100	3.98
2425.0	89.0	507.8	0.209	10.50	−41.4	98.3	4.93
112.0	9.2	11.78	0.105		−66.8	1.7	3.86
5560.0	1.8						
1029	100	159	0.155	11.04	−53.8	100	3.93
947.1	56.6	151.3	0.160	11.41	−54.9	100	4.39
136.0	43.4	23.11	0.170				
1030.0	70	245.9	0.238	12.6	−67.1	100	4.86
142.0	30	31.74	0.223				

from 442 to 385 FTU by increasing average particle diameter. It follows from Figure 3.4b on the distribution of charges on the surface of the poly-electrolyte ZETAG 9014® that at pH 7.32, the z-potential on the surface of the polymer particles is 41.4±4.93 mV. Bimodality occurred around the 50 nm size by increasing the pH to 11.04. This can be explained by the destruction of some of the polyelectrolyte molecules at high pH values.

There is a change in the surface charge of −41.4 to −53.8mV in the transition from pH 10.5 to 11.04 as seen from Figure 3.5b. A further increase in pH leads to uncontrolled deagglomeration of particles and it is observed a pronounced bimodal structuring. However, this system appears to have some labile nature and no reproducibility of measurements as observed at pH 11.04 (see Figure 3.5a) and at pH 12.16. The system is more stable due to an average particle size that becomes larger (1030 nm and 142 nm). Moreover, there is a decrease in the value of turbidity to 303 FTU as seen from Table 3.1.

With the increase in pH from 9.76 to 10.5, the z-potential continues to decrease almost linearly and reaches from −14.8 to −41, 4 mV. The reproducibility of the measurements is deteriorating as shown in Figure 3.4b. The

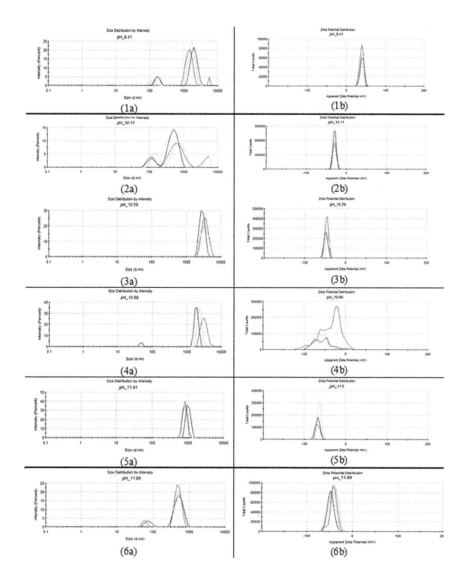

FIGURE 3.4 Particle size distribution of the aqueous solution(0.3 g/L) and charge distribution on the surface of the particles of cationic polyelectrolyte ZETAG9014®at different values of pH. The pH was adjusted by means of NaOH.

z-potential continues to decrease with a further increase in pH (from 10.50 to 11.04) reaching values from −41.4 to −53.8 mV as seen from Figure 3.5b and Table 3.2.

In the pH range from 11.04 to 11.41 (Figures 3.5 and 3.6), the z-potential continues to decrease to a value difference of −1.1 at a difference of pH of

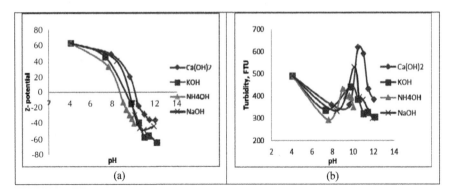

FIGURE 3.5   (a) Variation of the z-potential of the polymer suspension in water in all the pH range investigated for each base;(b) Changes in turbidity of the polymer suspension in water in all the pH range investigated for each base.

0.37 as seen in Table 3.2. The difference of z-potential is 11.9 mV when the difference of pH is 10.5–11.04 = 0.9. This means that at a higher pH changes, the z-potential becomes lower and tends to stabilize.

3.3.2   pH ADJUSTMENT WITH 0.1 N NH₄OH

As compared with the starting polymer PSD(Figure 3.1a), aggregates can be formed due to uneven distribution of surface charge on the particles. The degree of structuring of the polyelectrolyte or agglomerated particles in the initial state is 72.58%, whereas at higher NH_4OH doses to a pH 7.61, the degree of structuring increases and the background of the surface charge decreases to 32.2 mV and 83.5% as seen from Figure 3.1a. With further increase in pH from 7.61 to 9.01, the polyelectrolyte becomes fully structured as seen from Figure 3.2a. Thereby, the z-potential is reduced to −13.8 mV as shown in Figure 3.2b and hence this pH range is within the isoelectric point.

Some fluctuation pattern is observed at pH 9.32 where there is the most labile state of the polyelectrolyte. There is a dipolar structuring but PSD is very broad. A redistribution of charge occurs most likely, on the surface and as a result of this phenomena there is a broadening of the PSD as seen from Figure 3.3a and the z-potential is reduced to −21.2 mV as shown in Figure 3.3b.

Further increase in pH 9.62 leads to the complete re-structuring of the polyelectrolyte having an average particle diameter of 1102±214.7 nm as seen from Figure 3.4a. De-structuring occurs when the polyelectrolyte is exposed to ammonia and thus by increasing the pH. Apparently, this behavior occurs due to the fact that NH_4OH is a weak base.

The greatest structuring is observed at pH 9.83. In this case, the particle size is up to 3813 ± 1058 nm(Figure 3.5a). In this case, the z-potential is reduced to −33.3mV as shown in Figure 3.5b. Afterwards, in the transition from pH 9.83 to pH 10.04 (Figures 3.6 and 3.7), a sharp de-structuring can be seen most likely due to secondary structures and the particle size obtained is 1230 ± 244.8 nm(see Figure 3.7a). Thereby, the z-potential decreases to −40.4mV (see Figure 3.7b). There is a narrowing of PSD and the value obtained is 1023 ± 106.1 nm by increasing the pH to 10.68 as shown in Figure 3.8a. In this pH level, the z-potential decreases to −44.2mV as seen from Figure 3.8b.

3.3.3   pH ADJUSTMENT WITH 0.1 N CA(OH)$_2$

A trimodalPSD is achieved by increasing of the pH of the system by means of a slurry of lime to 7.9 as shown in Figure 3.1a. The intensity of the third peak is 11%. The intensity of the PSD starting polymer is about 26%. The greatest structuring of the system is observed at pH 9.66 as shown in Figure 3.2. The average size is 2809 ± 1108 nm. Some shift is observed in the initial polymer particles to larger sizes (instead of 115 nm these have 150.8 nm). Apparently, such change occurs as a result of the redistribution of particle aggregates as seen from Figure 3.2a. Such structuring shown in Figure 3.2b can be a result most likely of the occurrence of the isoelectric point at that pH. It is important to note that there is a bimodal distribution z- potential with 1.12 mV and 15.3 mV with an intensity of 68% and 32% respectively. It is assumed that the reason for this structuring shows a bimodal PSD, since it is at this pH that the system provides the polymer particles with such different z-potential values. Therefore, in these circumstances such structuring may occur. In other words, this behavior of the suspension is most probably closer to the isoelectric point rather than to the formation of structures with relatively large aggregates.

A further increase of the pH to 10.50 leads to a monomodalPSD in which the average diameter is 989.5 ± 200.6 nm. Moreover, the standard deviation of the particle size is reduced more than 5 times (from 1108 nm to 200.6 nm) as seen from Figure 3.3a. Furthermore, the PSD becomes narrower. The system has already passed through the isoelectric point of the surface charge is equal to −16.2 mV, the particles responsible for the second peak PSD again shift to 989.5 nm, almost like at the pH 7.90 (955.9 nm).

Increasing the pH to 11.03, leads to de-structurization of agglomerates with an observed peak area of 31.2% intensity and 155±31.4 nm of average particle size. Apparently there is a partial de-agglomeration to form stable dimers (155±31.4/2 = 98 nm, which is in accordance with the size of the original polymer).

A further increase in the pH of 11.03–11.49 (Figures 3.3–3.4 and 3.3–3.5) leads to some decrease in the average particle size of 1164±217.5 nm to 816.7±130.6 nm (about 1.5 times). There is also a reduction in the size of fine particles of 155±31.2 nm to 41.21±4.82 nm. Hence, the latter indicates the de-aggregation of the polymer molecules.

Since the polymer is a polyelectrolyte, it is possible to change the size of electrolytes via coil particles. For instance, if the initial size of the polymer is about 100 nm, then the increase in pH can have a denser deformation of the polymer chains and the size of the coil can be reduced. The question that arises is if 63.38 nm (peak with the lower intensity in Figure 3.6a) results from this deformation or partially it is the outcome of de-structuring by bases.

3.3.4   pH ADJUSTMENT WITH 0.1 N NAOH

In the Figure 3.1a is represented the PSD of the polyelectrolyte at pH 8.41 in which the degree of structuring is 79.0% of intensity. A further increase leads to a sharp pH of 10.11 and hence de-structuring of the particles occurs. The 1452 nm particle size is reduced to 666.3 nm as presented in Figure 3.2a. Such behavior of the polyelectrolyte in suspension is not observed with other bases including KOH. It is assumed that the structure of the already formed secondary particle size at pH 10.11, 663±285 nm formed larger particles (3535±740.4 nm) at a pH of 10.98 as seen from Figure 3.4a. The initial particles increased 5 times approximately. When the difference of pH is 0.48 (from pH 10.11 to 10.59) there is not tendency of de-agglomeration as shown in Figures 3.3 and 3.4. On the contrary, by increasing the pH in this range, enlargement in size holds the particles rather than releasing them to the water as shown in Figure 3.3a. From Figure 3.3a, it also follows that, after the previous de-agglomeration it takes place unusual structuring of bigger particles (average size grows up to 3535 nm). Moreover, this structuring behavior is observed when using ammonia at pH 9.83 (average particle size increases to 3813 ± 1058 nm).

The de-structuring process begins at a pH of 10.96 (Figure 3.4a). As a sign of de-structuring, associated polyelectrolyte molecules appear with a PSD around 50 nm. It is most likely at such pH level that a partial de-agglomeration of polyelectrolyte molecules occurred. The degree of de-agglomeration is 4% in terms of intensity as seen from Figure 3.3a. The enhancement of the secondary de-agglomeration occurs at a pH of 11.41, for instance the particle size decreases from 3134 ± 675 nm to 823 ± 110.4 nm (Figure 3.5) and further on to 499.5±95.42 nm at the pH level of 11.89 as seen from Figure 3.6a.

3.3.5   VARIATION OF THE Z-POTENTIAL

The variation of the z-potential is shown in Figure 3.5a. The agglomeration occurs when the system is tends to stability for all the pH adjustments between pH 4 and 10. However, there is a cationic effect associated to such trend. Stability and hence agglomeration is achieved by increasing the pH to the respective isoelectric points and such tendency is reflected in Figure 3.5b. The turbidity of the system for each alkali has a minimum when there is stability and aggregation of the polymer particles occurs. Otherwise, an increase of the turbidity takes place at high pH values for all the bases as an indicator of de-agglomeration.

3.4   CONCLUSIONS

In an alkaline environment, cationic polyelectrolyte ZETAG® 9014 is relatively more stable, and only when using calcium hydroxide has been some degradation of the polymer molecule. By using all types of alkalis, there is an increase of the agglomerate size with respect to the initial particles. Decomposition of the base polymer particles is random and thus a change in the z-potential and its standard deviation is more moderate.

The zero charge depends on the type of cation. The lowest pH in which the zero charge on the surface is observed was with NH_4OH whereas the highest value observed was with $Ca(OH)_2$.

There was an increase in turbidity of the solution of the cationic polyelectrolyte ZETAG® 9014 when getting away from the isoelectric points at high pH values. This fact is attributed to de-agglomeration of the polymer particles that destabilize the suspension with a decrease of the average particle size.

ACKNOWLEDGMENTS

The financial support of the Norwegian Research Council Grant # ES 459248/0 is gratefully acknowledged.

KEYWORDS

- **flocculation**
- **cationic polymer**
- **particle size**
- **particle size distribution**
- **turbidity**
- **z-potential**

REFERENCES

1. Bolton, B., (2006). Interface Science in Drinking Water Treatment: Theory and Applications, Chapter 5: Coagulation and Flocculation with Organic Polyelectrolytes, in: Newcombe, G., Dixon, D., (Eds.), Elsevier Ltd. pp. 63.
2. Jones, L. D., (1986). Wastewater Flocculating Agent, US Patent 4631132 A, USA.
3. Bolto, B., Gregory, J., (2007). Organic polyelectrolytes in water treatment. Review. *Water Research 41*, 2301–2324.
4. Rout, D., Verma, R., Agarwal, S. K., (1999). Polyelectrolyte treatment—an approach for water quality improvement.*Water Sci. Technol. 40*(2), 137–141.
5. Nozaic, D. J., Freese, S. D., Thompson, P., (2001). Long term experience in the use of polymeric coagulants at Umgeni Water. *Water Sci. Technol.: Water Supply 1*(1), 43–50. NSF International. Certified product listings. ANSI/NSF Standard 60. Washington, NSF, 2001.
6. Napper, D. H., (1983). *Polymeric Stabilization of Colloidal Dispersions.*Academic Press, London.
7. Scott, J. P., Fawell, P. D., Ralph, D. E., Farrow, J. B. (1996). The shear degradation of high-molecular-weight flocculant solutions.*J. Appl. Polym. Sci. 62*(12), 2097–2106.

8. Kam, S. K., Gregory, J. (2001). The interaction of humic substances with cationic poly-electrolytes.*Water Res. 35*(15), 3557–3566.

9. Bolto, B. A. (1995). Soluble polymers in water-purification.*Progr. Polym. Sci. 20*(6), 987–1041.

10. Baade, W., Hunkeler, D., Hamielec, A. E. (1989). Copolymerization of acrylamide with cationic monomers in solution and inverse micro suspension.*J. Appl. Polym. Sci. 38*(1), 185–205.

11. Ekberg, R., Wagberg, L., (1989). Hydrolysis of cationic polyacrylamides.*J. Appl. Polym. Sci. 38*(2), 297–304.

12. Smith-Palmer, T., Campbell, N., Bowman, J. L., Dewar, P., (1994). Flocculation behavior of some cationic polyelectrolytes.*J. Appl. Polym. Sci. 52*(9), 1317–1325.

13. Lafuma, F., Durand, G., (1989). C-13 NMR-spectroscopy of cationic copolymers of acrylamide.*Polym. Bull. 21*(3), 315–318.

14. Van de Wetering, P., Zuidam, N. J., van Steenbergen, M. J., van der Houwen, O. A. G. J., Underberg, W. J. M., Hennink, W. E., (1998). A mechanistic study of the hydrolytic stability of poly(2-(dimethylamino)ethyl methacrylate). *Macromolecules 31*(23), 8063–8068.

15. Jang, W., Nikolov, A., Wasan, D. T., (2004). Effect of depletion force on the stability of food emulsions.*J. Dispers. Sci. Technol. 25*(6), 817–821.

CHAPTER 4

CHAPTER 4

AROMATIC AZIDES AND PENTAZENES IN MULTISTEP RADICAL POLYMERIZATION

V. G. SYROMYATNIKOV, D. G. VYSHNEVSKY, and A.YU. KOLENDO

Department of Macromolecular Chemistry, Taras Shevchenko National University of Kyiv, 60 Volodymyrska Street, 01033, Kyiv, Ukraine, E-mail: svg@univ.kiev.ua

CONTENTS

ABSTRACT

Products of transformation of nitrene radicals formed during decomposition of aromatic azides can initiate radical polymerization. This fact has been found firstly for carbonyl-containing azides, such as derivatives of acetophenone, xanthone, fluorane, anthraquinone, phenylphthalimides, etc. Diazides are more effective than monoazides as a rule. For some diazides double step mechanism of photoinitiation due to the direct energy supply (intramolecular transfer) for differently positioned azidogroups with different values of photolysis quantum yields have been recently found.

Reaction of alkyl diamines with diazonium salts passes through two steps. Monopentazenes were found to be obtained as the first step of the reaction and after as the second step new type of organic compounds – bis-pentazadienes, can be obtained. Mono- and bis-pentazadienes can produce free radicals of several kinds due to cleavage of chromophore pentazadiene group. Therefore, they can initiate multistep polymerization processes.

4.1 INTRODUCTION

Such unstable light-sensitive organic compounds as triazenes, azides, pentazadienes etc., contain chains of nitrogen atoms. As well as diazonium salts they have bonds type Xe-O, formed owing to the interaction of completely occupied atom orbital of xenon and partially occupied orbital of triplet oxygen. Such molecules are characterized by considerable activity in lower excited states or on the transfer of an additional electron into them. It provides the necessary conditions for the dissociation of the molecule [1]. This hypothesis was approved by the results of numerous quantum calculations.

4.2 RESULTS AND DISCUSSION

Nitrogene atoms can form chains containing from 2 to 6 atoms. Such compounds are commonly unstable, high energetic and reasonably explosive. But some of them are more stable and can be used in physical experiments and practice due to their sensitivity to light exposure and heating [2]. We had synthesized and studied series of aromatic azides (3 nitrogen atoms in the chain) with such equilibrium of bonds in azido groups:

$$-N=\overset{+}{N}=N \quad \longleftrightarrow \quad -N-\overset{+}{N}\equiv N \quad \longleftrightarrow \quad -N=N=\overset{+}{N}$$

$$\text{I} \qquad\qquad\qquad \text{II} \qquad\qquad\qquad \text{III}$$

Aromatic azides can be obtained by the action of sodium azide on corresponding diazonium salts. They are photosensitive and form biradicals – nitrenes (singlet or triplet) [3].

$$ArN_3 \rightarrow ArN: + N_2$$

The last ones can be transformed into active free radicals, which can cause the radical polymerization. This fact we had found out firstly [2, 4–9] for carbonyl-contained azides, such as derivatives of acetophenone, benzophenone, xanthone, phthalimide, fluorine, anthraquinone, phenylphthalimides[4–7]. All synthesized azides are able to initiate a radical polymerization of vinyl monomers. Their mixtures with ordinary photoinitiators were found to be more efficient than only last ones [8].

Diazides are more effective than monoazides as a rule [4]. For some diazides (ex. 4-azido-N-4'-azidophenylphthalimide [7]) we have recently found double-step mechanism of photo initiation due to the direct energy supply (intramolecular energy transfer) [9] for differently positioned azido groups with different values of photolysis quantum yields. For MMA the polymerization degree here is about 1000 (Figure 4.1).

Two steps of polymerization are obviously seen on the curve (2) in Figure 4.1. From the curve 1 we can observe that the polymerization kinetics has a complex character, because of simultaneous formation of several products at the nitrene decomposition under irradiation. Some of them inhibit the polymerization process thus resulting in low conversions. Using as a monomer 1,6-hexandiol diacrylate (HDODA) we had found that azides have energies of activation lower than well-known cross-linker IRGACURE-1700 and can be applied in very small concentrations [10]. Experiments were made by method of differential scanning photocalorimetry(Photo-DSC) at

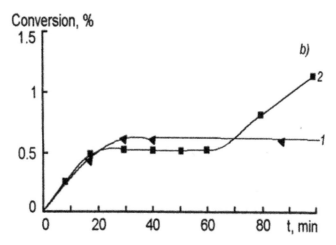

FIGURE 4.1 Polymerization kinetics curves for 15% DMF solutions of MMA in the presence of 2% of 4-azidophenylphthalimide (1),4-azido-N-4'-azidophenylphthalimide (2).

the University Montpellier 2 [11]. Investigated azides appeared themselves as good photo initiators of radical polymerization of vinyl monomers and their efficiency can be significantly increased by addition of electron donors.

Monopentazenes or pentazadienes can be obtained by action of monoal-kylamines on diazonium salts [12]. They are also photosensitive and during photolysis reaction can form free radicals [13]:

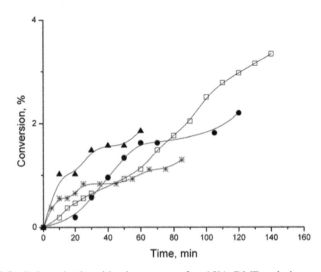

We had firstly discovered that such compounds can initiate the radical polymerization (Figure 4.2) of vinyl monomers. Mechanisms of correspond-ing processes were proposed and discussed. Polymerization activity was found to be correlated with values of their photolysis quantum yields [14, 15].

All polymerization kinetic curves support the common idea on multistep radical polymerization of MMA in a presence of all kinds of pentazadienes with different structures.

We also compared activity of obtained pentazenes with one of commonly used cross-linkers IRGACURE-1700.

FIGURE 4.2 Polymerization kinetics curves for 15% DMF solutions of MMA in the presence of 1% of 1,5-bisphenyl-3-(ethan-2-ol)pentazadiene-1,4)(▲), 1,5-bis(4-butyloxycarbonylphenyl) – (*), 1,5-bis-(4-methoxyphenyl)—3-methyl pentazadiene-1,4 (□), 1,5-bisphenyl-3-methyl pentazadiene-1,4 (●).

We can see that IRGACURE-1700 has activation energy higher than pentazenes therefore last ones are more effective cross-linkers (Table 4.1).

Recently, as prolongation of these investigations we have studied an interaction of diazonium salts with diaminoalkanes. Monopentazenes were found to be obtained as the first stage of the synthesis and after as the second stage we can obtain new compounds – bis-pentazadienes with the common structure as below:

TABLE 4.1 Activation Energies for Polymerization of HDODA in Presence of Pentazenes and IRGACURE 1700

Compound	E_a, kJ/mol	E_a, kcal/mol	Coefficient of correlation
	9.6	2.3	0.952
	5.6	1.3	0.999
	8.5	2.0	0.978
IRGACURE 1700	10.1	2.4	0.982

Structures of all products were proved by analytical and spectral methods. Photochemical studies and experiments on applications of obtained products as photocrosslinkers and photoinitiators have been carried out. During the photolysis they can produce free radicals of several kinds, due to cleavage of one (or both) pentazadiene groups and radicals transformations. Therefore, they can involve multistep polymerization processes.

In adsorption spectra of bis-pentazadienes we can see two adsorption bands at ~300 nm (corresponds to n→π* **transition**) and ~360–380 nm (corresponds to π→π* **transition**). During photolysis reaction in UV light (wavelength 354 nm) significant decreasing of adsorption in both bands can be observed (Figure 4.3).

The perspectives of their applications in the industry were evaluated as very high ones.

4.3 CONCLUSIONS

Some aromatic azides, diazides and pentazadienes were synthesized and their photochemical properties have been investigated. All of them are able to initiate radical polymerization of vinyl monomers. It was discovered that in case of diazides and pentazadienes multistep radical polymerization kinetics are

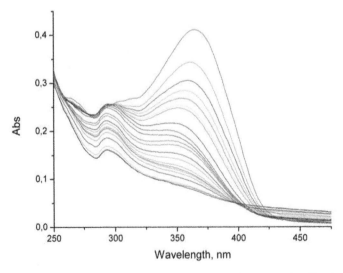

FIGURE 4.3 Adsorption spectra of 1,2-bis(1,5-di-m-tolylpentaaza-1,4-dien-3-yl) ethane in CHCl₃ recorded every 1 second under irradiation of UV-luminescent lamp (λ = 354 nm).

observed. Polymerization rate on each step depends on quantum yield of photolysis of differently positioned azidogroups (in case of diazides) and activity of initiating radical formed after photolysis (in case of pentazadienes).

KEYWORDS

- **aromatic azides**
- **bis-pentazadienes**
- **diazides**
- **multi step radical polymerization**
- **pentazadienes**
- **photocrosslinking**
- **photolysis**

REFERENCES

1. Kondratenko, P., Lopatkin, Yu. Yu, Kondratenko, N. P. (2002). Molecules with bonds such as Xe-O between moieties and their application. *Materials Science., 4*, 93–99.
2. Novikova, O. O., Syromyatnikov, V. G., Avramenko, L. F., Kondratenko, N. P., Kolisnichenko, T. M., Abadie Marc, J. M. (2002). Photo initiation ability of some pentaaza-1, 4-dienes. *Materials Science, 4,* 19–28.
3. Gritsan, N. P., Pritchina, E. A. (1992). The mechanism of photolysis of aromatic azides. *Russian Chemical Reviews, 61,* 500–516.
4. Novikova, E., Kolendo, A., Syromyatnikov, V., Avramenko, L., (1997). Azides of xanthone as photo initiators of methyl methacrylate radical polymerization. *Ukr. Khim. Zhurn., 63*(9), 65–68.
5. Avramenko, L., Yeshchenko, N., Kondratenko, P., Novikova, E., Syromyatnikov, V., (2005). Photochemical properties and mechanisms of photodissotiation of phthalimide azides. *Ukr. Khim. Zhurn., 71*(3), 64–70.
6. Novikova, E., Kolendo, A., Syromyatnikov, V., Avramenko, L., Prot, T., Golec, K., (2001). Azidoxanthone and azidofluorane photo initiated polymerizations of methyl methacrylate. *Polimer, 46*, 406–413.
7. Kolendo, A., Avramenko, L., Grigorenko, T., Novikova, E., Olkhovik, L., Kolisnichenko, T., Mogilevich, R., Serov, V., Prot, T., Golec, K., (2000). Reactivity of carbonyl-containing azides in radical polymerization of methyl methacrylate. *Functional Materials, 7,* 667–670.
8. Brzozowski, Z., K, Zadrozna, I., Kowalczyk, T., Syromyatnikov, V. G. (1995). Studies on the synthesis of polyarylates with allylic side groups. Pt.2. Photo cross-linking of polyarylates. *Polymers and Polymer Composites, 3*, 421–426.

9. Syromyatnikov, V. G., Yashchuk, V. N., Ogulchansky, T. Y., Savchenko, I. A., Kolendo, A. Y. (1996). Some light-sensitive imide molecular systems with the determined functional properties. *Mol. Cryst. Liq. Cryst., 283*, 293–298.

10. Abadie, M. J. M., Novikova, O. O., Voytekunas, V., Yu, Syromyatnikov, V. G., Kolendo, A., Yu. (2003). Differential scanning photocalorimetry studies of 1, 6-hexanedioldiacrylate photo polymerization initiated by some organic azides. *J. Appl. Polym. Sci., 90*, 1096–1101.

11. Abadie, M. J. M., Appelt, B. K. (1989). Photocalorimetry of light-cured dental composites. *Dental materials., 5*(1), 6–9.

12. Baindl, A., Lang, A., Nuyken, O., (1996). High and low molar mass 3-alkyl-l, 4-pentazadienes. Synthesis and photolysis. *Macromol. Chem. Phys., 197*, 4155–4171.

13. Kondratenko, N. P., Avramenko, L. F., Kondratenko, P. A., Syromyatnikov, V. G. (2000). Photodissociation mechanism of pentazene molecules. *Functional Materials, 7*, 613–618.

14. Syromyatnikov, V. G., Kondratenko, N. P., Novikova, E. A., Avramenko, L. F. (2003). Photo initiation of vinyl polymerization with aromatic pentazenes. *Polymer Science, 45*, 32–37.

15. Yeshchenko, N. P., Syromyatnikov, V. G. (2005). Pentazadienes as new photo initiators in the development of new materials. *Mol. Cryst. Liq. Cryst., 427*, 181–189.

CHAPTER 5

SYNTHESIS AND ADSORPTION PROPERTIES OF 4-AMINOSTYRENE AND METHACRYLIC ACID COPOLYMER, IMMOBILIZED IN SITU ON SILICA SURFACE

E. YANOVSKA,[1] L. VRETIK,[1] D. STERNIK,[2] O. KYCHKIRUK,[3] O. NIKOLAEVA,[1] and M. MELNYK[1]

[1]*Chemistry Faculty, Kyiv National Taras Shevchenko University, 12 L. Tolstogo Str., 01033 Kyiv, Ukraine, E-mail: lvretik@gmail.com*

[2]*Maria Curie-Skłodowska University, pl. Maria Curie-Skłodowskiej 3, 20-031 Lublin, Poland*

[3]*Department of Sciences, Ivan Franko Zhytomyr State University, 40 Velyka Berdychivska Str., 10008 Zhytomyr, Ukraine*

CONTENTS

ABSTRACT

In situ immobilization of 4-aminostyrene and methacrylic acid copolymer has beenperformed on silica gel surface. ^{1}H NMR, IR and mass spectroscopy as well as TG analysis have been used to elucidate the structure of immobilized copolymer. An adsorption capacity of the synthesized composite towards Cu(II), Zn(II), Pb(II), Mn(II), Fe(III), Co(II), Ni(II) ions has been estimated. Adsorption activity to microquantity of Pb(II), Mn(II) and Fe(III) in a neutral aqueous medium has been observed.

5.1 INTRODUCTION

In the last two decades significant progress has been made in the study of complex formation processes on solid surfaces, in the compilation methods of purposeful synthesis of major groups of complex-forming adsorbents, including complex-forming chemically modified silica. To date, many examples of proven efficiency of complexes on solid carriers in catalytic processes have come to exist, these examples include the development of hybrid and combined methods in analytical chemistry that enhance the scope and accuracy of defining chemical elements, reduce the time of analysis, and are important for treatment technologies.

In particular, research efforts are aimed at finding methods of targeted one-step synthesis of adsorbents with monofunctional hard surface coating agents, capable of complex formation. To increase the adsorption capacity of adsorbents, adsorption or chemical attachment to inorganic surfaces and nitrogen–oxygen containing polymeric materials is increasingly used; the latter have inherent both complex-forming and ion exchange properties (polyions, *polyhexamethylene guanidine* and its derivatives, polyaniline and polyacrylic acid) [1–4]. To synthesize such materials it is appropriate to use in situ formation of immobilized polymer layer in the presence of inorganic carrier particles as a promising way of creating nanocomposites with new valuable properties.

The most universal method of polymer synthesis is radical polymerization. It allows you to implement sedimentary polymerization, for example, the process where polymer is formed in the polymerization solution as sediment precipitation. In the case of such polymerization polymerization rate of the polymer increases, moreover, the latter has a high purity [5]; the process conducted in the presence of particles of inorganic nature is a simple method to create polymer composites – inorganic carriers.

4-aminostyrene, as well as methacrylic acid, is polymerized in an inert atmosphere with radical initiators. However, in contrast with methacrylic acid, products with low molecular weight are formed under direct polymerization of 4-aminostyrene[6–10]. For high molecular weight polymer formation in conditions of radical polymerization, aminostyrene with blocked amino group is typically used (e.g., tert-butoxycarbonyl"protection") [11]. Copolymerization of aminostyrenes with such active monomer as methacrylic acid, gives reason to expect the formation of sufficiently high molecular weight products.

This chapter deals with the in situ immobilization of 4-aminostyrene with methacrylic acid copolymer on the surface of silica gel, it concerns the research of the immobilized polymer structure and its absorption properties for ions of Cu (II), Zn (II), Pb(II), Mn(II), Fe (III), and Co (II).

5.2 EXPERIMENTAL PART

4-aminostyrene and methacrylic acid copolymerization in the presence of silica (fraction of particles with a diameter of 0.1–02 mm, specific surface 428.61 m^2/g, Merck) has been carried out under the argon atmosphere. A solution of 1.18 g (0. 0.006 mol) methacrylic acid, 0.82 g (0.006 mol) 4-aminostyrene and 0. 02 g of AIBN in 16 ml CCl4 (12.5 wt. %) were poured into a flask containing 10 g of silica gel. When argon blowing was finished after 25 min, the reaction mixture was heated to 70°C; polymerization continued for 5 hours with stirring. The reaction was stopped by cooling the reaction mixture. The resulting suspension was poured into a porcelain cup and left overnight to evaporate the solvent; the synthesized composite was washed 3 times with isopropyl alcohol, filtered and air-dried for 24 hours at room temperature.

A model sample of 4-aminostyrene and methacrylic acid copolymer obtained under identical conditions in the absence of silica gel was used for the estimation of copolymer composition by ^{1}H NMR-spectroscopy.

^{1}H NMR: (400 MHz, DMSO-d$_6$, TMS) δ (ppm): 6.99 (2H, Ar); 6.5 (2H, Ar); 1.89 (3H, CH$_3$); 1.46–1.14 (5H, CH$_2$, CH).

The fact of a heterophase polymerization was additionally confirmed by infrared spectroscopy. IR spectra of the composite and silica output were recorded with an infrared spectrophotometer Thermo Nicolet Nexus FT-IR, USA.

The amount of copolymer on the surface of silica gel was evaluated by thermo gravimetric analysis with TG/DTA analyzer "Shimadzu DTG-60 H" (Japan) in the 15–1000°C temperature range. The heating rate of the samples was 10°/min.

To investigate composite surface, the BET method (low-temperature nitrogen adsorption-desorption) at the boiling point of liquid nitrogen was used with ASAP 2420 V1.01 (Micromeritics, USA) sorptometer. Before measurements, samples were degassed at 60°C. According to the computer processing results of the adsorption-desorption isotherms of nitrogen, the surface area of the composite and pore distribution with regard to diameter was determined.

Adsorption characteristics of synthesized composite as to ions of Cu(II), Zn(II), Pb(II), Mn(II), Fe(III), Ni(II) Co(II) were studied in static mode. Thus 0.1 g of the composite contacted 25–100 ml of nitrates solutions of corresponding metals under the permanent stirring with a mechanical vibrator at room temperature. Determination of the equilibrium concentration of the metals was performed by atomic absorption using a "Saturn" (Ukraine) atomic absorption spectrophotometer with flame atomizer in "air – propane – butane" flame mixture.

The adsorption degree (R) was calculated by the formula:

$$R = (m_{ads}/m_o) \cdot 100\% = (m_o - m) / m_o \cdot 100\%,$$

where m_o – mass of metal in the output solution, m_{ads} – mass of adsorbed metal, m – mass of metal in equilibrium solution after adsorption, which was calculated as

$$m = [M] \cdot V,$$

where $[M] \cdot$ – is equilibrium concentration of metal and V is volume of equilibrium solution.

Nitrate solutions of Cu (II), Zn (II), Pb (II), Mn (II), Fe (III), Ni (II) Co (II) were prepared using sets "standard sample solutions" of these salts against the background of 1 M HNO_3 (manufactured by $A. V. Bogatsky$ FHI Co, in Odesa) with 1 and 10 mg/ml concentrations. To create the appropriate pH environment, standard buffer solutions (ISO 8.135:2009, manufacture red by RIAP JSC in Kyiv) were used.

Determination of the adsorption capacity of synthesized composite for the above mentioned transition metal ions included determining of the optimal range of pH environment, medium required phase contact time to achieve adsorption equilibrium in static mode and building sorption isotherms of the appropriate metal ions to establish the adsorption capacity.

5.3 RESULTS AND DISCUSSIONS

Intergel intensities of proton signals near 6.99 ppm (2H, Ar, styrene) and 1.89 ppm (3H, CH$_3$ methacrylic acid) were estimated as 3.8:1. This suggests that the ratio of 4-aminostyrene and methacrylic acid contained in the copolymer immobilized on the surface of silica is about 4:1. Then the *chemical structure* of *in situ* copolymer immobilization of 4-aminostyrene and methacrylic acid on the surface of silica could be presented as follows:

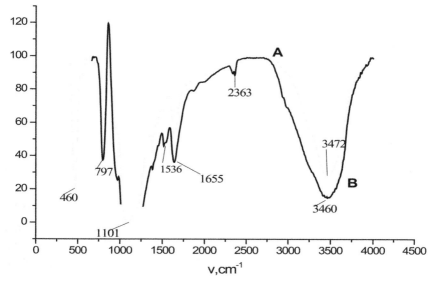

Chemical structure diagram of *in situ* immobilization of 4-aminostyrene-methacrylic acid copolymer on the silica surface ($m = 4$, $n = 3$).

Comparative analysis of IR spectra of synthesized composite and original silica gel (Figure 5.1) shows that absorption bands are present at 3460

FIGURE 5.1 The FTIR-spectra of the original silica gel (A) and synthesized composite (B).

cm⁻¹in the spectrum of the immobilized copolymer unlike the original carrier, which can be identified as valence vibrations of υ (N-H) – aminostyrene bonds. Absorption bands which can be attributed to the stretching vibration of aminostyrene aromatic systems are also present in the spectrum of the composite in the range of 1500 to 1600 cm⁻¹, and so are bands of ca. 800 cm⁻¹ deformation vibrations of CH-polymer bonds [12, 13].

To determine the mass of immobilized copolymer thermo gravimetric analysis was performed (Figure 5.2). It can be seen from the thermo gram presented that most of the copolymer decomposes in the temperature range from 300 to 600°C. Approximately, 10.5% of the composite weight is lost, which suggests that this is exactly the mass of copolymer to be found on the surface of silica gel.

Mass spectral analysis showed that thermal destruction of copolymer occurs in two stages. When temperatures are close to 100°C (118°C) carboxyl group water is produced, according to Ref. [14] accompanied by the formation of anhydride fragments. Further destruction occurs at temperatures above 300°C. The major products of the relative weight of 44 Da may be CO_2 and N_2O (Figure 5.3).

As seen in Figure 5.4, the forms adsorption-desorption isotherms of source silica gel nitrogen and composite are similar and belong to type IV

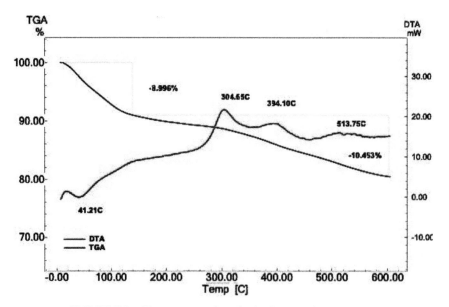

FIGURE 5.2   Thermogram of synthesized composite.

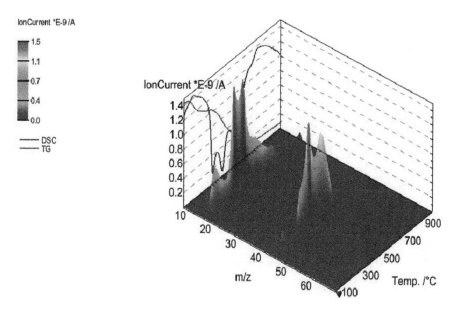

FIGURE 5.3 TG-MS-3D of synthesized composite.

isotherms according to IUPAC classification. The received data indicate that immobilized copolymer has virtually no effect on the structure of the surface layer of silica gel. The value of the specific surface area of silica gel following immobilization with polymer, calculated using the BET method, is 341.7 m^2/g. That is to say that the surface area decreases slightly after immobilization, which is a logical result of the consolidation on the surface of small pore polymer.

For a more detailed study of changes in the structure of the silica gel surface after modification with copolymer its surface pore size distribution diagrams were built, calculated using the BET method (Figure 5.5). As seen in Figure 5.5, modification virtually no effect on the silica gel pore size which is before and after the modification is predominantly of macro porous nature. Such pore distribution is usually characteristic of "islet" placement of the polymer on the surface of silica gel matrix, which was confirmed by the data of scanning electron microscopy.

The study of synthesized composite adsorption capacity for ions Cu (II), Zn (II), Pb(II), Mn(II), Fe(III), Ni(II) and Co(II) at different values of pH environments has shown in Table 5.1. Data presented have shown an adsorption activity of the synthesized composite against all studied trace metals in neutral and slightly alkaline environment (In the absence of initial hydrolysis

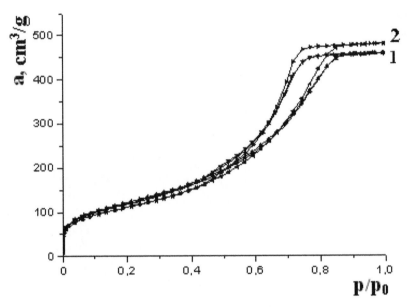

FIGURE 5.4 Adsorption–desorption isotherms of nitrogen of original silica (1) and composite (2).

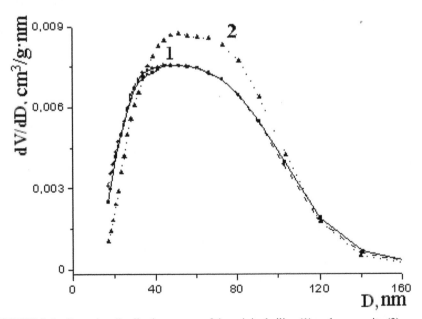

FIGURE 5.5 Pore size distribution curves of the original silica (1) and composite (2).

TABLE 5.1 Dependence of the Adsorption Degree of Metal Cations on Silica Gel with Immobilized Copolymer of 4-Aminostyrene and Methacrylic Acid

pH	Degree of sorption, %						
	Cu^{2+}	CO_2^+	Ni^{2+}	Mn^{2+}	Pb^{2+}	Fe^{3+}	Zn^{2+}
1	0.0	0.0	20.0	0.0	17.2	36.8	12.9
4.01	14.1	2.3	26.5	0.0	56.3	27.3	1.1
6.86	0	9.5	4.0	10.1	98.3	47.3	40.9
Distilled water	80.5	0.0	64.7	40.2	99.9	64.7	71.8
8.40	75.0	67.3	81.3	*	*	*	83.2

* – precipitation by hydrolysis of the salts of the corresponding metals was observed.

Experimental Conditions: 0.1 g Sorbent Mass, Volume of Solutions – 25 mL, m^0_{Me} – 100 mcg.

salts). But, as one can see, the quantitative adsorption recorded only for ions Pb(II) in neutral aqueous medium. In an acidic environment (0.1 M HCl) the synthesized composite was not stable and partial flushing of the immobilized polymer into solution was observed.

Adsorption isotherms of all metals studied have the same shape (Figure 5.6.), which is typical for chelating modified silica [15].One can make a preliminary conclusion that the adsorption capacityfor research synthesized composite transition metal ion complexation of due process aminostyrene nitrogen and oxygen atoms methacrylic acid.

An investigation of the degree dependence of adsorption of transition metals ions upon contact time in static mode (Figure 5.7) shows that all ions are adsorbed most instantly thus indirectly confirming the complexation mechanism of adsorption.

However, calculations of adsorption capacity of the synthesized composite made on the basis of adsorption isotherms, suggest (Table 5.2) that only for ions of Pb(II), Mn(II) and Fe(III) adsorption capacity of the synthesized composite is several times more than the adsorption capacity of source silica gel.

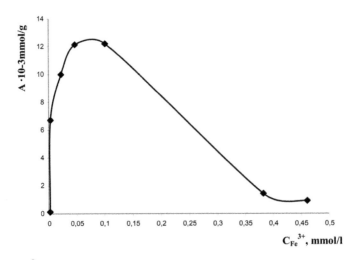

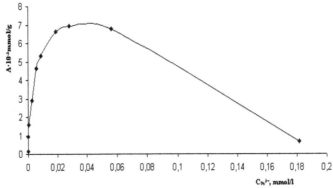

FIGURE 5.6   Adsorption isotherms of ions Fe(III) (a) and Pb(II) (b) on silica gel immobilized with copolymer of 4-aminostyrene and methacrylic acid from aqueous solutions of their nitrates (without addition of buffer solutions).

TABLE 5.2   Adsorption Capacity of the Synthesized Composite

Cation	Sorption capacity			
	Source silica gel		Composite	
	mmol/g	mg/g	mmol/g	mg/g
Mn(II)	0.010	0.55	0.020	1.10
Fe(III)	0.008	0.45	0.012	0.70
Pb(II)	0.002	0.41	0.007	1.45
Ni(II)	0.008	0.47	0.008	0.47

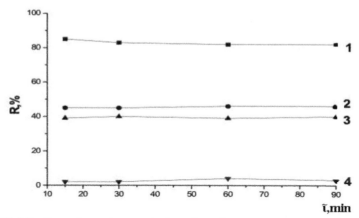

FIGURE 5.7 Dependence degree of adsorption of aqueous complexes $Pb^{2+}(1)$, $Cd^{2+}(2)$, $Mn^{2+}(3)$, $CO_2^+(4)$ (Terms of experiment: sorbent mass, 0.1 g volume of solutions, -20 mL, $m^0Me - 80$ mg).

5.4 CONCLUSIONS

A new organic composite material has been synthesized by *in situ* immobilization of 4 aminostyrene and methacrylic acid copolymer on the surface of silica. The fact of heterophase polymerization has been confirmed by H^1NMR, IR and mass spectrometry. According to H^1 NMR it has been found that the ratio of 4-aminostyrene and methacrylic acid containedin the copolymer immobilized on the surface of silica is about 4:1. It has been found due to the results of thermo graphic analysis that in the synthesized composite is 10.5wt.% of copolymer. A comparison of the adsorption-desorption isotherms of nitrogen and source silica gel and composite shows that immobilized copolymer has virtually no effect on the structure of the surface layer of silica gel. Adsorption activity of the synthesized composite for micro-ions of Pb(II), Mn(II) and Fe (III) in a neutral aqueous medium has been observed.

KEYWORDS

- **4-aminostyrene and methacrylic acid copolymer**
- **adsorption properties**
- **in situ immobilization**
- **metal complexes**
- **modified silica**

REFERENCES

1. Tertykh, V., Polishchuk, L., Yanishpolskii, V., Yanovska, E., Dadashev, A., Karmanov, V., Kichkiruk, O., (2008). Adsorption properties of functional silicas towards some toxic metal ions in water solutions, In book: Recent Advances in Adsorption Processes for Environmental Protection and Security, J. P. Mota, S. Lyubchik (eds.), NATO Science for Peace and Security Series – C: Environmental Security, Dordrecht: Springer, 119–132.
2. Yanovska, E. S., Dadashev, A. D., Tertykh, V. A. (2009). Inorganic anion-exchanger based on silica with grafted polyhexamethyleneguanidine hydrochloride. *Functional Materials. 16*(1), 105–109.
3. Dadashev, A. D., Tertykh, V. A., Yanovska, E. S., Yanova, K. V. (2012). Ion-exchange properties of modified silicas with bound amides of polyhexamethylene guanidine and maleicoro-phthalicacids with respect to metal-containing anions. *Chemistry, Physics and Technology of Surface, 3*(4), 419–428.
4. Yanovska, E. S., Ryabchenko, K. V., Tertykh, V. A., Kichkiruk, O., Yu. (2012). Complexing properties of silica gel-polyaniline composites with grafted heterocyclic azo reagents. *Chemistry, Physics and Technology of Surface, 3*(4), 439–447.
5. Bunyakan, C., Hunkeler, D., (1999). Precipitation polymerization of acrylic acid in toluene. I: synthesis, characterization and kinetics. *Polymer, 40*, 6213–6224.
6. Schildknecht, C. E. Vinyl and Related Polymers, J. Wiley & Sons, Inc., p. 156. (1952).
7. Covolan, V. L., D'Antone, S., Ruggeri, G., Chiellini, E., (2000). Preparation of aminated polystyrene latexes by dispersion polymerization. *Macromolecules, 33*, 6685–6692.
8. Covolan, V. L., Ruggeri, G., Chiellini, E., (2000). Synthesis and characterization of styrene/Boc-*p*-aminostyrene copolymers.*J. Polym. Sci, Part A: Polym. Chem., 38*, 2910–2918.
9. Ober, C. K., Lok, K. P., Hair, M. L. (1985). Monodispersed, micron-sized polystyrene particles by dispersion polymerization. *J. Polym. Sci, Polym. Lett. Ed., 23*, 103–108.
10. Ober, C., K, Hair, M. L. (1987). The effect of temperature and initiator levels on the dispersion polymerization of polystyrene.*J. Polym. Sci, Part A: Polym. Chem,25*, 1395–1407.
11. Goikhman, M., Ya, Subbotina, L. I., Martynenkov, A. A., Smirnov, M. A., Smyslov, R., Yu., Popova, E. N., &Yakimanskii, A. V. (2011). *Russian Chemical Bulletin, International Edition, 60*(2), 295–303.
12. Shnayner, Q., Fyuzon, R., Kertyn, D., Morrill, T., (1983). Identification of organic compounds/M.: Mir, 704 pp.
13. Sobolev, V. A., Chuyko, A., A, Tertykh, V. A. (1972). The spectral determination of some organic groups chemisorbed surface of silica. *Zh. J. Spectroscopy, 17*(3), 477–483.
14. Giordanengo, R., Viel, S., Allard-Breton, B., Thévand, A., Charles, L., (2009). Tandem Mass Spectrometry of Poly (Methacrylic Acid) Oligomers Produced by Negative Mode Electro spray Ionization. *J. Am. Soc. Mass. Spectrom., 20*, 25–33.
15. Trohimchuk, A. K. (1997). The processes of complex formation on the surface modified silicas and their use in inorganic analysis: Author. Thesis. Doctor. Chem. Sciences: 02.00.01. KNU. Taras Shevchenko. Kyiv. 38 p.

CHAPTER 6

SYNTHESIS OF POLYARYLATES CONTAINING ETHER BONDS IN MACROMOLECULES

G. PAPAVA, N. DOKHTURISHVILI, N. GELASHVILI,
M. GURGENISHVILI, K. PAPAVA, I. CHITREKASHVILI,
and Z. TABUKASHVILI

Petre Melikishvili Institute of Physical and Organic Chemistry of Ivane. Javakhishvili Tbilisi, State University, I. Chavchavadze Ave. 1, 0179, Tbilisi, Georgia, E-mail: marina.gurgenishvili@yahoo.com

CONTENTS

ABSTRACT

Polyesters which contain ether bonds in the main polymer chain have been synthesized. Diphenyloxidedicarbonic acid was used as acid component. Polyesters are characterized by high heat and thermal stability. The presence of an oxygen atom between the phenyl nucleus of diphenyloxidedicarbonic acid causes reduction of softening temperature and increase of elasticity.

6.1 INTRODUCTION

Dependency of some properties of polymers from their structure has been proved on the example of various class polymers which enables us to resolve successfully the tasks of synthesis of polymers of preliminarily focused properties. Knowing in advance such relations enabled us, in our case, by inculcation of aromatic nuclei in ether macromolecule, to synthesize new class of thermally stable polymers – polyarylates, which are distinguished by a set of high physical-mechanical and other interesting properties.

Endeavors of researchers to receive more thermally stable polymers than the existing ones, lead them to the creation of rigid structure polymers. In most cases softening temperature of polymers possessing such rigid macro-molecules is higher than their destruction temperature. This is why repro-cessing of such polymers from the melt becomes impossible, while their poor solubility in organic solvents makes difficult their reprocessing from the polymer solutions. Very often such situation greatly limits the possibili-ties of at first sight very interesting practical application of polymers.

With this in view it was interesting to create polymers, in which it would be possible to fuse two as if reciprocally excluding properties, rigid structure of a macromolecule and good solubility of a polymer. But sometimes, some factors which determine increase of polymer softening temperature causes deterioration of its solubility. With this in view, the factor which would simultaneously condition increase of polymer softening temperature and improvement of its solubility seems very interesting. Such factor is inculca-tion of card type groups of side cycle structure in a macromolecule.

This chapter pursued synthesis of polyarylateson the base of cyclic bisphenols containing norbornene-type non-coplanar substitutes at the cen-tral carbon atom. Synthesis of polyarylates on the base of such bisphenols was interesting since presence of groups with volumetric non-coplanar structure at the central carbon atom in bisphenols molecules enabled us to hope that it would be possible to obtain the polymers which would be char-acterized by precious properties, in particular by good solubility in organic solvents and at the same time, by high thermal stability. Polyarylates are namely such polymers.

Due to the fact that properties of polyarylates greatly depend on their chemical structure, it is possible to alter polymer properties in desired direction by adequate selection of starting components.

Polyarylates obtained on the base of aliphatic dicarboxylic acid are characterized by lower softening temperature [1–4].If we replace aliphatic dicarboxylic acid by aromatic dicarboxylic acid, softening temperature of polyarylates sharply is increased because of increase of rigidity of a macromolecule [4–6].

By the increase of a number of aromatic nuclei in aromatic dicarboxylic acid softening temperature of polyarylates suffers sharp increase [7–9]. At the infringement of symmetry of spreading of carboxyl groups in aromatic dicarboxylic acid the polyarylate softening temperature suffers significant decrease since at this moment macromolecules arrangement density decreases [2, 4, 5, 10].

At the insertion of various groups between aromatic nuclei of biatomoc aromatic dicarboxylic acid flexibility of polymer chains increases resulting in decrease of polymer softening temperature and increase of polymer solubility [11–13].

Polyarylates on the base of the bisphenols the central carbon atom of which simultaneously is in the main polymer chain and in the side cyclic groups, are characterized by especially interesting properties.Thus, the properties of polyarylates depend both on the structure of bisphenol as well as dicarboxylic acid.

Earlier conducted researches, dealing with polycarbonates [14] showed that for the synthesis of polycarbonates, application of bisphenols containing norbornene-type substitutes at the central carbon atom, leads to significant increase of thermal stability of polycarbonates. Considering this fact and also the above-referred data about chemical structure of bisphenols, as well as the impact of dicarboxylic acid on polymer properties, we had the right to suppose that polyarylates obtained on the base of norbornene-type groups-containing polycyclic bisphenols and biatomic aromatic dicarboxylic acid would possess interesting properties.

6.2 MATERIALS AND METHODS

The following bisphenols were used for the synthesis of polymers:4,4¹-diphenyloxide carboxylix acid dichloroanhydride was used as an acid component.

Polymer synthesis was performed by the high-temperature polycondensation method in ditolyl methane. Reaction of formation of homogeneous polyarylate by high-temperature polycondensation can be imagined as follows:

$$nHOAOH + nClOCBCOCl = 2nHCl + [-OAOOCBCO-]_n,$$

where A is for remainder of bisphenol molecule, Bis for remainder of dicarboxylic acid molecule.

6.3   RESULTS AND DISCUSSIONS

IR spectral study of polymers showed absorption band characteristic for ester groups (1770 cm^{-1}) and absorption band in 1460 cm^{-1} region characteristic for bisphenol CH$_2$ group.

Table 6.1 offers the properties of polyarylates obtained on the base of polycyclic bisphenols, their phenyl substituted derivatives and biatomic aromatic dicarboxylic acid – 4,4^1-diphenyloxide dicarboxylic acid.

All homo-polyarylates are characterized by high softening temperature. High softening temperature is conditioned by rigid structure of macromolecules. Because of presence of volumetric norbornene-type substitutes of non-coplanar structure, free rotation of phenol nuclei bound to norbornene carbon atom is complicated. Due the same reason, migration of separate sections of macromolecules towards each other is hindered too. Besides, due to the fact that norbornene-type substitutes have non-coplanar structure, these cycles have significant dimensions in all directions. This is why such group is deprived of orientation in any definite direction, and hence, the ability of migration of separate fragments of macromolecules in small space, similar to the extent possessed for example by allyl or phenyl groups, which have insignificant size at least in one direction.

Due to the fact that polymer softening and melting is connected with movement of macromolecules, presence of non-coplanar polycyclic structure in a molecule contributes also to the increase of polymer softening temperature. The size of a substitute linked at the central carbon atom is also important. Polymer softening temperature increases at the increase of a substitute size, which is well seen from the data of the Table 6.1. Presence of oxygen atom among the phenyl nuclei in diphenyloxide dicarboxylic acid

TABLE 6.1 Properties of Aromatic Polyesters Obtained on the Base of Card-Type Bisphenols and Diphenyloxidedicarboxylic Acid by High Temperature Polycondensation

S. No.	D Structure of polymer recurring ring	Decrease of polymer mass at heating on air			Yield%	0.5% polymer solution in chloroform $\eta^{20}_{finished}$ dl./g	Softening temperature, °C		Polymer structure according to X-ray diffraction analysis
		Start	By 10%	By 5%			In capillary	According to thermo mechanical curves	
		Start	By 10%	By 5%			In capillary	According to thermo mechanical curves	
1		340	395	550	85	0.90	320–345	320	Amorphous
2		310	400	570	90	0.60	300–325	290	Amorphous
3		320	370	510	75	0.80	290–295	270	Amorphous

TABLE 6.1 (continued)

S. No.	D Structure of polymer recurring ring	Decrease of polymer mass at heating on air			Yield%	0.5% polymer solution in chloroform $\eta^{20}_{finished}$' dl./g	Softening temperature, °C		Polymer structure according to X-ray diffraction analysis
		Start	By 10%	By 5%			In capillary	According to thermo mechanical curves	
4		330	365	580	65	0.40	305–335	275	Amorphous
5		305	400	560	80	0.70	325–340	280	Amorphous
6		300	375	540	85	0,65	295–300	270	Amorphous
7		290	355	480	70	0.40	300–310	285	Amorphous

S. No.	D Structure of polymer recurring ring	Decrease of polymer mass at heating on air			Yield%	0.5% polymer solution in chloroform η^{20} finished' dl./g	Softening temperature, °C		Polymer structure according to X-ray diffraction analysis
		Start	By 10%	By 5%			In capillary	According to thermo mechanical curves	
8		275	390	510	60	0.50	290–330	270	Amorphous
9		270	330	385	90	0.90	260–280	270	Amorphous
10		280	340	500	95	0.88	290–310	285	Amorphous
11		275	410	480	95	0.65	290–300	290	Amorphous

TABLE 6.1 (continued)

S. No.	D Structure of polymer recurring ring	Decrease of polymer mass at heating on air			Yield%	0.5% polymer solution in chloroform $\eta^{20}_{finished}$, dl./g	Softening temperature, °C		Polymer structure according to X-ray diffraction analysis
		Start	By 10%	By 5%			In capillary	According to thermo mechanical curves	
12		260	400	495	78	0.80	270–320	300	Amorphous
13		370	400	570	90	0.80	310–315	285	Amorphous
14		365	440	580	95	0.65	310–320	280	Amorphous

S. No.	D Structure of polymer recurring ring	Decrease of polymer mass at heating on air			Yield%	0.5% polymer solution in chloroform $\eta^{20}_{finished}$ dl./g	Softening temperature, °C		Polymer structure according to X-ray diffraction analysis
		Start	By 10%	By 5%			In capillary	According to thermo mechanical curves	
15		335	410	495	75	0.77	300–310	270	Amorphous
16		300	360	405	55	0.65	305–325	290	Amorphous
17		330	340	480	70	0.90	320–325	280	Amorphous

TABLE 6.1 (continued)

S. No.	D Structure of polymer recurring ring	Decrease of polymer mass at heating on air			Yield%	0.5% polymer solution in chloroform $\eta^{20}_{finished}$, dL/g	Softening temperature, °C		Polymer structure according to X-ray diffraction analysis
		Start	By 10%	By 5%			In capillary	According to thermo mechanical curves	
18		340	350	505	80	0.96	310–325	275	Amorphous
19		315	340	475	60	0.70	280–290	270	Amorphous
20		300	450	560	75	0.70	300–320	285	Amorphous

results in decrease of polymer softening temperature. Thus, for example, if softening temperature of a polymer obtained on the base of norbornene cycle-containing bisphenols and diphenyl dicarboxylic acid is 345–350°C, at the substitution of diphenyldicarboxylic acid by diphenyloxidedicarboxylic acid, polymer softening temperature decreases to 260–280°C. Softening temperature of a polymer obtained on the base of indangroup containing bisphenol and diphenyldicarboxylic acid is 355–360°C. At the substitution of diphenyl-dicarboxylic acid by diphenyloxidedicarboxylic acid, polymer softening temperature falls to 310–315°C. Substitution of methyl group as well as chlorine atoms in phenyl nucleus also decreases softening temperature of a polymer.

Polymer thermal stability on air (heating rate 3°/min) was studied by the method of thermo gravimetric analysis. The results are given in Table 6.1. The Table data show that thermal stability of a polymer prepared on the base of diphenyl oxide dicarboxulic acid is lower than thermal stability of a polymer prepared on the base of diphenyl dicarboxylic acid, although these polymers too retain rather high thermal stability. All polymers obtained on the base of diphenyloxide dicarboxylic acid are characterized by amorphous structure.

In organic solvent solutions, thus, for example,in chloroform, polymers made on the base of diphenyloxide dicarboxylic acid and polycyclic (card type) bisphenols form transparent, solid films, which possess good mechanical and dielectric properties. Thus, for example, tangent of angle of dielectric losses for #9 and #14 polymers of the Table 6.1 approximately equals to 6×10^{-3} and 5×10^{-3}, correspondingly. Tenacity is approximately 400 kg/cm².

The advantage of the polymers is that they retail high mechanical and dielectric properties not only at common ambient temperature but also at high temperatures. Thus, for example, when heated at 200°C, they retain more than 50% of their strength. Specific volumetric resistance of polymers is within 10^{17} ohms/cm. When heated at 200°C it falls only to 10^{-13} ohms/cm, which refers to the fact that card type polyarylates which contain norbornene-type substitutes are good dielectrics and might be used successfully as insulation materials.

6.4   CONCLUSIONS

Polyesters containing ether bonds in the main polymer chain have been synthesized. Diphenyl oxidedicarboxylic acid was used as acid component. The presence of an oxygen atom between the phenyl nucleus of diphenyloxide dicarboxylic acid causes reduction of softening temperature and increase of elasticity. From the solutions polymers form transparent films, having good mechanical and dielectric properties which are preserved at high temperature.

KEYWORDS

- aromatic
- bisphenol
- dicarboxylic
- polyarylates
- polycondensation
- polycyclic
- polyester

REFERENCES

1. YamagutiEakayanogiKuryata. (1957). *J. Chem. Soc. Japan, Industr. Chem. Sect., 58*, 358.
2. Korshak, V. V., Vinogradova, S. V., (1959). *High Molecular Compounds, 1*, 834 (Rus.).
3. Korshak, V. V., Vinogradova, S. V., (1959). *High Molecular Compounds, 1*, 1442 (Rus.).
4. Eareckson, W. M., (1959). *J. Polymer Sci., 40*, 399.
5. Korshak, V. V., Vinogradova, S. V. (1958). Proceedings of the Academy of Sciences of USSR, Dept. Org. Chemistry, 637.
6. .Korshak, V. V., Vinogradova, S. V., Salazkin, S. N. (1962). *High Molecular Compounds, 4,* 339. (Rus.).
7. Pankratov, V. A. Candidate's Thesis, (1965), D. I. Mendeleev Moscow Institute of Chemical Technology, Moscow Russian Federation.
8. Ingold, K. K. (1959). Mechanism of reaction and structure of organic compounds "Izdatinlit" M.
9. Bender, M. L. (1964). Mechanism of catalysis of nucleophilic reactions of carboxylic acid derivatives. Ed. "Mir." M. (Rus.).
10. Korshak, V. V., Vinogradova, S. V., Lebedeva, A. S. (1960). *High Molecular Compounds, 2,* 61 (Russ.).
11. Conix, A., (1958). *Ind. Eng. Chem., 51,* 147.
12. Fomina, Z., Ya. Candidate's Thesis, (1965) D. I. Mendeleev Moscow Institute of Chemical Technology, Moscow, Russian Federation.
13. Korshak, V. V., Vinogradova, S. V. (1964). Polyarylates. Ed. "Nauka," M. (Rus.).
14. Jackson, W. J, Caldwell, I.R, (1963).*Ind. Eng. Chem, Prod. Res. Develop, 2,* 246.

CHAPTER 7

SYNTHESIS OF POLYARYLATES ON THE BASE OF NUCLEUS BISPHENOLS IN HIGH BOILING SOLVENTS

G. PAPAVA, N. MAISURADZE, E. GAVASHELIDZE, Sh. PAPAVA, M. GURGENISHVILI, I. CHITREKASHVILI, and V. SHEROZIA

Petre Melikishvili Institute of Physical and Organic Chemistry of Ivane, Javakhishvili Tbilisi State University, I. Chavchavadze Ave. 1, 0179 Tbilisi, Georgia, E-mail: marina.gurgenishvili@yahoo.com

CONTENTS

ABSTRACT

Card group-containing heterochain polyesthers were synthesized on the base of norbornene-type polycyclic bisphenols and diphenyldicarboxylic acid dichloroanhydride by the method of high temperature polycondensation. The obtained polymers are characterized by high heat- and thermal stability, are well soluble in chlorinated hydrocarbons and they form transparent films in solutions which are characterized by good mechanical and dielectric properties.

7.1   INTRODUCTION

Alongside with the progress of modern technologies demand on polymer materials has been increased sharply. Development of a series of branches of new technology, such as space engineering, aviation, rocket technology, radio electronics and others ask for the creation of new type materials, which should preserve high strength at strict conditions of exploitation at high temperature, at high mechanical charges. With this in view the polymers, which are characterized by high heat- and thermal resistance, high strength and other significant indices are most interesting.

As has been shown by the researches, with the view of heat- and thermal stability, the polymers with macromolecules which contain aromatic and heterocyclic rings are most perspective materials, but because of high rigidity of recurrent cycles of such polymers, often it becomes impossible to use them for obtaining films, fibers and other articles, where materials must possess high plasticity, flexibility, high dielectric and other specific properties. General method to obtain such polymers is creation of the polymers, macromolecules of which, together with aromatic and heterocyclic rings would contain relatively more flexible groups such as ester, ether and the like bonds; creation of such macromolecules can be achieved either in the process of polymer synthesis, when such bonds are formed in the reaction process, for example, polyarylates, polyimides, epoxy, silicon-containing or other polymers, or by inculcation of such fragments into polymer chain by the use of the monomers, which contain such groups. With this in view the so-called card-type polymers are perspective polymers, which in their recurrent ring contain side cyclic groups, one of the atoms of which simultaneously is in the composition of a macromolecule structure. Presence of such groups grants specific properties to the polymers, when high thermal stability of a polymer is fused with its good solubility [1–4]. It is especially important for aromatic heterocyclic polymers of rigid structure, when softening temperature of such polymers is close, or even is higher than polymer destruction temperature, which, of course complicates the process of obtaining articles from such materials.

Considering the above stated the card-type polymers, aromatic polyimides, polyamides, polyoxazoles, which in their polymer chain contain phthalide, norbornene and the like card-type groups [5, 6].

In special literature there are rather scarce data about the polymers, which contain norbornene-type card groups, which are very interesting because of their original spatial structure, creating closed three dimensional, volumetric cycles. Such structure distinguishes them from other card groups. This is why it was expected that for building macromolecules, it would be interesting and perspective to use the monomers, which as substituted groups contain norbornene-type and other cyclic groups characterized by non -coplanar structure.

7.2   MATERIALS AND METHODS

For obtaining of heterochain polymers we used the norbornene type monomers, which contain biatomic phenols in which phenol hydroxyl groups were in various nuclei of benzol. In their synthesis we used the easily accessible material such as wastes of gas and petroleum treatment.

For the synthesis of polymers we used the following norbornenc-type polycyclic bisphenols:4,4^1-(1-cyclopentyl)diphenol; 4,4^1-(1-cyclopentyl) di-ortho cresol; 4,4^1-(cyclopentyl)-2-chloro phenol; 4,4^1-(cyclopentyl) bis-2,6 dichlorophenol;4,4^1-(cyclohexyl)diphenol; 4,4^1-(cyclohexyl) di-ortho cresol; 4,4^1-(cyclohexyl) bis-2-chlorophenol; 4,4^1-(cyclohexyl) bis-2,6-dichlorophenol; 4,4^1-(2-norbornyliden)diphenol; 4,4^1-(2-norbornyliden) di-ortho-cresol; 4,4^1-(2-norbornyliden) bis-2,6 dichloro phenol; 4,4^1-(hexahydro-4,7-methylenindan-5-yliden)diphenol;4,4^1-(hexahydro-4,7-methylenindan-5-yliden)diortho-cresol; 4,4^1-(hexahydro-4,7-methylenindan-5-yliden)bis-2-chlorophenol; 4,4^1-(hexahydro-4,7-methylenindan-5-yliden)bis-2,6-dichlorophenol; 4,4^1-(decahydro-1,4,5,8-dimethylennaft-2-yliden) di ortho cresol; 4,4^1-(decahydro-1,4,5,8-dimethylennaft-2-yliden) bis-2-chlorophenol and 4,4^1-(decahydro-1,4,5,8-dimethylennaft-2-yliden) bis 2,6-dichlor phenol.

As an acid component we used 4,4^1-(diphenyl dicarboxylic acid dichloroanhydride. Synthesis of polymers was performed by the method of high temperature polycondensation in ditolyl methane. Properties of the synthesized polymers are given in Table 7.1.

7.3   RESULTS AND DISCUSSION

IR spectroscopic studies of the synthesized polymers showed an absorption band in the 1770 cm^{-1} zone in IR spectrum, which is characteristic for

Chemical Engineering of Polymers

TABLE 7.1 Properties of Aromatic Polyesters Obtained on the Base of Card-Type Bisphenols and Diphenyldicarboxylic Acid by High Temperature Polycondensation

S. No.	D-P-P Structure of polymer recurring ring	Decrease of polymer mass at heating on air			Yield %	0.5% polymer solution in chloroform $\eta^{20}_{finished}$, dL/g	Softening temperature, °C		Polymer structure according to X-ray diffraction analysis
		Start	By 10%	By 5%			In capillary	According to thermo mechanical curves	
1		325	30	340	90	0.45	345–360	320	Crystalline
2		330	325	370	75	0.60	305–315	290	Crystalline
3		365	375	400	65	0.40	310–320	280	Amorphous
4		360	400	430	60	0.30	350–365	335	Amorphous

S. No.	D-P-PStructure of polymer recurring ring	Decrease of polymer mass at heating on air			Yield %	0.5% polymer solution in chloroform $\eta^{20}_{finished}$, dL/g	Softening temperature, °C		Polymer structure according to X-ray diffraction analysis
		Start	By 10%	By 5%			In capillary	According to thermo mechanical curves	
5	*(chemical structure)*	340	460	540	90	0.85	360–365	315	crystalline
6	*(chemical structure)*	330	400	475	90	0.80	315–335	315	amorphous
7	*(chemical structure)*	320	425	510	80	0.65	320–330	280	amorphous
8	*(chemical structure)*	450	460	530	75	0.50	350–365	325	amorphous

TABLE 7.1 (Continued)

S. No.	D-P-PStructure of polymer recurring ring	Decrease of polymer mass at heating on air			Yield %	0.5% polymer solution in chloroform $\eta^{20}_{finished}$, dL/g	Softening temperature, °C		Polymer structure according to X-ray diffraction analysis
		Start	By 10%	By 5%			In capillary	According to thermo mechanical curves	
9		370	415	540	90	0,52	345–350	250	amorphous
10		370	400	515	70	0,55	320–330	290	amorphous
11		365	410	500	65	0,35	310–315	280	amorphous
12		360	440	550	70	0,40	330–335	215	amorphous

S. No.	D-P-P Structure of polymer recurring ring	Decrease of polymer mass at heating on air			Yield %	0.5% polymer solution in chloroform $\eta^{20}_{finished}$, dL/g	Softening temperature, °C		Polymer structure according to X-ray diffraction analysis
		Start	By 10%	By 5%			In capillary	According to thermo mechanical curves	
13		450	480	560	85	0,48	355–360	305	amorphous
14		435	440	500	90	0,88	345–355	300	amorphous
15		440	520	540	65	0,40	3315–325	280	amorphous

TABLE 7.1 (Continued)

S. No.	D-P-P Structure of polymer recurring ring	Decrease of polymer mass at heating on air			Yield %	0.5% polymer solution in chloroform $\eta^{20}_{finished}$, dL/g	Softening temperature, °C		Polymer structure according to X-ray diffraction analysis
		Start	By 10%	By 5%			In capillary	According to thermo mechanical curves	
16		470	540	615	55	0.55	3340–350	335	amorphous
17		450	510	580	94	0.40	360–365	325	amorphous
18		445	490	500	80	0.88	340–345	320	amorphous

S. No.	D-P-PStructure of polymer recurring ring	Decrease of polymer mass at heating on air			Yield %	0.5% polymer solution in chloroform $\eta^{20}_{finished}$, dl./g	Softening temperature, °C		Polymer structure according to X-ray diffraction analysis
		Start	By 10%	By 5%			In capillary	According to thermo mechanical curves	
19		450	480	500	75	0.65	320–325	290	amorphous
20		480	510	600	60	0.30	355–360	320	amorphous

polymer ester groups and an absorption band in the 1460 cm-[1] zone, which is characteristic for bisphenol CH_2 groups.

As is seen from the data of Table 7.1 the polymers obtained on the base of polycyclic bisphenols and aromatic dicarboxylic acid are characterized by high thermal stability. High softening temperature of the polymers is explained by high rigidity of polymer chain, which is conditioned by non-coplanarity of norbornene-type cycles. These cycles have significant dimensions in all directions. This is why these groups are not capable to orientate in any definite direction, hence they can't move in a small space between macromolecules similar to that taking place in case of alkyl or phenyl groups. By the same reason free circulation of two phenyl groups which are linked to norbornene cycle carbon atom is complicated. Due to the fact that the polymer softening and melting temperatures are associated with movement of certain definite zones of macromolecules towards each other, presence of non-coplanar polycyclic structure in polymer recurring ring conditions high softening temperature of polymers.

Polymer thermal stability was investigated by the method of thermo gravimetric analysis, at heating on air (sample heating velocity – 4.5°/min). As is seen from the data of Table 7.1, polymers are characterized by high thermal stability. Decrease of their mass at heating on air commences above 300–400°C. Especially high heat resistance is inherent to the polymers obtained on the base of chlorine substituted bisphenols, destruction of which (e.g., polymers # 16 and 20) commences only at 470–480°C.

As has been shown by the studies polyarylates which contain the poly-cyclic bisphenols are characterized by high resistance to water. Films prepared on the base of these polymers are characterized also by high resistance to 20% sodium alkaline solution and to ultraviolet rays. At the long-term impact of ultraviolet rays polymers factually remain unchanged. Chlorine containing polymers are characterized also by refractory properties.

7.4 CONCLUSION

Card group-containing heterochain polyesters were synthesized on the base of norbornene-type polycyclic bishenols and diphenyl dicarboxylic acid dichloroanhydride by the use of high temperature polycondensation method.

The obtained polymers are characterized by high heat- and thermal-stability, are well soluble in chlorinated hydrocarbons. In solutions they form transparent films, which are characterized by good mechanical and dielectric properties.

KEYWORDS

- **aromatic hydrocarbons**
- **card type**
- **dicarboxylic acid**
- **heterochain**
- **polycyclic bisphenol**
- **polyester**

REFERENCES

1. Vinogradova, S. V., Salazkin, S. N., Chelidze, G., Sh, Slonimskii, G. L., Askadskii, A. A., Bychko, K. A., Komarova, L. I., Zhuravleva, I. V., Korshak, V. V. (1971). Methyl-ydenphthalyde copolymers. *Plastic Masses, 8*, 10 (Rus.).
2. Korshak, V. V., Vinagradova, S. V., Salaskin, S. N. (1962). "Hetero chain Polyesters. Polyarylates on the basis of Phenolphthalein. *Vysokomol. Soed., 4*(3), 339, (Rus.).
3. Korshak, V. V., Vinagradova, S. V., Slonimskii, G. L., Salaskin, S. N., Askadskii, A. A. (1966). Polyarylates with side phtalide in polymer chain based on diphenylphtalide carboxylic acid. *Vysokomol. Soed., 8*(3), 548. (Rus.).
4. Vinogradova, S. V., Vygodskii,Ya. S. (1973). Card Polymers. *Achievements in Chemistry, 42*, 1225. (Rus.)
5. Korshak, V. V. (1970). Chemical structure and temperature characteristics of polymers. Publ. House "Nauka," M. (Rus.).
6. Korshal, V. V., Vinogradova, S. V. (1964). Polyarylates. Publ. House "Nauka," M. (Rus.).

PART II

COMPOSITES AND NANOPARTICLES

CHAPTER 8

STRUCTURE AND PROPERTIES OF THE POLYSTYRENE/FULLERENE COMPOSITE FILMS

O. ALEKSEEVA, N. BAGROVSKAYA, and A. NOSKOV

G. A. Krestov Institute of Solution Chemistry, Russian Academy of Sciences, Akademicheskaya Str., 1, Ivanovo, 153045, Russia, E-mail: avn@isc-ras.ru

CONTENTS

ABSTRACT

This chapter includes the detailed study on the structural properties and biological activity of both polystyrene films and polystyrene films filled with fullerene. The structural characteristics of polystyrene films modified by the incorporation of fullerenes were researched by the infrared spectroscopy and X-ray diffraction technique. According to IR-spectroscopy data it was suggested that intermolecular interaction between polymer's phenyl ring and fullerene molecule occurs in the composite material. Semi quantitative analysis of IR spectra of the studied films with application of a method of a base line and internal standard was carried out. By the X-ray diffraction technique it is shown that there is the intermediate-range order in these materials. Incorporation of fullerenes into polymer matrix does not change the value of the repetitive characteristic distance in structural arrangement, but increases a correlation length (a scale of the intermediate-range ordering) in the disordered phase.

The glass transition temperatures, T_g, in range of 293–423 K have been determined for the films by DSC technique. The plasticization of the polymer is observed in thermal properties of the films under influence of small fullerene additions. The value of T_g decreases as fullerene content increases up to 0.02 wt.%. The effect of interaction between polymer and fullerene molecules on thermal properties becomes evident at higher fullerene content in range from 0.02 to 0.1 wt.%.

To assess bioactivity of pure polystyrene and filled composite with fullerene the laboratory tests were conducted in biologic fluids (blood serum). We have studied the influence of polymer materials researched on free-radical processes in blood serum in vitro. Lipid per oxidation was evaluated by induced chemiluminescence. We supposed that fullerene-containing polystyrene films may inhibit peroxidation processes.

Antimicrobial activity of the polystyrene-fullerene composites was tested against gram-positives (*Staphylococcus aureus*) and gram-negative (*Escherichia coli, Pseudomonas aeruginosa*) microorganisms and mushrooms of the type *Candida albicans*. The test results showed absolute death of the microorganisms under the modified film. It should be noted that dynamics of the microorganism inactivation persists during a month.

8.1   INTRODUCTION

Polymer-matrix composites have attracted much attention of researchers because of the perspective using of these materials in instrument engineering, medicine, biology, etc. [1]. Insertion of fillers, such as fullerenes, results to modify the original polymer matrix, which can lead to creation of materials with improved physical and chemical properties and the main service characteristics (mechanical and electrical). There are numerous publications deals with the mechanical, optical, tribological, thermo chemical properties and structure of polymers containing fullerene [2–8].

Modification by carbon nanoparticles results to occurrence of new properties of the polymer, for example, biological activity. Biological activity of fullerenes are due to, firstly, lipophilic properties, so that they can penetrate into the cell membrane, secondly, electron deficit, promoting to react with free radicals, and, thirdly, capacity of excited C_{60} to generate active oxygen species [9].

However biological activity of polymer/fullerene composites studied not enough. It can be assumed that the insertion of fullerenes into a polymer matrix will result in creation of biocomposites, which may be used as agents for drug delivery, antiseptic preparations [9].

Polystyrene (PS) is well-known film-forming polymer often used for different modifications with low molecular compounds of special properties, including fullerenes. PS is well dissolved in benzene, toluene, o-xylene, etc., which are also solvents for fullerenes. This dissolution peculiarity allows to employ widespread procedure of polymer/fullerene composite formation that consists in casting of the solution containing polystyrene and fullerene followed by the evaporation of solvent. That such a procedure we used.

This chapter includes the study of the physicochemical properties, structure and antioxidant effect of both polystyrene films and polystyrene/fullerene composite films. We also report results on intermolecular interactions of macromolecules of polystyrene and fullerene in polystyrene/fullerene composite films. Findings have very important practical significance for materials science, because polystyrene is used often for various modifications with low molecular compounds, including fullerenes.

8.2   EXPERIMENTAL PART

8.2.1   MATERIALS

Atactic polystyrene purchased from Aldrich, US, with molecular mass of 140,000, polydispersity index of 1.64, and melt mass flow index of 6÷9 g/10 min was used as a polymer matrix. Fullerenes C_{60} (NeoTechProduct Ltd, Russia) and fullerenes mix $C_{60}+C_{70}$ (Fullerene Technologies Ltd., Russia) were used as filler agents. For fabrication of films a solvent casting of perspective components from solutions was employed.

To produce the polystyrene films, a polymer batch was dissolved in o-xylene (17 wt. % of PS) and the solution was stirred for about 1 day. After casting onto a glass substrate, the solvent was slowly evaporated at room temperature over several days until the thin film formation.

Polystyrene/fullerene composition films have been fabricated as follows. Fullerene batches were dissolved in solvent at various concentrations. Then polystyrene batches were dissolved in all obtained solutions (17 wt. % of PS) and the mixed solutions were stirred for about 1 day before being cast into thin films. After casting the solvent was slowly evaporated at room temperature over several days to produce the composite films. By this technique, we prepared some of composites samples with various fullerene percentages in the form of film.

The samples obtained have been examined by optical microscope "Boetius" (Germany). We found both polystyrene films and fullerene-polystyrene composite films transparent, that is, the films are homogeneous on the optical level. Unmodified polystyrene samples were colorless, whereas the fullerene-polystyrene composite films were light purple. The intensity of color depended on the content of fullerene in the composite. The thickness of the films was in the range from 60 to 80 μm.

8.2.2   METHODS

Structure of both PS films and PS/C_{60} composite films was evaluated by X-ray diffraction (XRD) measurements on the base of Debay-Sherrer method. XRD patterns of film samples were obtained by X-ray diffractometer DRON-UM1 (Russia) equipped with MoK$_\alpha$ radiation that monochromate by the Zr-filter, $\lambda = 0.071$ nm. X-ray diffractometer was modernized for substances in condensed and polycrystalline state. The voltage and the current of the X-ray tubes were 40 kV and 40 mA, respectively. A scan rate of 0.04 degree/s was used. We investigated structure of the original polystyrene and fullerene-containing composites by the X-ray diffraction technique in wide angles from 2 to 15 degrees.

IR spectra of films were recorded by Avatar 360 FT-IR ESP spectrometer (Termo Nicolet, US).

DSC measurements were performed using DSC 204 F 1 apparatus (Netzsch, Germany) in argon atmosphere (15 $cm^3 \cdot min^{-1}$). A stack of films with a diameter of 5 mm was placed in a press-fitted aluminum crucible covered the pierced lid. The samples were undergone first heating up to 423 K with a scan rate of 10 $K \cdot min^{-1}$ to remove volatile substances from the polymer and cooled down to 283 K by means of liquid nitrogen. Second heating of the simples was carried out according to the following sequence: heating up to 293 K; isothermal mode at 293 K for10 minutes; and heating up to 423 K with a scan rate of 10 K·min^{-1}. The glass transition temperatures were determined from data of the second heating. The reference aluminum crucible was empty. All measurements were performed relative to the base line obtained with two empty crucibles. Three measurements required for each composite (base line, sample and standard) were carried out on the same day.

Influence of the PS/C_{60}+C_{70} nanocomposites on free-radical oxidation of lipids in biologic fluid (blood serum) was researched in vitro. Subject of research was native blood serum of 10 patients managed in V.N. Gorodkov Research Institute of Maternity and Childhood (Ivanovo, Russia). Specimen of pure PS or composite film (size 1.5 cm^2, weight 5 mg) was put into blood serum (1 ml). System was incubated for 1 hour at 4°C. Then, the film specimen was removed from biologic fluid.

The parameters of lipid peroxidation in serum after exposure to the film nanomaterials were determined by chemiluminescent analysis. The induced chemiluminescence (ChL) tests were performed by BChL-07 luminometer (Medozons, Russia). We used hydrogen peroxide and ferric sulfate as inductors of ChL. 0.1 ml of serum, 0.4 ml of phosphate buffer (pH 7.5), 0.4 ml of 0.01M ferric sulfate and 0.2 ml of 2% hydrogen peroxide were put into cuvette. Luminescence was registered for 40 s.

To estimate the intensity of lipid peroxidation we used following parameters:

- J_{max} is maximum intensity of ChL during the experiment. Value of J_{max} quantifies the level of free radicals, that is, gives an idea of the potential ability of the blood serum to free radical lipid peroxidation;
- *tanα* is tangent of maximum slope angle of ChL curve towards time axis. This value characterizes the decay rate of free radical oxidation, that is, quantifies an effectiveness of the antioxidant system;
- *A* is an area covered by intensity curve or total light sum. Value of *A* is inversely proportional to the antioxidant activity of the sample;
- $Z = AJ_{max}^{-1}$ is normalized light sum.

Free radical processes in serum have been studied after exposure to pure polystyrene films and fullerene-containing polystyrene films. The mean values of ChL parameters in native serum without film adding were used as controls. 8–10 measurements required for each film were carried out on the same day. The results have been expressed as percentages relative to controls. A p-value of 0.05 was chosen as the significance limit.

Antimicrobial activity of both polystyrene films and polystyrene/fullerene composites was tested against gram-positives bacteria (*Staphylococcus aureus*), gram-negative bacteria (*Escherichia coli, Pseudomonas aeruginosa*) and fungi of the type *Candida albicans*. The essence of the tests is to storage the test samples under conditions optimal for growth and development of bacterial and fungal cultures.

8.3 RESULTS AND DISCUSSION

8.3.1 X-RAY DIFFRACTION ANALYSIS

XRD patterns of both PS films and PS/C_{60} composite films with 0.03, 0.1, 0.5, and 1 wt. % of C_{60} are given in Figure 8.1. XRD of solid C_{60} reported in Ref. [10]. Comparing these data it can be seen the reflexes associated with fullerene are absent in XRD patterns of composite films Figure 8.1. Apparently, this is due to the concentration of C_{60} in the studied films is inadequate for the occurrence of reflex. This finding correlates to the data of Ref. [11], in which there is a comparison of wide-angle diffraction patterns

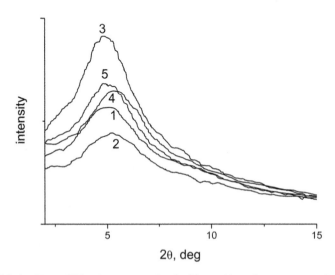

FIGURE 8.1 X-ray diffraction patterns for the films with various concentrations of C_{60}, wt. %: 0 (1); 0.03 (2); 0.1 (3); 0.5 (4); 1 (5).

of origin polymethylmethacrylate (PMMA), composites PMMA/C_{60} (1 wt. % of C_{60}), and PMMA/C_{60} (10 wt. % of C_{60}). It is found in Ref. [11] that diffraction pattern does not change under insertion of fullerene (up to 1 wt. %) into the polymer matrix. For opinion of the authors, the absence of these changes is due to the weak aggregation of C_{60}. But in diffraction pattern of PMMA/C_{60} composite (10 wt. % of C_{60}) there are additional peaks associated with fullerene aggregates [11].

It is shown in Figure 8.1, both PS films and PS/C_{60} composite films exhibit broad diffraction peak (halo). Its location does not depend on the film composition. Abscissa of maximum, $2\theta_m$, is equal to 5.2 degree. In the reciprocal space, the position of peak appears at 8.029 nm^{-1}. We calculated the repetitive characteristic distance in structural arrangement, h, using the relation [12]:

$$2h\sin\theta_m = K\lambda \tag{1}$$

where $K = 1.2 \div 1.3$. We obtained the value of h is equal to $0.94 \div 1.02$ nm for all studied composites.

The broad peak in the X-ray diffraction patterns of films is an indication of the existence of the intermediate-range order [13] in these materials. The full width at half maximum, Δs, makes it possible to specify a correlation length (a scale of the intermediate-range ordering) in the disordered phase, L, using the relation:

$$L = \frac{2\pi}{\Delta s} \tag{2}$$

where $s = 4\pi\sin\theta/\lambda$ the modulus of dispersion vector.

The values of L calculated by Eq. (2) are given in Table 8.1. The obtained data demonstrate the following. When the concentration of fullerenes in the

TABLE 8.1 X-Ray Diffraction Results of the PS and PS/C_{60} Films

C_{60} content, wt. %	Δs, nm^{-1}	L, nm
0	15.53	0.40
0.03	16.12	0.39
0.1	8.80	0.71
0.5	9.29	0.68
1.0	8.98	0.69

film does not exceed 0.03 wt. %, the value of L is 0.39÷0.42 nm. Greater correlation length (L = 0.68÷0.71 nm) occurs if concentration of C_{60} is 0.1÷1 wt. %. Thus insertion of fullerenes into polystyrene promotes "local ordering in the disordered phase." Note such conclusion is consistent with findings of Ref. [14]. It found in Ref. [14], if the content of fullerenes is more than 0.1 wt. %, the effects of the intermolecular interaction of fullerene with polystyrene dominate, which may result to cross-linking of the polymer chains.

8.3.2 OPTICAL PROPERTIES RESEARCH

Identification of interaction nature between fullerene and polymer in films was carried out by IR-spectroscopy. As an example, we analyzed the IR spectra of the original PS film and the modified PS films formed by casting from o-xylene (Figure 8.2). In the polystyrene IR spectrum the bands of the C = C stretching vibrations of the benzene ring appear as doublets in ranges of 1600–1585 cm^{-1} and 1500–1400 cm^{-1}. The bands corresponding to planar deformation vibrations of C-H bonds are in the region of 1300–1000 cm^{-1}. Out-of-plane deformation vibrations of C-H bonds are at 900–675 cm^{-1} [15]. Bands corresponding to valent vibrations in the phenyl rings are usually very sensitive to the possibility of conjugation π-electron density of the ring with

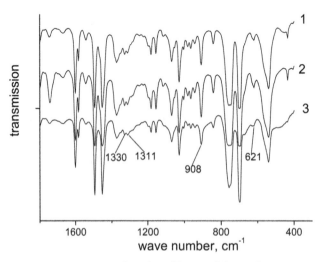

Figure 8.2. IR spectra for the films with various concentrations of $C_{60+}C_{70}$, wt. %: 0 (*1*); 0.02 (*2*); 0.035 (*3*).

reactive groups. In the IR spectrum of the fullerene molecule four vibrations with the absorption bands corresponding to 527, 577, 1183, and 1429 cm⁻¹ are active [16]. Absorption bands at 1429 cm⁻¹ are the most sensitive to the charge transfer.

When analysis of IR spectroscopy of the modified polystyrene samples we founded in spectra of the composites the new bands do not appear. However, in spectrum of film containing 0.035 wt. % of $C_{60}+C_{70}$ there are a change of contours of the absorption bands in the range of 700÷400 cm⁻¹ and displacement of the band maximum at 621 cm⁻¹ by 7 cm⁻¹ toward the higher frequencies. Quantitative interpretation of IR spectra is difficult due to overlapping of the bands corresponding to the vibrations of the phenyl ring and fullerene (1451 and 1429 cm⁻¹, respectively). Therefore, semi quantitative analysis of IR spectra of the studied films by the method of a base line and internal standard [17] was carried out. We determined the ratios of absorbance for characteristic bands to absorbance for the C-H bond at 908 cm⁻¹ selecting as the internal standard (Table 8.2).

Analysis showed in the IR spectra of the composite a significant change in modes relative intensities compared to the same parameters in the spectrum of polystyrene film is observed. This indicates the electronic structure of the phenyl rings varies considerably. We can assume non-covalent interaction of donor polystyrene macromolecule with fullerene acceptor molecule occurs.

It is known [16] that the molecular chains of polystyrene possess helical conformation that attributes by doublet at 1330 and 1311 cm⁻¹ in the vibrational spectrum of the polymer. In the spectra of fullerene-containing films, change in the ratio of intensities of these bands occurs. The observed effect is probably associates with a change in helical conformation of polystyrene macromolecules under the influence of the fullerene.

TABLE 8.2 The Relative Intensities of the Absorption Bands in the IR Spectra of PS and Composite Films

$C_{60}+C_{70}$ content, wt. %	D_{1452}/D_{908}	D_{1601}/D_{908}
0	2.44	2.13
0.01	2.19	2.08
0.035	3.16	2.28
0.10	2.81	2.31
1.0	3.45	2.40

8.3.3   DIFFERENTIAL SCANNING CALORIMETRY

Typical DSC curves for the films of original polystyrene and filled composite with fullerene ($C_{60}+C_{70}$) are shown in Figure 8.3. It can be seen that for all examined polymeric materials there is a reversible phase transition from the glassy state to elastic one, which manifests itself as a step of heat flow in endothermic direction.

We have characterized this phase transition by the following parameters: T_1 is the extrapolated temperature of the phase transition onset; T_2 is the extrapolated temperature of the phase transition end; $\overline{T}$ is the average temperature of the phase transition; T_g is the temperature of DSC curve inflection taken as the glass transition point; $\Delta T = T_2 - T_1$ is the temperature interval in which the phase transition occurs.

The average characteristics of the glass transition obtained from tree experiments for each polymer films are presented in Table 8.3. It was found that the $C_{60}+C_{70}$ addition affects all characteristic temperatures and temperature range of phase transition, ΔT, for the composite materials.

It can be seen in Table 8.3 that small additions of fullerene sharply reduce the glass transition temperature of the composite in comparison with original polymer film. The minimum value of T_g is observed when the concentration of fullerenes is equal to 0.01 wt %. Apparently, in this case, there is the

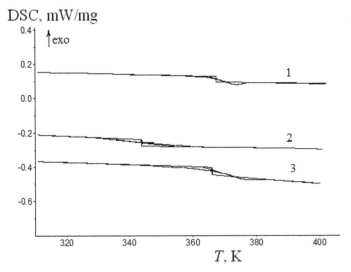

FIGURE 8.3   DSC curves of films: PS (*1*); PS/$C_{60}+C_{70}$ composite with 0.01 wt % of fullerene (*2*); PS/$C_{60}+C_{70}$ composite with 0.1 wt % of fullerene (*3*).

TABLE 8.3 Parameters of Phase Transition from Glassy State to Elastic One for the Pure and Filled Polystyrene Films with Various Contents of $C_{60}+C_{70}$

Fullerene content, wt %	T_1, K	$\overline{T}$, K	T_g, K	T_2, K	ΔT, K
0	357.0	361.5	363.8	365.8	8.8
0.01	328.1	343.8	335.8	355.1	27.0
0.02	341.6	348.6	345.0	359.3	17.7
0.035	347.0	358.4	357.4	366.5	19.5
0.1	365.3	369.1	371.6	373.5	8.2

plasticization of polystyrene with the fullerene [4, 14]. Molecules of $C_{60}+C_{70}$ embed between the chains and weaken interchain interactions in the polystyrene. It results in increasing mobility of polymer chain segments and reducing value of T_g. The noticeable extension of the temperature range for phase transition, $\Delta T = T_2-T_1$, at low concentration of fullerene (Table 8.3) confirms an evident plasticization effect of these additives on thermal properties of the composite.

The glass transition temperature increases with further increase in the $C_{60}+C_{70}$ content. When the fraction of fullerene is equal to 0.1 wt %, the T_g value for the composite exceeds that for pure polystyrene.

It is known that fullerene molecules are capable of strong intermolecular interaction due to the large number of conjugated double bonds [14]. Intermolecular interactions of polystyrene and fullerene are emerged on the values of T_g in the films with additions of more than 0.01 wt %. When the concentration of fullerenes is equal to 0.1 wt % and above, the effect of intermolecular interactions is dominant, the probability of interaction between fullerene molecules in the composite increases, that is manifested in increasing glass transition temperature as compared with the original polystyrene and, apparently, can lead to the physical cross-linking of polystyrene chains. Vice versa, at extremely low concentration (less than 0.01 wt %) the probability of interaction between fullerene and polystyrene molecules is very small because of there is one fullerene molecule per 55–60 polystyrene macromolecules (or one fullerene molecule per 70,000–80,000 monomeric unit) in the composite material. In this regime, weakening of interchain interactions in the polystyrene is dominant.

Early we researched the influence of the fullerene additives on structure of the polystyrene films formed by evaporation of the solvent [18, 19]. We concluded in films without fullerenes, packing of straightened chains parallel to each other predominated. This shows that a polystyrene molecule, which had the shape of a coil in solution, straightened and stretched itself along the surface of an aggregate when attached to it. When the concentration of fullerenes is equal to 0.035 wt % (one fullerene per 7–10 polystyrene macromolecules), attachment to aggregates occurred similarly. However, at such concentrations the intermolecular interactions between polystyrene and fullerene are appreciable, and under the influence of fullerene molecules, polystyrene molecules straightened with the formation of ordering elements in the arrangement of chains [18].

8.3.4 EVALUATION OF ANTIOXIDANT ACTIVITY

Figure 8.4 shows kinetics of chemiluminescence in serum after exposure of pure polystyrene and nanocomposite films. Peak of chemiluminescence due to free radical production was in 2 s of reaction. This can be explained by production of active oxygen species (HO_2^*, O_2^*, O_2, OH^-). The highest intensity, J_{max}, was registered in case of nanocomposites with 0.01 and 0.03 wt. % of fullerenes.

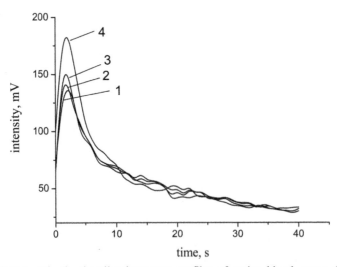

FIGURE 8.4 Kinetic chemiluminescence profiles of native blood serum (*1*) and after exposure to studied films (*2–4*): *2* – PS; *3* – PS/C$_{60}$+C$_{70}$ (1.0 wt. %); *4* – PS/C$_{60}$+C$_{70}$ (0.03 wt. %).

In Table 8.4 we represent the main ChL parameters for films prepared by casting of o-xylene solution. It can be seen in case of original polystyrene film the ChL parameters were approximate to controls. Value of J_{max} in case of PS/$C_{60}+C_{70}$composites is higher than for control serum samples. A light sum, A, was significantly increased only for films with 0.01 and 0.03 wt. % of fullerenes ($p<0.05$). For nanocomposite containing 1 wt. % of $C_{60}+C_{70}$ no significant changes in value of A was revealed. In addition we found both significant growth in value of tanα and reduction in value of Z in case of all fullerene-containing films (Table 8.4). So regardless of the fullerenes content the antioxidant activity of PS/$C_{60}+C_{70}$ composites is higher than for original polystyrene. It seems nanocomposites containing fullerenes were easy to react with oxygen species, preventing lipid peroxidation. We suggest the fullerene molecules can cross the external cellular membrane and they localize to the mitochondrions. Formation of oxygen free radicals occur due to the electrons leakage in the mitochondrial electron transport chain. Therefore, the localization of fullerenes near mitochondrions may contribute to their antioxidant action.

Note similar assumption has been proven for fullerene derivative $C_{61}(CO_2H)_2$ [20]. The authors of Ref. [20] researched distribution of the [^{14}C] radioactivity in various cellular compartments and concluded $C_{61}(CO_2H)_2$ molecules localize preferentially to the mitochondrions.

8.3.5 STUDY OF ANTIBACTERIAL ACTIVITY OF THE FILM MATERIALS

The results of research of antimicrobial action of both polystyrene films and polystyrene-fullerene composites are shown in Figure 8.5 and Table 8.5.

TABLE 8.4 Chemiluminescence Parameters in Blood Serum after Exposure to Original Polystyrene Film and Fullerene-Containing Composites

$C_{60}+C_{70}$ content, wt. %	A, %	J_{max}, %	tanα, %	Z, %
Controls	100.0	100.0	100.0	100.0
0	96.0	97.0	97.5	99.0
0.01	121.5 *	131.5 *	139.0 *	92.0 *
0.03	115.0 *	125.0 *	146.0 *	92.0 *
0.1	105.0	110.0 *	110.0 *	95.5
1.0	96.0	111.0 *	121.0 *	86.5 *

* – significant differences compared to control ($p<0.05$)

FIGURE 8.5   Effect of the pure PS film (*1*) and the modified PS film (*2*) on *Staphylococcus aureus Rosenbach.*

TABLE 8.5   Antibacterial Activity of the Film Materials

Fullerene content, wt %	R, mm		
	E. coli	*Staph. Aureus*	*E. coli+Staph. aureus*
0	0	0	0
0.03	2	3	1÷2
0.1	3	2	2

As can be seen there is a distinct zone of lysis (zone in which there is no microbial growth) around the sample of fullerene-polystyrene composite film against *Staphylococcus aureus* and *Escherichia coli*. But for the original polystyrene film there is no zone of lysis. Size of the lysis zone, R, that defines the inactivation degree of the material against bacteria, has been represented in Table 8.5.

From the data we can conclude that polystyrene becomes bacteriostatic action as a result of doping fullerenes. In addition we have found that polystyrene-fullerene composite films have been attributed by fungistatic action against fungi of the type *Candida albicans*. It is likely that one of the reasons

of microorganisms inactivation is interaction of the fullerenes with functional groups of the amino acids composing bacterial proteins. This results
in the cell membrane damage and destruction of the cell wall leading to their
death. It should be noted that dynamics of the bacteria inactivation persists
during a month.

ACKNOWLEDGMENTS

The study was supported by the Russian Foundation for Basic Research
(project No. 15–43–03034-a).

KEYWORDS

- **biological activity**
- **DSC**
- **fullerene**
- **IR spectroscopy**
- **polystyrene**
- **X-ray diffraction**

REFERENCES

1. Scholz, M., -S., Blanchfield, J. P., Bloom, L. D., Coburn, B. H., Elkington, M., Fuller, J.
D., Gilbert, M. E., Muflahi, S. A., Pernice, M. F., Rae, S. I., Trevarthen, J. A., White, S.
C., Weaver, P. M., Bond, I. P. (2011). The use of composite materials in modern orthope-
dic medicine and prosthetic devices: A review. *Compos. Sci. Technol.*, *71*, 1791–1803.
2. Badamshina, E. R., Gafurova, M. P. (2008). Characteristics of fullerene C_{60} doped polymers. *Polym. Sci. B*, *50*, 215–225.
3. Alekseeva, O. V., Barannikov, V. P., Bagrovskaya, N. A., Noskov, A. V. (2012). DSC
investigation of the polystyrene films filled with fullerene. *J. Therm. Anal. Calor.*, *109*,
1033–1038.
4. Weng, D., Lee, H. K., Levon, K., Mao, J., Scrivens, W. A., Stephens, E. B., Tour, J.
M. (1999). The influence of Buckminsterfullerenes and their derivatives on polymer
properties. *Eur. Polym. J.*, *35*, 867–878.
5. Jiang, Z., Zhang, H., Zhang, Z., Murayama, H., Okamoto, K., (2008). Improved bonding between PAN-based carbon fibers and fullerene-modified epoxy matrix. Compos.
Part A – Appl. Sci., *39*, 1762–1767.

6. Tajima, Y., Tezuka, Y., Yajima, H., Ishii, T., Takeuchi, K., (1997). Photo-cross-linking polymers by fullerene. *Polymer, 38,* 5255–5257.

7. Ogasawara, T., Ishida, Y., Kasai, T., (2009). Mechanical properties of carbon fiber/fullerene-dispersed epoxy composites. *Compos. Sci. Technol., 69,* 2002–2007.

8. Zhao, L., Guo, Z., Cao, Z., Zhang, T., Fang, Z., Peng, M., (2013). Thermal and thermo-oxidative degradation of high density polyethylene/fullerene composites. *Polym. Degrad. Stabil., 98,* 1953–1962.

9. Da Ros, T., (2008). Twenty years of promises: fullerene in medicinal chemistry. In: Cataldo, F., Da Ros, T., (eds.). Carbon Materials: Chemistry and Physics, vol. 1: Medicinal Chemistry and Pharmacological Potential of Fullerenes and Carbon Nanotubes, Springer. pp. 1–21.

10. Krätschmer, W., Lamb, L. D., Fostiropoulos, K., Huffman, D. R. (1990). Solid C_{60}: a new form of carbon. *Nature, 347,* 354–358.

11. Ginzburg, B. M., Pozdnyakov, A. O., Shepelevskij, A. A., Melenevskaya, E. Y., Novoselova, A. V., Shibaev, L. A., Pozdnyakov, O. F., Redkov, B. P., Smirnov, A. S., Shiryaeva, O. A. (2004). Structure of fullerene C_{60} in a poly(methyl methacrylate) matrix. *Polym. Sci. A, 46,* 169–175.

12. Guinier, A., (2001). X-ray Diffraction: In Crystals, Imperfect Crystals, and Amorphous Bodies, Dover Publications, New York.

13. N'Dri, K., Houphouet-Boigny, D., Jumas, J.-C. (2012). Study of first sharp diffraction peak in As_2S_3 glasses by X-ray powder diffraction method. *J. Non-Oxide Glasses, 3,* 29–37.

14. Gladchenko, S. V., Polotskaya, G. A., Gribanov, A. V., Zgonnik, V. N. (2002). The study of polystyrene-fullerene solid-phase composites. *Tech. Phys. Russ. J. Appl. Phys., 47,* 102–106.

15. Dechant, J., Danz, R., Kimmer, W., Schmolke, R., (1972). Ultraspectroscopische Untersuchungen an Polymeren, Akademie (Infrared Spectroscopy of Polymers) Verlag, Berlin, (in German).

16. Konarev, D. V., Lyubovskaya, G. N. (1999). Donor–acceptor complexes and radical ionic salts based on fullerenes. *Russ. Chem. Rev., 68,* 19–38.

17. Smith A. L. (1979). Applied Infrared Spectroscopy. Fundamentals, Techniques, and Analytical Problem-Solving, Wiley, New York.

18. Alekseeva, O. V., Bagrovskaya, N. A., Kuzmin, S. M., Noskov, A. V., Melikhov, I. V., Rudin, V. N. (2009). The influence of fullerene additives on the structure of polystyrene films. *Russ. J. Phys. Chem. A, 83,* 1170–1175.

19. Alekseeva, O. V., Rudin, V. N., Melikhov, I. V., Bagrovskaya, N. A., Kuzmin, S. M., Noskov, A. V. (2008). Kinetics of formation of hierarchical nanostructures in polystyrene films containing fullerene. *Dokl. Phys. Chem., 422,* 275–278.

20. Foley, S., Crowley, C., Smaihi, M., Bonfils, C., Erlanger, B. F., Seta, P., Larroque, C., (2002). Cellular localization of a water-soluble fullerene derivative. *Biochem. Biophys. Res. Commun., 294,* 116–119.

CHAPTER 9

COMPOSITES ON THE BASIS OF SAWDUST WITH SOME ORGANIC AND INORGANIC BINDERS

O. MUKBANIANI, J. ANELI, G. BUZALADZE, E. MARKARASHVILI, and T. TATRISHVILI

Department of Macromolecular Chemistry, Iv. Javakhishvili' Tbilisi State University, I. Chavchavadze Blvd., 1, Tbilisi 0179, Georgia, E-mail: omarimu@yahoo.com

CONTENTS

ABSTRACT

Novel composites on the basis of dry sawdust and different organic and inorganic binders have been obtained in the spatial molds. Phenylethoxysiloxane of two types (PhES-50 and PhES-80), polyethylene, colophony, wood glue and liquid glass were used as binders, concentrations of which varied in region 3–15 wt%. Spectroscopic investigations by method FTIR have shown a presence of some chemical bonds between components of the composites in result of reactions between active groups of the ingredients. These bonds may be the main reason of improving of physical mechanical and thermal properties of obtained composites and their water resistance. It is established that these properties in general depend on the concentration of the binders. It is shown that the maximal values of the noted parameters for the composites appearance at relatively low concentrations of binders. Especially in case of PhES-50 and PhES-80 improved properties of composites are reached already at their 3–5 wt%. With rather high values of these parameters are characterized the composites, containing polyethylene due to their good wet stability and composites containing 2 binders simultaneously. The experimental results show that at right selection of the used ingredients and their concentrations the materials with rather high exploitation properties may be obtained.

9.1 INTRODUCTION

Creation of wood polymer composites (WPC) presents very actual problem today. Obviously it is due to very large perspectives of application of these products in the life. At this time there are many scientific articles and inventions in the sphere of synthesis of WPC on the basis of different wood materials and binders – thermoplastic and thermosetting polymers [1–7]. For example, different type of wood was combined with different combinations of hexandiol diacrylate, hydroxyethyl metacrylate and maleic anhydride [2]. In this work treatment slowed the rates of water vapor and liquid water absorption. It gives the hardest and most dimensionally stable WPC. WPC prepared using hydroxyethyl metacrylate were hardened than specimens made without hydroxyethyl metacrylate and excluded water and moisture more effectively. It was established the effect maleic anhydride (50%) treatment on steam and water absorption of WPC prepared from wheat straw, cane bagasse and teak wood sawdust with sizes 425 mcm using novolac as matrix [3]. The values of shock viscosity and water absorption of the composite are 2–3 times better than that for analogous material obtained without treatment. WPC on the basis of cellulose, hemicelluloses and lignin impregnated with MMA, styrene, acrylonytril, vinyl chloride, acrylamide after polymerization initiated with gamma-radiation and electron beams, from one side, and polymerization with different catalysts, from the other one, are characterized with improved physical properties [4]. In the work [5] hydrophobisity of maritime pine wood surfaces was modified by silicone. A generic method for the introduction of a variety of silicones at the surface of pre-treated wood was developed. The initial treatment of wood with maleic anhydride and allyl glycidyl ether resulted in oligoesterified wood bearing terminal alkenes. The hydrosilylation of these groups, performed with hydride-terminated silicones, led to very hydrophobic surfaces, even after extensive soxhlet extraction with good solvents for silicones. The silicones appear to be attached to the wood by covalent bonds. The interest results were obtained in the work [6], where trembling aspen has been impregnated with the coupling agent γ-methaacryloxypropyltrimethoxysilane and the hygroscopicity and antishrink efficiency before and after water extraction were defined. There was a hygroscopicity depression with treatment, but not enough to account for the observed antishrink efficiencies. There was limited reversibility after water extraction. This indicated that the wood was interacting with part of the sorbed silane more strongly than with water and that was a diversity of sorption environments within the wood-silane complex. Wood plastic composite was prepared with simul and MMA in the presence of methanol used as swelling agent at different proportions [7]. In this WPC Sulfuric acid was incorporated into the impregnating solution to investigate its effect on the polymer loading and tensile strength of the composite. 1% acid solution induces the highest polymer loading.

In the work [8] it is established that PMMA does not form bonds with the hydroxyl groups of the cellulose fibers but simply bulks the void spaces within the wood structure Here it was used vinyl triacetoxysilane before impregnation with MMA in order to evaluate effects of the silane coupling agent on the physical properties of WPC. The composites were prepared using a catalyst accelerator method and polymerization process was initiated at room temperature. This method involves impregnation of wood samples using MMA, containing benzoil peroxide (1%), louril peroxide (0.5%) and accelerator N,N-dimethyl aniline (0.5%). This method requires no heating in the initiation process. The material obtained using this method has improved compression strength and hardness. Morphology of composite was examined with scan electron microscopy.

Various inorganic silicon compounds are used for the treatment of wood [9]. Among these compounds silico fluorides represent one of the long-known classes of wood preservatives based on silicon. Data on an organic fluoro silicon compound ("silafluofen") are additionally presented. The mode of action of these fluorides is based on their biocidal action. "Water glass," an alkali silicate, was able to enhance the durability of wood but showed some important drawbacks. Because of its high hygroscopicity and its high pH values, increased moisture absorption and strength loss of wood was frequently observed. Wood treated with tetra alkoxy silanes showed an enhanced dimensional stability, especially when the hydrolysis and the condensation of the silanes was controlled to react within the cell wall. Positive results are obtained in the works devoted to creation of WPC at using of silicon- organic compounds and sol-gel systems [10], where some compounds as organofunctional silanes which are mostly applied in combination with tetra alkoxy sylanes (sol-gel process) as well as chlorosilanes and trimethylsilyl derivatives were proposed for a full impregnation treatment

of wood. The effects related to the various treatments vary from an increase in dimensional stability, durability and fire resistance to an enhanced hydrophobization of wood.

9.2 EXPERIMENTAL PART

9.2.1 MATERIALS

There were prepared the composites based on dry sawdust with different binders as alcoxysilanes (PhES-50 and PhES-80) and polyethylene. By way of dry mix of ingredients there were prepared the composites with both binders. The composites were prepared by using of method of hot pressing of high-dispersed (50 mcm) components under pressures about 150 kg/cm^2 and 130°C in the spatial molds during 15 min. It was prepared 2 types samples: cylinder (for investigations of water absorption) and rectangle forms (for conducting of mechanical investigations).

9.2.2 MEASUREMENT

Following characteristics were studied for the composites with straw: Furie transformation infrared spectra, The KBr pellets of samples were prepared by mixing (1.5–2.00) mg of samples, finely grounded, with 200 mg KBr (FT-IR grade) in a vibratory ball mixer for 20 s.

Viscosity (by Sharpy), thermal stability (by Vicat), water absorption and investigations with use of optical microscope. The preliminary experiments have shown that the composite's characteristics were essentially depended on the external temperature and pressure. Consequently we obtained the samples under conditions of optimal temperature and pressure, which the best characteristics of composites were ensured at variation of the types and concentrations of binders.

The micro-structure of the wood composites was studied on the optical microscope of type NMM-800RF/TRF. With aim of investigation of the structure of the materials we have prepared sample by following manipulations: after polishing of the surface of sample with longevity about 1 cm we transferred it to the emery paper and continued this procedure during 1 h, after which the sample was polished on the coarse calico. After this the sample was transferred to the object glass of the optical microscope. There were studied the separate parts of the sample by use of the noted optical microscope. Scanning electron microscopic investigation was carried out on high-performance devices (SEM) JSM-6510LV with high resolution of 3.0 nm at 30 kV, for fast characterization and imaging of fine structures.

9.3 RESULTS AND DISCUSSION

The aim of our work is obtaining of new wood polymeric composites (WPC) on the basis of sawdust powders and ecologically friendly different organic binders.

It is known that the sawdust contains cellulose structural rings with hydroxyl groups. One of the used by us substances phenylethoxysilane (PhES-50and PhES-80) contains ethoxyl groups. The noted groups participate to the etherification reaction with a binder through the macromolecular and intra-molecular reactions. Processes that occur during the curing are complex and varied. A modern look at an overview of the curing liquid glass itself and in the various homogeneous and heterogeneous systems, the most widely encountered in practice, is presented in a number of reviews [1–3, 6]. Acting as an adhesive or binder liquid glass system goes from liquid to solid in many ways and may be divided into three types:

1. The loss of moisture by evaporation at ordinary temperatures;
2. The loss of moisture from the system, followed by heating above 100°C;

3. The transition to the solid state by introducing specific reagents, which is called hardeners. Naturally, these three types are used in combination.

In solution the polymerization degree depends on two factors – the silica modulus and the solution concentration. Each solution has a distribution of degree of polymerization of anions. Distribution is superimposed on the polymer distribution of anions on the charges, which is also determined by these two factors.

The processes occurring in the silicate solution, governed by two revers‑ible reactions:

$\equiv$SiOH + HO- —® $\equiv$SiO- + H_2O

$\equiv$SiOH + $\equiv$SiO- —® $\equiv$Si—O—Si$\equiv$ + HO-

On the intermediate stages it is possible the reactions between of $\equiv$Si-OH band contained liquid glass and cellulose molecule, which are take place with dehydration reactions. These reactions go through obtaining of three dimension structures. Formation of such structures takes place also at use of Na_2SiF_6, which accelerates the processes.

The colophony is in the structure of different plants main structure containing the isomeric acid rings. The noted structural ring can introduce to reaction both with PhES-50 or PhES-80 and with liquid glass. At presence of colophony it may be realized the donor-acceptor bond with straw matrix, which is connected with formation of additive intermolecular forces and leads to increasing of material strengthening.

9.3.1 FOURIER TRANSFORM INFRARED (FTIR) SPECTROSCOPY INVESTIGATION

For samples Fourier transform infrared spectroscopy investigation has been carried out in KBr.

We have investigated the FTIR spectra for pure sawdust, binders, two and threecomponents composites systems. It is interesting, that in composites based on sawdust one can see the absorption pick characteristic for asymmetric valence oscillation of bonds $\equiv$Si-O-Si$\equiv$ with maximum near 1066 cm^{-1}, which possess to siloxane bond in the fragment of cyclo-tetrasiloxane fragment. Besides of the sorption strips are coincided one to other. In the spectrum one can see the absorption strips 1267, 1373, 1429, 1515, 1600–1650, 1733, 2800–2950, 3363 typical for methyl groups, C-H bond absorption (-C/C-/CH$_3$), CH$_2$ cellulose –lignin, C = C aromatic, C = C alkene, (C = O etheric bond), C-H methyl, methylene and phenyl groups, O-H alcoholic group, respectively. In the spectrum of PhES-80 it is seen two maximums in the region of 1060 and 1132 cm^{-1} corresponding to asymmetrical valence oscillations of the fragment Si-O-Si and absorption stripe in the zone 1349 and 1483 cm^{-1} of Si-OC$_2$H$_5$.

9.3.2 MECHANICAL STRENGTH OF THE COMPOSITES

To testing were underwent the composites obtained at different temperatures and pressures. Testing on impact viscosity of the same samples was conducted on the apparatus of Sharpy type. Table 9.1 includes the numerical values of the mechanical strength on bending and impact viscosity.

Mechanical strength of the composites at bending deformation was investigated by mean of the standard method. On the Table 9.1, the numerical data of the bending strength for the obtained by us composites are presented.
[1]Average sizes of sawdust particles is no morethan 50 mcm.

On the basis of the table data it can be made the following conclusions:

1. The strength of composite on the basis of sawdust as filler (95 wt%) and PhES80 (5 wt%) as binder in general depends on the technological parameters – temperature and pressure. Namely the strength of composites obtained at increasing temperature in the range 90–120°C and fixed pressure (17 MPa) monotonically increases in the range 4.5 up to 21 MPa. The composites obtained at increasing pressure (8–15 MPa) and constant temperature (110°C) exhibit high value of this parameter at relatively low pressures (22 MPa at technological pressure 8 MPa). Probably more monolith structure in these materials creates at optimal conditions of the technological regime. In this case high values of the bending strength (20–22 MPa) have composites, obtained under two types of regimes: when temperature is about 120°C and pressure 17 Mpa or when T = 110°C and P = 12 MPa. The level of dispersion play also significant role in formation of mechanical properties – mechanical strength of these composites is the higher the lower is the average diameter of sawdust particles (compare samples N20 and N21).

2. The composites, where PhES-50 is used instead of PhES-80 the dependence of the bending strength practically is analogical, but the

TABLE 9.1 Bending Strength and Impact Viscosity of Composites Obtained at Different Temperatures and Pressures with Exposition Time 10 min (Average Size of the Sawdust Particles is Less than 1 mm)

S. No.	Ingredients (wt%)	Temperature of composite formation, °C	Pressure of composite formation, kg/cm²	Bending strength, MPa	Impact strength, kJ/M²
1	PhES80 (5) + sawdust (95)	90	17	4.5	16.3
2	PhES80 (5) + sawdust (95)	100	17	13.4	17.0
3	PhES80 (5) + sawdust (95)	110	17	12.6	19.4
4	PhES80 (5) + sawdust (95)	120	17	20.9	20.8
5	PhES80 (5) + sawdust (95)	110	8	12.6	18.8
6	PhES80 (5) + sawdust (95)	110	12	22.3	18.1
7	PhES80 (5) + sawdust (95)	110	13	12.6	17.2
8	PhES80 (5) + sawdust (95)	110	15	13.4	18.8
9	PhES50 (5) + sawdust (95)	110	8	13.4	21.6
10	PhES50 (5) + sawdust (95)	110	10	12.5	16.3
11	PhES50 (5) + sawdust (95)	110	12	14.6	18.9
12	PhES50 (5) + sawdust (95)	110	15	15.6	20.1
13	PhES50 (3) + sawdust(94)+ PE (3)	110	10	18.6	21.8
14	PhES50 (3) + sawdust(94)+ PE (3)	110	12	18.6	17.0
15	PhES50 (3) + sawdust(94)+ PE (3)	110	15	15.7	20.1
16	PhES50 (3) + sawdust(94)+ PE (3)	110	13	14.5	13.7
17	PhES50 (3) + sawdust(94)+ PE (3)	120	13	20.9	18.6
18	PhES50 (3) + sawdust(94)+ PE (3)	130	13	19.7	17.3
19	PhES50 (3) + sawdust(94)+ PE (3)	150	13	21.8	17.7
20	PhES50 (3) + sawdust(94)+ PE (3)	120	12	22.3	16.9
21	PhES50 (3) + sawdust(94)+ PE (3)	120	12	33.2	19.8

conditions of obtaining of materials with high values of the bending strength are changed to definite extent. This parameter is changed in the range 12–15 MPa (composites were obtained at constant temperature and different pressures) and 14–19 MPa (obtained at different temperatures and constant pressures).

3. The composites contained polyethylene (5 wt%) generally exhibit more high mechanical properties and depend on the regime weaker in comparison with ones don't contenting polyethylene. Obviously the polyethylene in this composite plays the role of additive binder. By more high indexes are characterized the composites including high dispersive sawdust and polyethylene. So for composites based on high dispersive filler, polyethylene and PhES-80 the strength reaches 33 MPa. This result may be described with high value of the sawdust particles surface and, consequently to creation of high number of bonds between filler particles and binder molecules. The experimental results show that the technological conditions essentially influence on the microstructure of composites. Obviously it is necessary to provide the optimization of thermo-dynamical conditions for obtaining the composites with best technical parameters.

4. The samples obtained were tested on the impact strength. The samples 1–4 containing the composites based on PhES80 and sawdust (the particles sizes up to 1 mm) is characterized by increasing of this parameter at increasing of obtaining temperature (90–120°C) at constant pressure (17 Mpa). This fact shows that at fixed pressure the conditions of formation of the solid composites with high impact factor are improved. It is possible the passing of some heterogeneous chemical reactions between active groups on the surface of sawdust and PhES, which may be reason of increasing of the impact strength of these composites to some extent. However, for the composites obtained at constant temperature (110°C) and variable pressures (8–15 MPa) the impact strength changes for the samples obtained at 15 MPa. These results show once more that formation of the composites with high impact strength may be at optimal pressures and temperatures. Analogical results we obtain for composites based on PhES50 and sawdust with same relation of components as in case of composites based on PhES80. We tried improve the mechanical properties of the composites by introduction to them polyethylene

of high density. However, obtained results show some insignificant improve of the properties if the particles size of sawdust are about 1 mm, while the composites containing more dispersive particles (up to 0.5 mm) the results are significantly better. In general obtaining of wood composites with high mechanical strength is due to true selection of the ingredients of composites and optimal technological conditions effectively influencing on the microstructure of materials.

9.3.2.1 Measurement of Thermal Stability by Method of Vicat

There were conducted the investigations on study of thermal stability of composites by method Vicat. Temperature dependence of the softening of composites has monotonic increasing character with approaching to definite limit values (Figures 9.1–9.3). Analysis of the curves of these figures allow make the following conclusions:

1. The composites based on PhES-80 with sawdust are characterized with improved thermal stability, when the "technological" pressure increases till definite level. This result is obvious, because of two reasons: (a) the intergel volume of communicated micro-empties distributed chaotically in the composite body reduced at relatively

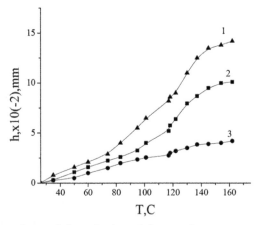

FIGURE 9.1 Dependence of the softening of the sample on temperature for composite PhES80 (5) + sawdust (95), obtained under conditions of constant temperature 110°C and pressures: 10 (1), 12 (2), and 15(3) Mpa.

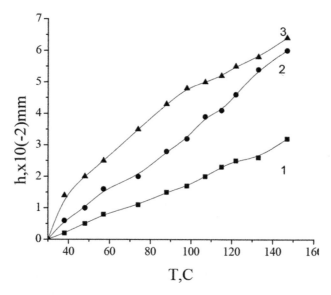

FIGURE 9.2 Dependence of the softening of the sample on temperature for composites PhES50 (5)+ sawdust (95), obtained under conditions of constant temperature 110°C and pressures: 15(1), 12 (2), and 10 (3) Mpa.

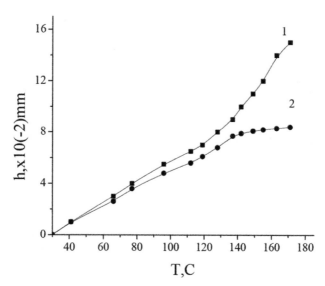

FIGURE 9.3 Dependence of the softening of the sample on temperature for composites PhES80 (3 wt%) + sawdust (97 wt%) with particles sizes up to 1 mm + PE (3 wt%) (1) and the same composite containing sawdust with particles sizes up to 0.5 mm (2) obtained under conditions of constant temperature 120°C and pressure 12 MPa.

high pressures and, consequently, the rigidity and thermal stability of the material increases; (b) PhES80 creates intensive heterogeneous chemical bonds in composites.

2. The composites based on PhES-50 with sawdust are characterized with rather high thermal stability in comparison with analogs based on PhES-80. Probably the composites based on PhES-50 contain non-great numbers of the micro-empties. Dependence of the composite's softening on the technological pressure is same as for composites with PhES-80 (the higher is this pressure the higher thermal stability). So the molecular segments of PhES-50 have less mobility in comparison with PhES-80, which is characterized with high branching of the molecular fragments.

3. The effect of the particles sizes in the softening and thermal stability of the composites clearly seen from Figure 9.3. The composite containing the particles with sizes up to 1 mm is less stable than the analog with same sawdust but with particles up to 0.5 mm. This result is interpreted in the same terms as in case of considered above (the intergel surface of the heterogeneous interaction is more in case of small particles and respective composite is more rigid than that containing big sawdust particles).

9.3.2.2   The Measurement of Water Absorption

Water absorption is one of the important characteristics of such materials as wood composites, because they are often used in conditions of damp climate. Products included in the wood structure in result of evaporation of the secondary products of these reactions. This phenomenon may be connected with formation of the associates on the basis of binder molecules and water ones.

With low water absorption is characterized the composites containing colophony. Here it is exhibited the ability of this binder to fill well of this empties. The composites containing binary binders are characterized with different water absorption, the amount of which depends on the character of interactions of them with water molecules. The experiment shows that the water of the wood composites may be changed by selection of type and concentration of the binders.

Another situation realized in case of the composites with LG. Here the weight of the composites after 24 h exposition in the water increases. In

TABLE 9.2 Water Absorption of the Composites Based on Saw Dust and Organic and Inorganic Binders (the Samples are Obtained under Conditions of t = 120°C and p = 12 MPa)

S. No.	Composite (wt. %)	Weight, G	Volume, cm	Density, g/cm³	Weight after exposition in the water during 3 h, G	Water absorption after exposition in the water during 24 h, %
1.	Sawdust (97), ES(3)	7.1	5.1	1.4	7.0	−1.1
2	Sawdust (94), ES (3), PE(3)	6.7	4.9	1.4	6.7	−0.8
3	Sawdust (90), ES(5), PE(5)	5.2	3.7	1.4	6.3	−0.7
4	Sawdust (92), PhES-80(3), PE(5)	4.0	3.4	1.4	3.9	−1.7
5	Sawdust (90), PhES-80(5), PE(5)	6.4	4.8	1.4	6.4	−0.5
6	Sawdust (95), PhES-80(5)	4.8	3.5	1.4		
7	Sawdust (95), LG(5)	6.2	4.4	1.4	6.0	−2.46
8	Sawdust (94), PhES-50(3), PE(3)	5.3	3.8	1.4		
9	Sawdust (92), PhES-50(3), PE(5)	5.5	3.9	1.4		
10	Sawdust (92), PhES-50(5), PE(5)	5.1	3.9	1.3	5.1	0.1
11	Sawdust (90), PhES-50(5), PE(5)	4.8	3.7	1.3	4.7	−1.0
12	Sawdust (90), PhES-50(5), PE(5)	5.3	4.2	1.3	5.3	−0.6
13	Sawdust (95), PhES-50(5)	5.0	3.9	1.3	5.0	1.1

this case it may be supposed that LG adsorbs the water molecules easily at relatively low concentrations. At increasing of concentrations of LG the micro-empties of the composites saturate and consequently the diffusion of the water molecules to all composites decreases.

With low water absorption is characterized the composites containing colophony. Here it is exhibited the ability of this binder to fill well of this empties.

The composites containing binary binders are characterized with different water absorption, the amount of which depends on the character of interactions of them with water molecules. The experiment shows that the water absorptionof the wood composites may be changed by selection of type and concentration of the binders.

9.4   CONCLUSIONS

Composites on the basis of dry sawdust and some organic (PhES, polyethylene) binders under conditions of optimal pressure and temperature in the molds with special forms have been obtained. Best results are achieved at relatively low concentrations of the binders. Structural investigations by means of FTIR show that with rather high physical–mechanical and thermal properties are characterized the materials, in which a formation of new chemical bonds takes place in result of chemical reactions between active groups of composite's ingredients. With more good properties are characterized the composites containing small particles of sawdust and both PhES and PE binders simultaneously.

ACKNOWLEDGMENTS

The financial support of the Georgian National Science Foundation Grant #STCU 5892 is gratefully acknowledged.

KEYWORDS

- binders
- composites
- phenylethoxysiloxane
- physical properties
- sawdust

REFERENCES

1. Kliosov, A. A. (2007). *Wood-Plastic Composites*. John Wiley & Sons: Hoboken, New Jersey, 698 pp.
2. Ellis, W. D., & O'Dell, J. L. (1999). *J. Appl. Polym. Sci., 73*(12), 2493–2505.
3. Patil, Y. P., Gajre, B., Dusane, D., & Chavan, S. (2000). *J. Appl. Polym. Sci., 77*(13), 2963–2967.
4. Li, Y. (2002). *Wood Polymer Composites*, Northeast Forestry University: Harbin, China.
5. Sebe, G., & Brook, M. A. (2001). *Wood Sci. and Technology, 35*, 269–282.
6. Brember, K. I., & Shneider, M. H. (1985). *Wood Sci. and Technology, 19*, 75–81.
7. Husain, M. M., Khan, M. A., & Idriss, K. M. (1996). *Radiat. Phys. Chem., 47*(1), 149–153.
8. Elvy, S. B., & Dennis, G. R. J. (1995). *Material Processing and Technology, 48*(1–4), 365–371.
9. Mai, C., & Militz, H. (2004). *Wood Science and Technology, 37*(5), 339–348.
10. Mai, C., & Militz, H. (2004). *Wood Science and Technology, 37*(6), 453–461.

CHAPTER 10

AGING PROCESS OF GOLD NANOPARTICLES SYNTHESIZED IN SITU IN AQUEOUS SOLUTIONS OF POLYACRYLAMIDES

NATALIYA KUTSEVOL,[1] VASYL CHUMACHENKO,[1] YULIIA HARAHUTS,[1] and ANDRIY MARININ[2]

[1]*Taras Shevchenko National University, Faculty of Chemistry, 60 Volodymyrska Str., Kyiv, 0160, Ukraine, E-mail: kutsevol@ukr.net*

[2]*Problem Research Laboratory, National University of Food Technology, 68, Volodymyrska Street, 01601, Kyiv, Ukraine*

CONTENTS

ABSTRACT

Gold sols were synthesized in polymer matrices. The star-like copolymers Dextran-graft-Polyacrylamide and linear Polyacrylamide were used. The influence of macromolecular structure of polymer matrix on size distribution gold nanoparticles in sols and their aging effect were studied by Dynamic

Light Scattering (DLS). It was shown that storage stability and nanoparticles size distribution are affected by molecular architecture of polymer matrix. Branched structure of polymer matrix provides higher sol stability in comparison with linear one.

10.1   INTRODUCTION

In recent years nanosystems based on gold nanoparticles (Au NPs) were proposed to be used as carriers for antitumor drugs for photothermal[1, 2] and photodynamic therapy (PDT) [3, 4]. Moreover, researches maintain for Au NPs to be not only passive carriers but allow to improve their activity. Targeted delivery offers the opportunity to increase the PDT efficiency to the cancer cells and minimize potential side effects to healthy tissues. For practical application it is important to use the sols having high storage stability and high concentrations. Well-defined nanoparticles size distribution in the sols becomes significant because of quantum-sized effects determining nanosystem properties. Moreover, it is necessary to know what changes take place in high concentrated sols during its storage. Aggregation stability of sols can be reached by using water-soluble polymers [5].

In this chapter, the influence of macromolecular structure of polymer matrix on size distribution of Au NPs in sols were studied by Dynamic Light Scattering (DLS) during 6 months of sol storage. As it was shown in Ref. [6], DLS allows to measuresize distribution of multicomponent sols.

10.1.1   MATERIALS

Thetrachloroauric acid (HAuCl$_4$), Sodium Borohydride(NaBH$_4$) were purchased from Aldrich and used without further purification. Copolymers Dextran-graft-polyacrylamide (D-g-PAA), its anionic derivative D-g-PAA (PE) and linear polyacrylamide (PAA) were used as polymer matrix for Au NPs synthesis. The stable sol was not synthesized in PAA(PE) linear matrix. In our previous work D-g-PAA polymers had been successfully used for Ag NPs synthesis [7].

Molecular parameters of the polymer matrices are represented in the Table 10.1. Polymer characteristics were determined by SEC and potentiometer: M$_w$, the weight average molecular weight; I = M$_w$/M$_n$, the polydispersity index; R$_g$, the radius of gyration; A, the chemical charge fraction of polyelectrolyte's obtained by alkaline hydrolysis of polyacrylamide.

TABLE 10.1   Molecular Parameters of Polymer Matrices

Sample	$M_w \times 10^{-6} [g \cdot mol^{-1}]$	$I = M_w/M_n$	R_g, nm	A [%]
PAA	1.43	2.4	68	—
D-g-PAA	1.57	1.81	67	—
D-g-PAA(PE)	1.57	1.81	—	37

10.1.2   METHODS

10.1.2.1   Au NPs Synthesis

Au NPs were synthesized by in situ by reduction of $HAuCl_4$ in aqueous polymer solution using $NaBH_4$ as reducing agent.

10.1.2.2   Transmission Electron Microscopy (TEM)

For the sample preparation 400 mesh Cu grids with plain carbon film were rendered hydrophilic by a glow discharge treatment (Elmo, Cordouan Technologies, Bordeaux, France). A 5µl drop was deposited and let adsorbed for 1 min then the excess of solution was removed with a piece of filter paper. The observations of the AgNPs were carried out on two TEMs, TecnaiG2 or CM12(FEI, Eindhoven Netherlands) and the images were acquired with anCCD Eagle camera on the Tecnai and a Megaview SIS Camera on the CM12.

10.1.2.3   Quasy Elastic Light Scattering (QELS)

DLS measurements were carried out using Zetasizer Nano ZS90(Malvern Instruments Ltd., UK). The apparatus contains a 4 mW He-Ne laser with a wavelength 632.8 nm and the scattered light is detected at an angle 60°.

10.1.2.4   UV-Visible Spectroscopy

UV-visible absorption spectra of silver sols were recorded by Varian Cary 50 scan UV-visible spectrophotometer (Palo Alto, CA, USA) from 200 to 800 nm.

10.2   RESULTS AND DISCUSSIONS

Images of aqueous Au NPs stabilized by polymers of different structures and UV vis spectra of these sols are shown in Figure 10.1a. A color of sols

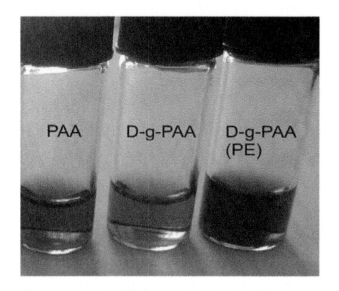

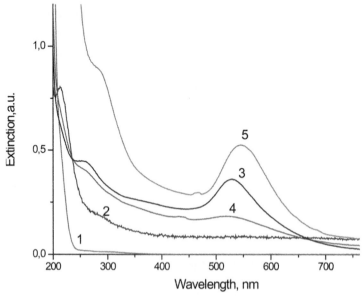

FIGURE 10.1   Image (a) and UV-vis spectra (b) of Au sols synthesized in different polymer matrices. 1 – polymer solution; 2 – HAuCl₄ in polymer solution; 3 – Au NPs in linear PAA; 4 – Au NPs in D-g-PAA; 5 – Au NPs in D-g-PAA(PE).

obviously depends on polymer matrix type, because other synthesis conditions were the same. UV-visible spectroscopy was used for primary characteristic of Au sols. Typical Surface Plasmon resonance for gold nanoparticles was observed (Figure 10.1b).

It is seen, the nature of polymer matrix influences on the form of extinction spectra and position of maxima. For linear PAA maximum position at 530 nm may correspond to spherical particles of 10–12 nm in size [8, 9]. For branched nonionic matrix, plasmon peak at 518 nm and peak at 438 nm have been observed. The presence of two maxima may dedicate anisotropic shaped particles [10, 11]. It should be noted, that branched polyelectrolyte matrix allows to obtain lager particles (20 nm, maxima at 545 nm) in comparison with their nonionic polymer analog.

*DLS analysis*has been chosen to study the aging process. Nanoparticles hydrodynamic radii distribution in Au sols was analyzed during 2 months. Multimodal size distribution for all synthesized Au sols was observed. Dependences of nanoparticles number versus their radii in the sols stabilized by various polymers are shown in Figure 10.2. The first one maximum corresponds to the smallest distinguished Au NPs, the second oneto larger specie of Au NPs, and the third one to Au NPs aggregates and macromolecules of polymer matrix. In the Table 10.2, the size analysis characteristics measured during 2 months are represented (peak 1, peak 2, and peak 3, respectively).

Just after sol synthesis the size distribution of Au NPs obtained in linear PAA included two types of scattering particles (12 nm and 26 nm for single particles and aggregates respectively). It should be noted that the dependence of intensity of a scattering versus particles size revealed also the peak at 60–100 nm corresponding to macromolecules of polymer matrix (Figure 10.3). The results represented in the Table 10.2 testified that in 7 days of storage the peak corresponding to Au NPs had not shifted but number of the particles have reduced and one more maximum at 3.2 nm have been observed. The appearance of this fraction may be a result of dispersion process. On the other hand, broad peak with maximum at 235 nm may correspond to polymer intermolecular aggregates. Finely, in 60 days the aggregation process becomes more significant, small particle peak disappeared.

The Au sol synthesized in branched D-g-PAA matrix appeared to be more stable during the observation time. Anionic derivative of D-g-PAA allows to synthesize larger Au NPs (approximately 20 nm), but aging process for this nanosystems includes the stages similar to D-g-PAA(Figure 10.2, Table 10.2). Intermolecular aggregation of polymer molecules hasn't

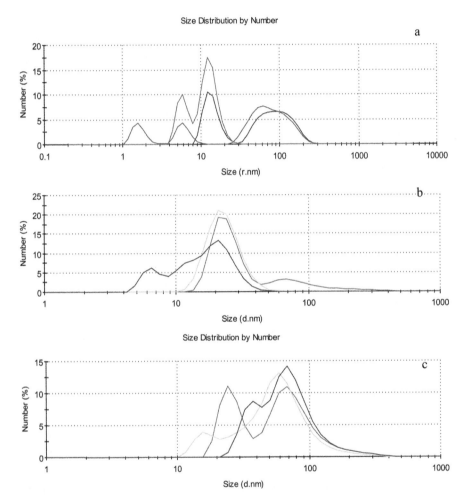

FIGURE 10.2 Evolution in time of nanoparticles hydrodynamic radii in Au sols stabilized by linear PAA(a), D-g-PAA(b), D-g-PAA(PE)(c). Green line – 1st day, red line – 8th day, and blue line – 60th day.

taken place only for the systems obtained in branched polymer matrices (Table 10.2).

For both nanosystems synthesized in branched uncharged and polyelectrolyte matrices the aggregates were revealed in TEM images in 2 months (Figure 10.3). For Au sols obtained in D-g-PAA(PE) matrix the rod-like particles were observed along with spherical ones (Figure 10.4).

TABLE 10.2 Size Characteristics Analysis of Au Sols Measured During Months

Sample	Time, days	Peak 1	Peak 2	Peak 3
PAA	1	—	12.1	26.2
	8	3.2	11.9	235.1
	60	—	27.2	247.4
D-g-PAA	1	—	23.4	72
	8	—	23	—
	60	—	7.02	22
D-g-PAA(PE)	1	—	25,9	71.8
	8	—	17,3	62.7
	60	—	35,9	71.2

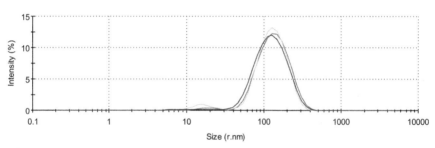

FIGURE 10.3 Intensity of scattering versus hydrodynamic radii of nanoparticles in Au sol stabilized by linear PAA.

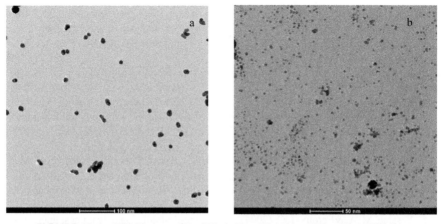

FIGURE 10.4 TEM image of Au NPs in D-g-PAA (a) and D-g-PAA (PE) (b).

10.3 CONCLUSIONS

High aggregation stability of Au sols can be reached by using branched water-soluble polymers as matrices for gold nanoparticles synthesis. The linear uncharged polymer matrix is less efficient for this aim. The stable sol was not obtained in linear polyelectrolyte matrix. During Au sols storage the dispersion process as well as aggregation process was revealed, but the nanosystems were stable without any precipitation in 60 days after synthesis.

KEYWORDS

- aging
- branched polymer
- gold
- nanoparticle
- nanosystem
- sol

REFERENCES

1. Aravind Kumar Rengan, Amirali B. Bukhari, ArpanPranhan, RenuMalhotra, Rinti Banerjee, &RohitSrivastava, et al. (2015). In vivo analysis of biodegradable liposome gold nanoparticles as efficient agents for photothermal therapy of cancer.*Nano Lett.,15*(2), 842–848.
2. Jing-Liang Li & Min Gu (2010). Gold-nanoparticle-enhanced cancer photothermal therapy. *IEEE Journal of Selected Topics in Quantum Electronics, 16*(4), July/August.
3. Joseph D. Meyers, Yu Cheng, Ann-Marie Broome, Richard S. Agnes, Mark D. Schluchter, &SeungheeMargevicius, et al. (2014). Peptide-targeted gold nanoparticles for photodynamic therapy of brain cancer: Part. *Syst. Charact.,*doi: 10.1002/ppsc.201400119.
4. Tanya Stuchinskaya, Miguel Moreno, Michael J. Cook, A. Dylan, R. Edwards, & David A. Russell. (2011). Targeted photodynamic therapy of breast cancer cells using antibody-phthalocyanine-gold nanoparticle conjugates. *Photochemical &Photobiological Sciences, 10*, 822.
5. Yessine, M. A, &Leroux, J. C. (2004). Membrane-destabilizing polyamines: interaction with lipid bilayers and endosomal escape of biomacromolecules. *Adv. Drug Deliv. Rev., 56*, 999–1021.

6. Emilia Tomaszewska, KatarzynaSoliwoda, KingaKadziola, BeataTkacz-Szczesna, GrzegorzCelichowski, Michal Cichomski, et al. (2013). Detection Limits of DLS and UV-vis spectroscopy in characterization of polydisperse.*Nanoparticles Colloids*, Article ID 313081, 10 pp. http://dx.doi.org/10.1155/2013/313081.

7. VasylChumachenko, NataliyaKutsevol, Michel Rawiso, Marc Schmutz, & Christian Blanck. (2014). *In situ* formation of silver nanoparticles in linear and branched polyelectrolyte matrices using various reducing agents, *Nanoscale Research Letters, 9*, 164.

8. Wolfgang Haiss, Nguyen T. K. Thanh, Jenny Aveyard, & David G. Fernig (2007). Determination of size and concentration of gold nanoparticles from UV-vis spectra. *Anal. Chem., 79*(11), 4215–4221.

9. Vincenzo Amendola, &Moreno Meneghetti(2009). Size evaluation of gold nanoparticles by UV-vis spectroscopy.*J. Phys. Chem. C,113*(11), 4277–4285.

10. Franklin Kim, Jae Hee Song, &Peidong Yang (2002). Photochemical synthesis of gold nanorods.*J. Am. Chem. Soc., 124*, 14316–14317.

11. Na Li, Peng Xiang Zhao, &Didier Astruc(2014). Anisotropic Gold Nanoparticles: Synthesis, Properties, Applications, Toxicity,*Angew. Chem. Int.Ed.,53*(7), 1756–1789.

CHAPTER 11

MICROSCALE TEMPERATURE VISUALIZATION IN SILVER NANOPARTICLE DOPED POLYMER NANOCOMPOSITE

N. PONJAVIDZE[1] and G. PETRIASHVILI[2]

[1] Tbilisi State University, 1 Ilia Chavchavadze Ave, Tbilisi 0179, Georgia

[2] Georgian Technical University, 68 Merab Kostava Street, Tbilisi, Georgia Tbilisi, 0175, Georgia, E-mail: fonjo@rambler.ru

CONTENTS

ABSTRACT

The visualization and control of optical to thermal energy conversion in nano and microstructures is a key challenge in many fields of science with applications to areas as nanofluidics, nanocatalysis, photothermal cancer therapy, drug delivery, imaging and spectroscopy, information storage and processing, nanoscale patterning and solar energy harvesting. In the presented work a polymer nanocomposite incorporated with silver nanoparticles and organic luminescence dye has been fabricated, which exhibits thermo chromic properties. It was shown that this kind of composite dramatically changes its color when changes an environmental temperature and acts as a thermo chromic material with such improved parameters as temperature controlled fine-tuning of absorption. This combination of silver nanoparticles and organic luminescent dyes can find versatile application in the areas such as: fabrication thermo addressable luminescence displays and thermo optical printers, environmentally friendly thermo chromic clothes. In medicine: modeling, simulation and visualization of heat spreading to the surrounding biological medium, new possibilities for measuring the heat release at the nanoscale, fabricating nano sized storage media for quantum information devices, thermo controllable filters, windows and smart coatings, infrared image converters. A simple technology promises to fabricate thermo chromic material based cost-effective devices.

11.1 INTRODUCTION

The development of nanoscience and nanotechnology has allowed to create new nano-sized materials having distinguished electronic and optical properties quite different from those of their bulk state [1]. An example of these materials is nanoparticles (NPs) that have unique electronic, magnetic, optical and photonic properties. NPs have many potential applications in electro-optics, as in ultrafast data communications, optical data storage [2], and in modern bio-nano-technology, as markers [3], biosensors [4] and actuators [5]. Recently, heat generation by NPs has attracted a lot of interest. As an example, in nano-medicine this could be a viable way to address the heat in specific parts of the human body to kill cancer cells [6]. The enhancement in the photothermal properties of metal NPs arises from the resonant oscillation of their free electrons in the presence of the oscillating electromagnetic field of the impinging light, also known as localized surface plasmon resonance. The surface plasmon oscillation decays by radiating its energy resulting in light scattering or decays nonradiatively as a result of conversion of absorbed light into heat [7]. The heating effect is especially strong for metal NPs since they have many mobile electrons. Interestingly, noble metal NPs, gold and silver, exhibit plasmon resonances in the visible wavelength range. One of the major challenges is to measure the temperature in the medium surrounding the NPs, which is a crucial parameter for applications in nano-medicine. Govorov et al. [8] attempted to measure the temperature on the surface of gold NPs by embedding them into a matrix and exciting them optically. Using ice as a matrix, for its well-known thermodynamic proper-ties, and observing the melting effect, the amount of heat generated by nano-heaters was determined. At small light powers, it was possible to estimate an average number of Au NPs within the laser spot and with gradually increas-ing power, the threshold melting intensity was determined which could be used to estimate the heat power generated by optically driven NP complexes. In another work, the same authors used a matrix composed of water, poly-mers and a CdTe quantum dot surrounding a gold NPs. The detachment of the quantum dot from the NPs was obtained when the polymer shell was melted due to the heat released by the optical driven NPs [9]. In a more recent study, Carlson et al. used a thin film of Al0.94/Ga 0.06/N embedded with $Er3^+$ ions as an optical temperature sensor to image the temperature profile around optically excited gold nanostructures [10].

Further, van de Broek et al. [11] showed how the NP temperature could be measured through the thermal influence on the gold lattice parameters (using X-ray absorption spectroscopy). In 2009, Baffou et al. [12], reported on a technique based on fluorescence polarization anisotropy measure-ments to map the temperature around nanometer sized heaters. The tech-nique consisted in introducing fluorescent molecules inside a medium and mapping the emitted fluorescence polarization anisotropy, that is directly related to rotational diffusion induced by molecular Brownian dynamics. The key point of this technique was that an increase of the temperature contributed to lower the anisotropy since it gave rise to a faster rotation of the molecules. The technique has a temperature accuracy of $0.1°C$ and a spatial resolution of 300 nm. More recently, Baffou et al. [13] proposed a thermal microscopy technique that was sensitive to the thermal-induced variation of the refractive index of the medium. This technique used an optical microscope equipped with a CCD camera in combination with a modified Hartman diffraction grating. Their technique allowed to detect temperature variations of the order of 1 K and offered the advantage that no sample modification to the fluorescent target was needed. As noted above, heat generation becomes especially strong in the case of plasmon resonance. In the absence of phase transformations, temperature distribu-tion around optically stimulated NPs is described by the usual heat transfer equation:

$$P(r)\ C(r) = \partial T(r,\ t)/\ \partial t = \Delta K(r) \times \Delta T(r,\ t) + Q(r,\ t) \tag{1}$$

where **r** and **t** are the coordinate and time, $T(r,\ t)$ is the local temperature, and the material parameters $P(r)$, $c(r)$, and $k(r)$ are the mass density, specific heat, and thermal conductivity, respectively. The function $Q(r,\ t)$ represents an energy source coming from light dissipation in NPs:

$$Q(r,\ t) = J(r,\ t) \times E(r,\ t) \tag{2}$$

where $J(r,\ t)$ is the current density and $E(r,\ t)$ is the stimulating electric field in the system. The heating effect can be strongly enhanced in the presence of number of NPs. The accumulative effect comes from the addition of heat fluxes generated by single NPs and is described by the thermal diffusion equation (1). The energy source in this equation should be written as a sum over all NPs:

$$Q(\mathbf{r},\ t) = \sum_k Q_k(\mathbf{r},\ t) \tag{3}$$

where $Q_k(\mathbf{r},\ t)$ describes the heat generation by the kth NPs. The more NPs, the stronger the temperature increases that appears in the system. Here, we propose a novel method to map the temperature profile around silver NPs using thermo chromic compounds. These kinds of NPs, in which polymers are mixed with NPs, are of particular interest as they combine the properties of two or more different materials to obtain novel mechanical, electronic or chemical properties. A key challenge, however, remains the homogenous dispersion of the metal NPs and their stability within the matrix. Metal NPs are prone to aggregation because of their high surface free energy and oxidation by air or moisture. These factors can dramatically change the properties of the composite materials. The presented materials consist basically in homogeneously dispersed silver NPs in a polyvinylpyrrolidone (PVP) matrix and an organic dye, used in the form of thin film. The mixture acquires thermo chromic properties when nitric acid is added to it. Actually thermo chromic properties already exist in polymer matrix doped with luminescent dye. This behavior is increased significantly after doping small quantity of nitric acid. When the nanocomposite is irradiated with visible light, it shows a color variation, that gives an instantaneous information about the temperature change around the silver NPs. This provides a tool to estimate the temperature reached by the medium surrounding the nanoscale heaters.

11.2 EXPERIMENTAL PART

11.2.1 MATERIALS, RESULTS, AND DISCUSSIONS

Thermo chromic materials, as liquid crystals or leuco dyes, are prone to change their color when changes in temperature occur. For the preparation of the thermo chromic mixture we used Oxazine-17 (Ox-17, NIOPIK) (Figure 11.1), PVP, and nitric acid (HNO_3 67, 5%) (Sigma Aldrich). Ox-17 and PVP, both soluble in ethanol (EtOH), were mixed according to the following proportions in weight: 99.98% {90% EtOH + 10% PVP} + 0.02% Ox-17.

The mixture was stirred for 1 hour at room temperature, the resulting solution was then deposited by drop coating onto a glass slide and stored at room temperature for 24 hours to evaporate the solvent. A homogeneous film of about 20 μm was obtained. A heating stage for optical bench (Series F, Calctec, Italy), and a multi-channel fiber optic spectrophotometer (AVS-S2000 and DH-2000, Avantes) were used to investigate the optical properties of the sample at increasing temperatures. In Figure 11.2, the absorption spectra of the polymer film, varying the temperature from 25°C (black curve) to 150°C (green curve), are reported. The spectra showed that the thermal sensitivity of the mixture was weak, and it was observed a small blue shift of the absorption peak of about 30 nm in a temperature range from 25°C to 150°C.

The Nitric acid (HNO_3) was added to the mixture, according to the following proportions in weight: 89% {99.98% {90% EtOH + 10% PVP} +0.02% Ox-17} + 11% HNO_3. As in the previous case, the solution was stirred for 1 hour at room temperature, and then drop coated onto a glass slide and left 24 h to dry.

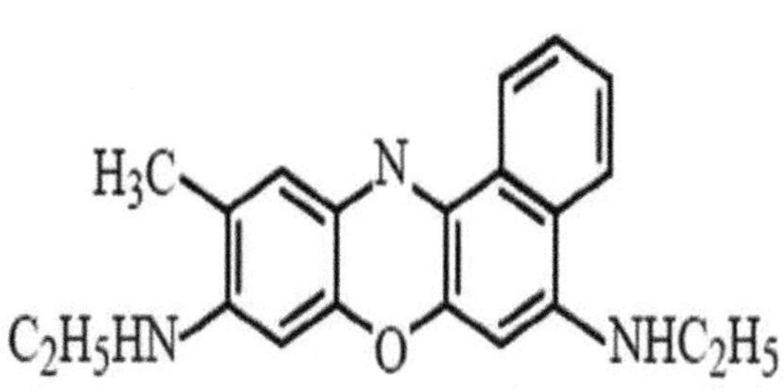

FIGURE 11.1   Structural formula of luminescence dye Ox-17.

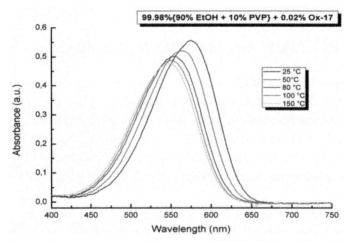

FIGURE 11.2 Absorption spectra of the mixture 99.98% {90% EtOH + 10% PVP} + 0.02% Ox-17 acquired at different temperatures.

After the preparation of polymer film, it was heated up and using a spectrometer was measured spectral tuning of absorption band. Figure 11.3 shows the absorption spectra of the material at different temperatures going from 25°C (black curve) to 120°C (yellow curve). Measurements were stopped at 120°C, when the blue shift of the absorption peak was already enough to cover the visible spectrum. The presence of nitric acid in the polymer/dye mixture substantially improved the sensitivity of the material to the temperature and a considerable blue shift of about 300 nm was observed in the above-mentioned

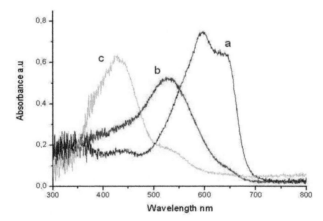

FIGURE 11.3 Absorption spectra for the mixture 89% {99.98% {90% EtOH + 10% PVP} + 0.02% Ox-17} + 11% HNO₃ as a function of temperature.

range. Nitric acid acts as an oxidizing agent in the mixture, the dye molecule becomes protonated and a shift in its absorbance is observed. A blue shift is also observed varying temperature in a wider wavelength range.

A visualization of the temperature/color dependence of the thermo chromic film upon heating is shown in Figure 11.4. Blue color corresponds to room temperature (25°C), red corresponds to 80°C, and yellow corresponds to 120°C. We observed that the color reversibility after the heating was depended mostly on factors as the maximum reached temperature, the film thickness and the concentration of the organic dye. Silver NPs were added to the thermo chromic mixture. Colloidal silver NPs were dispersed in PVP in an 8 mg/1 ml proportion, using water as a solvent, and the solution was stirred for 3 hours at room temperature. We found that a PVP matrix allows to obtain a good dispersion of the silver NPs.

In Figure 11.5, the absorbance of the water solution of PVP and silver NPs is shown. The surface plasmon resonance at about 420 nm, as for single particles, suggested that silver NPs were homogeneously dispersed in the medium, and we could infer that the formation of NP aggregates was minimized. The solution containing silver NPs was mixed with the 89% {99.98% {90% EtOH + 10% PVP} + 0.02% Ox-17} + 11% HNO_3 solution in 1/1 ratio by volume and was stirred in a closed container for 24 hours at room temperature. The final blend was then drop coated onto a glass slide as previously described. To estimate the temperature around NPs we have irradiated samples with a 100 W mercury lamp HG 100 AS, (Jelosil), equipped with a 390–450 nm band-pass filter. The distance from the lamp to the sample was

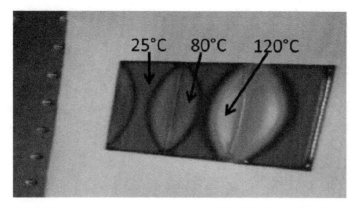

FIGURE 11.4 Temperature distribution of the thermo chromic film. Blue, red and yellow colors correspond to 25°C, 80°C, and 120°C, respectively.

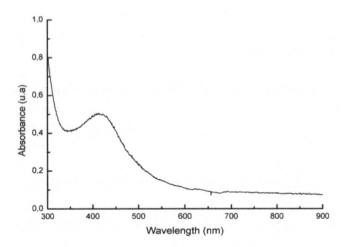

FIGURE 11.5   Surface plasmon resonance of silver NP doped PVP.

about 15 cm. The intensity of the light at the sample surface was about 0.2 mW/cm² and the total exposure time was 12 min. When the light hits the sample, if thermo chromic molecules are located in the close proximity of the silver NPs, the temperature variation causes a change in the color of the nanocomposite. Figure 11.6 shows the color distribution on the surface of the thermo chromic nanocomposite induced by light irradiation at different exposure times. At t = 0 min the sample appeared blue, after 6 minutes it became red-brown, and at t = 12 min it becomes orange-yellow. The color distribution on the film was uniform and it could easily be approximated to the temperature. As an example, referring to the color of the sample of the central image in Figure 11.6 and to the absorption spectra in Figure 11.2 and the visualization of the temperature/color dependence of the thermo chromic film upon heating in Figure 11.3, we can conclude that the temperature of the matrix surrounding the NPs was about 80°C, after 6 minutes of exposure time.

It must be pointed out that we have examined different species of organic luminescent dyes, such as coumarin, stilbene and rhodamine based ones, but the best thermo chromic sensitivities are obtained using oxazine based dyes.

11.3   CONCLUSIONS

Photothermal effects in metal NPs are important for many applications. For example, in biomedicine NPs may be used for hyperthermic destruction of

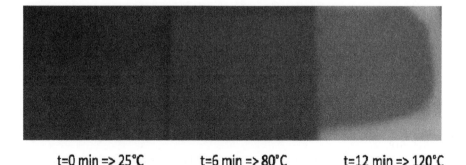

t=0 min => 25°C   t=6 min => 80°C   t=12 min => 120°C

FIGURE 11.6 Temperature distribution of the polymer nanocomposite film. Blue corresponds to the not irradiated sample, red-brown corresponds to 6 minutes of irradiation, and orange-yellow corresponds to 12 minutes of irradiation.

cells, targeted protein denaturation and photothermal interference contrast imaging. In optics they can be used in smart windows and thermo optical printers. The temperature increase is the most important parameter for applications of heated NPs and this demands the development of sensors for temperature detection. Here, an approach for measuring temperature at micro-scales was investigated. We have obtained for the first time a mapping of the temperature around silver NPs using a thermo chromic mixture. The color change is induced by the temperature variation of the medium due to the light interaction with silver NPs and the consequent heat release. In general, thermo chromic mixtures, even prepared using commercial thermo chromic materials, can be considered as suitable candidates for the visualization of the temperature distribution at the micro- and nano-scales.

KEYWORDS

- nanoscience and nanotechnology
- polymer nanocomposite
- silver nanoparticles
- thermal energy conversion
- thermo chromic dye
- visualization

REFERENCES

1. Alivisatos, A. P. (1996). Semiconductor clusters, nanocrystals and quantum dots. *Science, 271*, 933–937.
2. Ditlbacher, H., Krenn, J. R., Lamprecht, B., Leitner, A., & Aussenegg, F. R. (2000). Spectrally coded optical data storage by metal nanoparticles. *Optics Letters, 25*, 563–565.
3. Berciaud, S., Cognet, L., Tamarat, P., & Lounis B. (2005). Observation of intrinsic size effects in the optical response of individual gold nanoparticles. *Nano Letters, 5*, 515–518.
4. El-Sayed, M. A. (2001). Some interesting properties of metals confined in time and nanometer space of different shapes. *Accounts of Chemical Research, 34*, 257–264.
5. Kelly, K. L., Coronado, E., Zhao, L. L., & Schatz, G. C. (2003). The optical properties of metal nanoparticles: The influence of size, shape, and dielectric environment. *Journal of Physical Chemistry B, 107*, 668–677.
6. Jain, P. K., El-Sayed, I. H., & El-Sayed M. A. (2007). Au nanoparticles target cancer. *Nano Today, 2*, 18–29.
7. Jain, P. K., Huang, X., El-Sayed, I. H., & El-Sayed, M. A. (2008). Noble metals on the nanoscale: Optical and photothermal properties and some applications in imaging, sensing, biology, and medicine. *Accounts of Chemical Research, 41*, 1578–1586.
8. Govorov, A. O., Mand Richardson, H. (2007). Generating heat with metal nanoparticles. *Nano Today, 2*, 30–38.
9. Govorov, A. O., Zhang, W., Skeini, T., Richardson, H., Lee, J., & Kotov, N. A. (2006). Gold nanoparticle ensembles as heaters and actuators: Melting and collective plasmon resonances. *Nanoscale Research Letters, 1*, 84–90.
10. Richardson, H. H., Carlson, M. T., Tandler, P. J., Hernandez, P., & Govorov, A. O. (2009). Experimental and Theoretical studies of light-to-heat conversion and collective heating effects in metal nanoparticle solutions. *Nano Letters, 9*, 1139–1146.
11. Van de Broek, B., Grandjean, D. J., Trekker, J. Ye., Verstreken, K., Maes, G., Borghs, G., et al. (2011). Temperature determination of resonantly excited plasmonic branched gold nanoparticles by X-ray absorption spectroscopy. *Small, 7*, 2498–2506.
12. Baffou, G., Kreuzer, M., Kulzer, F. & Quidant, R. (2009). Temperature mapping near plasmonic nanostructures using fluorescence polarization anisotropy. *Opt. Express, 17*, 3291–3298.
13. Baffou, G., Bon, P., Savatier, J., Polleux, J., Zhu, M., Merlin, M., et al. (2012). Thermal imaging of nanostructures by quantitative optical phase analysis. *ACS Nano, 6*, 2452–2458.

CHAPTER 12

POLYANILINE-CO$_3$O$_4$ NANOCOMPOSITES: SYNTHESIS, STRUCTURAL, MORPHOLOGICAL, OPTICAL, AND ELECTRICAL PROPERTIES

Z. HESARI,[1] and B. SHIRKAVAND HADAVAND[2]

[1]*Department of Chemistry, Faculty of Science, Science and Research of Tehran Branch Islamic Azad University, Tehran, Iran, E-mail: zeinab.hesari@yahoo.com*

[2]*Department of Resin and Additives, Institute for Color Science and Technology, Tehran, Iran*

CONTENTS

ABSTRACT

In this study, polyaniline-Co$_3$O$_4$ nanocomposite with different contents of Co$_3$O$_4$ nanoparticles (1, 2.5, 5 wt%) were prepared by in situ chemical polymerization of aniline using ammonium persulfate as an oxidizing agent. Optical, electrical properties and morphological of the products were characterized by FTIR spectroscopy, X-Ray diffraction (XRD), electron microscopy (SEM), UV-vis spectroscopy and electrical conductivity measurements. UV-Vis and FT-IR spectra and X-Ray diffraction which offered the information about the chemical structure of the PANi and PANi-Co_3O_4 nanocomposite. The SEM images of the nanocomposites showed uniform distribution of Co_3O_4 nanoparticles in PANi matrix. Electrical conductivity measurements of PANi-Co_3O_4 nanocomposites revealed that the conductivity of PANi increased with increasing content of Co_3O_4 nanoparticles.

12.1 INTRODUCTION

In recent years, intrinsic conducting polymers with conjugated double bonds attracted attention as functional materials, due to low cost, high environ-mental stability, facility of processing, adaptability with flexible substrates and wide variety of applications in batteries, corrosion inhibition, super capacitors, chemical sensor, separation membrane and electronic devices [1–3].

Among the existing conducting polymers, polyaniline has been extensively studied due to good environmental stability, facile synthesis, and high conductivity, low cost, unique electrochemical and physicochemical properties [4, 5].

Recently, the development of inorganic/organic composites have been receiving attention due to appropriate mechanical, electrical and optical properties and a wide range of potential applications in chemistry, medicine material science and biology [6].

Polymer nanocomposites synthesized from polymers and organic/inorganic fillers at the nanometer scale that used in the different industrial fields such as microwave absorption layers, membranes, aerospace, coatings, adhesives, fire-retardants, medical devices, gas sensors, super capacitors, etc. [7]. Among of nanocomposites, PANi-inorganic nanocomposites are considered because applications in many industrial fields [8]. In this study, polyaniline-Co_3O_4 nanocomposite with different contents of Co_3O_4 nanoparticles was prepared [9] by in situ chemical polymerization of aniline using ammonium persulfate (Scheme 1) and then Optical, electrical and morphological properties of the products were characterized.

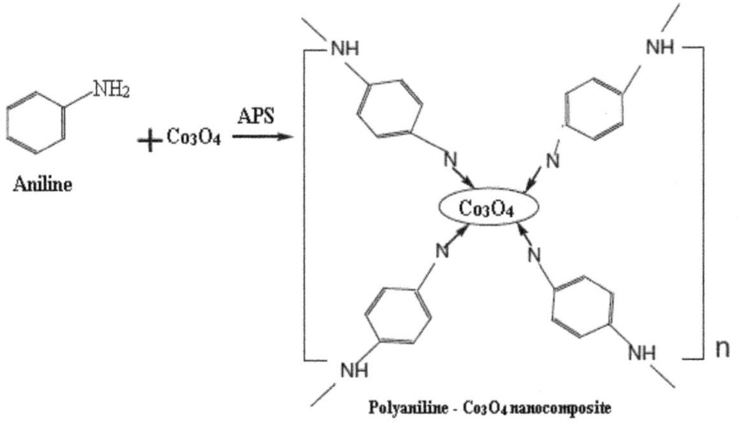

SCHEME 1   Synthesis of PANi-Co$_3$O$_4$ nanocomposite.

12.2   EXPERIMENTAL PART

12.2.1   SYNTHESIS OF POLYANILINE (PANI)

Polyaniline (PANi) was synthesized by oxidative polymerization in the presence of ammonium per-sulfate ((NH_4)$_2$S$_2$O$_8$) as an oxidant. First aniline (2 ml) in 100 ml of 1 M HCl solution at room temperature was stirred for ½ h afterward 1M (molar) APS solution (125 ml) was added at the rate of 15 drops/min. After adding complete solution of APS the stirring was continued for 4.5 h and temperature was maintained 0°C and then mixture was kept overnight in the for complete reaction. The green precipitate was collected by filter and repeatedly rinsed with water and HCl solution. The product was dried in an oven for 10 h. Thus, finally obtained green precipitate of conducting form of polyanilines, emeraldine salt (ES).

12.2.2   SYNTHESIS OF PANI-CO$_3$O$_4$ NANOCOMPOSITES

PANI-Co$_3$O$_4$ nanocomposites was synthesized with different weight ratios (1, 2.5, and 5%) of Coo nanoparticles into PANi matrix. Co$_3$O$_4$ nanoparticles (0.025, 0.05 and 0.1 g) was added to 0.24 mol (2 g) of aniline monomer in HCl (1 M) solution and then APS solution (125 ml) was added slowly to mixture. The reaction mixture was stirred for 5 h at 0°C in ice bath and then filtration. Obtained green precipitate washed several times with distilled water and HCl solution.

12.3   RESULTS AND DISCUSSIONS

12.3.1   UV-VIS SPECTROSCOPY ANALYSIS

The optical absorption spectrum is an important method to achieve optical energy band gap of crystalline and amorphous materials. At present, Co_3O_4 nanoparticles as a filler semiconductor in PANI polymer have been used to make a red shift to the absorbed light.

The UV-visible absorption spectra of PANi (ES) and PANi-Co_3O_4 hybrid nanocomposites were recorded at room temperature are given in Figure 12.1.

Figure 12.1a shows that two distinctive peaks of polyaniline appear at about 285 nm, 390 nm which are attributed to $\pi-\pi^*$-conjugated ring systems

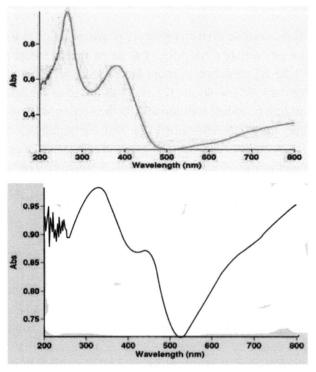

FIGURE 12.1   UV-Vis spectra of (a) PANI and (b) PANI-Co_3O_4 nanocomposite.

FIGURE 12.2   SEM images of a: PANi (ES), b: PANi- Co$_3$O$_4$ (5%) nanocomposite.

and π–π*transition of the benzenoid ring. Figure 12.1b shows UV-visible spectra of PANi-Co$_3$O$_4$(5%) hybrid that shows two distinctive peaks at 330 nm, 450 nm that shifting in the bands is noticed.

12.3.2 SCANNING ELECTRON MICROSCOPY ANALYSIS

The morphologies of PANi (ES) and PANi-Co$_3$O$_4$ (5 wt%) hybrid nanocomposite are as shown in Figure 12.2. The SEM image of polyaniline shows clearly nanofibers structure with many pores and gaps among the fibers in Figure 12.2a. Figure 12.2b shows SEM images of PANi-Co$_3$O$_4$ hybrid nanocomposites (5%). SEM images shows that the loading of Co$_3$O$_4$ nanoparticles has a powerful effect on the polyaniline morphology and with the increase of the amount of Co_3O_4 nanoparticles, a change in morphology of PANi is observed.

12.3.3 ELECTRICAL CONDUCTIVITY

Table 12.1 shows the variation of conductivity of pure PANi, and PANI-Co_3O_4 (1–5wt%) hybrid composites.

Generally, the conductivity increase with the increase of Co_3O_4 nanoparticles in the PANi matrix. The conductivity of these nanocomposites was found to be in the semiconducting range due to conductivity properties of PANi and Co_3O_4 nanoparticles.

12.3.4 XRD ANALYSIS

XRD analysis was used to examine the structure of the PANI and PANi-Co_3O_4 nanocomposites and investigate the effect of Co_3O_4 nanoparticles on the PANi structure. Figure 12.3. The X-ray diffraction pattern of PANi shows amorphous nature in partially crystalline state with broad diffraction peaks of at about 25.5.XRD patterns of PANi-Co_3O_4 nanocomposite show diffraction peaks almost similar to the free PANi (Figure 12.3b), due to low concentr

12.4 CONCLUSIONS

PANi-CuO nanocoposite nanocomposite have been synthesized with different molar ratios of aniline:Co_3O_4 by in situ chemical polymerization of aniline using ammonium persulfate as an oxidizing agent. The

TABLE 12.1   Conductivity of Pure PANi, and PANI-Co_3O_4 (1–5wt%) Hybrid Nanocomposites Pellet

Sample	σ (at 300 K) S cm^{-1}
Pure PANI	1.5
PANI-5 wt% Co_3O_4	0.11
PANI-2.5 wt% Co_3O_4	0.9
PANI-1 wt%Co_3O_4	0.3

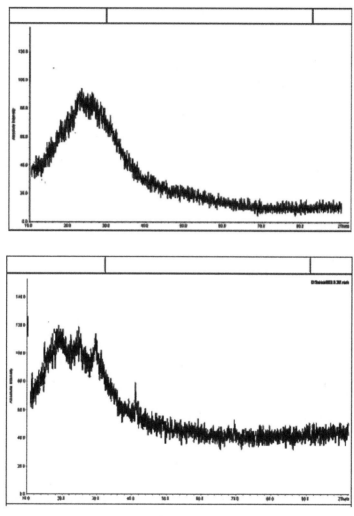

FIGURE 12.3   X-ray diffraction pattern of a: PANi (ES) and b: PANi-Co$_3$O$_4$(5%) nanocomposites.

morphological studies (SEM) showed uniform distribution of the Co$_3$O$_4$ nanoparticles in the PANi matrix. XRD, UV-Vis and FT-IR studies confirm that there is strong chemical interaction between PANI and Co$_3$O$_4$ nanoparticles, which cause the red shift of UV and FT-IR characteristic bands. Electrical conductivity measurements of PANI-Co$_3$O$_4$ nanocomposites revealed that the conductivity of PANi increases with increasing content of Co$_3$O$_4$ NPs.

KEYWORDS

- Co_3O_4 nanoparticles
- electrical conductivity
- nanocomposite
- polyaniline
- chemical polymerization
- optical properties

REFERENCES

1. Zhang, X., Ji, L., Zhang, S., & Yang, W. (2007). *Journal of Power Sources, 173,* 1017–1023.
2. Akel, T. M., Pienimaa, S, Taka, T, Jussila, S, & Isotalo, H. (1997). *Synth. Met., 85,* 1335–1336.
3. Lu, X., Ng, H. Y., Xu, J., & Chaobin, H. (2002). *Synth. Met., 128,* 167–178.
4. Han, M. G., Cho, S. K., Oh, S. G., & Im, S. G. (2002). *Synth. Met., 126,* 53–60.
5. Gupta, K., Jana, P. C., & Meikap, A. K. (2012). *Solid State Sci., 14,* 324–329.
6. Sui, X. M., Shao, C. L., & Liu, Y. C. (2005). *Appl. Phys. Lett., 87,*113.
7. Song, F., Shen, X., Liu, M., & Xiang, J. (2011). Preparation and magnetic. *Mater. Chem. Phys., 126,* 791–796.
8. Khairy, M., & Gouda, M. E. (2013). *Journal of Advanced Research, 6,* 341–345.
9. Umare, S. S., Borkar, A. D., & Gupta, M. C. (2002). *Mater. Sci., 25,* 235–239.

CHAPTER 13

STRATEGY FOR NANOHYBRIDIZED SYNTHESIS OF $M_A M_B O_X$ SYSTEMS

M. DONADZE and T. AGLADZE

Georgian Technical University, Kostava Str., 77, 0175 Tbilisi, Georgia, E-mail: tamazagladze@emd.ge

CONTENTS

ABSTRACT

Bottom-up design of hybrid nanomaterials requires individual nanosized building blocks to fuse together usually one building block to nuclate on the surface of another [1]. Hybridization process is frequently complicated by presence of a surfactant shell adjusted to a surface of nanocomponent cores to prevent agglomeration or corrosion. Maintenance of balance between reactivity and stability of nanocomponent cores in the process of heterogeneous nucleation frequently becomes critical for nanohybridized synthesis. Here, we describe alternative strategy for synthesis of $M_A M_B O_X$ nanohybrids which envisages surface fictionalization of electrochemically synthesized free standing core/shell metal (A) nanoparticles provided by direct chemical interaction of shell molecules with metal (B) oxide precursor. The strategy is implemented to synthesis Ag-MnO$_2$ nanohybrid particles displayed diverse functions.

13.1 INTRODUCTION

Metal based core-shell nanoparticles (NPs) have attracted attention owing to their unique properties compared to their bulk counter parts. Integration of various nanoscale materials into a single nanoparticleyields multifunctional hybrid materials (MHM) with the novel or improved functions of the constituents [2, 3].Colloidal metal-metal oxide hybrids were recently intensively investigated owing to their potential application in a wide range of fields [4]. Coupling metal nanoparticles with metal oxides enhances electro catalytic activity of the metal center toward electrode reactions involved in energy conversion and storage as well ascatalytic conversion of various organic matters. The common strategy for the synthesis of MHM materials involves preparation of individual NPs stabilized by an organic shell and further use as seeds to deposit the other component. Though a number of successful synthesis of such heterostructures has been reported [5], synthesis of hybrid nanoparticles (HNPs) consisting of components with strongly differing structural and chemical properties is still a challenging task; the desirable level of control over a particle size and homogeneity was achieved only in a limited number of cases because of complexity of heterodimers growth [6]. Among critical factors is a controversial role of a shell material (usually molecules of organic surfactant), which prevents aggregation of a seed material but frequently blocks surface active sites and thus inhibits deposition of another component.

In this chapter, we focus on a novel bottom up strategy for the synthesis of HNPs which involves an electrochemical formation of metal (core)-surfactant (shell) NPs undergoing subsequent chemical coupling with metal oxide via interfacial reduction of a metal oxide precursor by surfactant molecules. The strategy provides a facile preparation of various $M_AM_BO_x$-type hybrid nanomaterials.

A sequential electrochemical-chemical approach was applied to synthesis of silver-manganese dioxide nanocomposite. We chose Ag-MnO$_2$ hybrid as a model system owing to useful properties of individual nanocomponents of the system (strong biocide action of nanosilver NPs [7,8], catalytic/electrocatalytic activity of nanostructured manganese oxides) and to potential synergy of properties provided by coupling of the components [9,10]. Ag-MnO$_2$ NPs can be considered as promising material for application in the field of water treatment, heterogeneous catalysis, electrode materials, sensors, etc. [11–16]. Ag-MnO$_2$ nanohybrids two-step synthesis is schematically illustrated in Figure 13.1.

13.1.1 ELECTROCHEMICAL STEP

The fundamentals of electrochemical fabrication of the individual metal core-surfactant shell nanoparticles are given in Ref. [17]. Here we present brief description of the basic principles of electro synthesis in bi-layer cell equipped by the rotated ring cathode. Upon rotation ring electrode crosses the layers of the immiscible liquid-no polar organic solvent (toluene, hexane) with dissolved surfactant (oleic acid) and aqueous solution of metal salts (Figure 13.1a). The metal ions formed at the anode are discharged at the poisoned by adsorbed surfactant surface of the cathode. Molecules of oleic

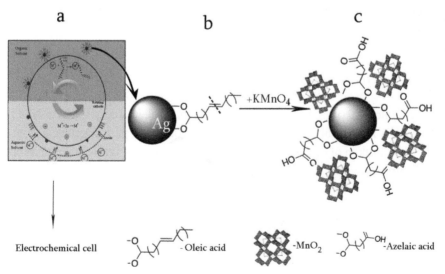

FIGURE 13.1 Schematic image for preparation of Ag-MnO$_2$ NPs. (a) Sole of electro synthesized Ag-OA core-shell NPs; (b) Reduction of potassium permanganate to manganese dioxide via shell oxidation reaction; and (c) Formation of hybrid nanoparticles.

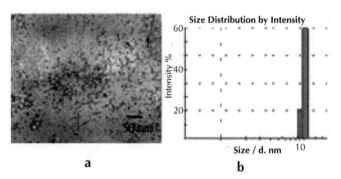

a   b

FIGURE 13.2 (a) TEM images of Ag-OA nanoparticles. (b) Histogram of particle size distribution.

acid predominantly occupy favorable sites for formation of the metal and atoms. Adsorbed surfactant molecules inhibit grows of nanoclasters, which being weakly adsorbed at poisoned surface and strongly bounded to amphiphile surfactant easily washed out from the surface upon rotation forming a solution of the free standing metal NPs in organic solvent. The size control of nanozoles is achieved by variation of residence time τ_R during which a segment of the ring electrode is in contact with each liquid phase. The linear relationship between τ_R and the position of liquid-liquid boundary (z) provide base for easy tuning of the size of metal NPsby variation in partial volumes of immiscible solvents. In present study we used 10–12 nm size silver building blocks (Figure 13.2).

13.1.2 CHEMICAL STEP

The chemical step involves partial substitution of oleic acid shell by a MnO_2 formed via interfacial reduction of precursor–potassium permanganate accompanied by broken of double bond in oleic acid molecule (Figure 13.1c) and formation of pelargonic and azelaic acids according to the following reaction [18]:

$$
\begin{aligned}
&3CH_3-(CH_2)_7-CH{=}CH-(CH_2)_7 \quad \rightarrow \quad 3CH_3-(CH_2)_7-COOH \\
&-COOH + 4\ KMnO_4 + 2H_2O \qquad\qquad + 3HOOC-(CH_2)_7-COOH \\
&\qquad\qquad\qquad\qquad\qquad\qquad\qquad\qquad + 4\ KOH
\end{aligned}
\tag{1}
$$

$\qquad\qquad$ (*Pelargonic Acid*) $\qquad\qquad\qquad\qquad$ (*Azelaic Acid*)

The hybrid properties of Ag-MnO_2 NPs were assessed by probing bactericidal activity of the material toward gram-positive and gram-negative bacteria and catalytic ability to convert carbon monoxide to a carbon dioxide.

13.2 EXPERIMENTAL PART

13.2.1 CHEMICALS AND INSTRUMENTS

The reagents used in this study were purchased from Sigma-Aldrich unless otherwise specified and used without purification. Optical properties of a silver and Ag-MnO$_2$NPs were characterized by UV spectroscopy in a range 200–600 nm (Carry 100, Varian). Chemical interactions of surfactant with NPs and permanganate ions were studied by FT-IR spectroscopy in a range 400–4000 cm^{-1} with the resolution of 0.5 cm^{-1} (Thermo Nicolet, Avatar 370) using KBr technique. Size, shape and chemical composition of isolated NPs were estimated from the TEM (Tesla BS 500) and SEM (JSM-6510LV) images. The samples were prepared by placing small drops of a sole onto the carbon coated copper grid. The size distribution of NPs in a sole was evaluated from laser beam dynamic light scattering (DLS) data (Zetasizer-Nano, Malvern). Prior to sample placing in to a cuvette the as prepared soles were diluted with hexane 1:8. The structural and thermal stability properties of the nanomaterials were characterized by XRD (Russian Dron 4) and the thermo-gravimetric analysis technique (TGA, Derivatograph Q-1500D).In both cases soles were initially evaporated at ambient temperature during 5 hours and 1 g powder was subjected to analysis. Concentration of Ag and K in liquid and solid phases was measured by the inductively coupled plasma optical emission spectrometer (ICP-OES, Varian). Concentration of MnO$_2$ in the reaction mixture was determined by the redox titration method [19]. The consumption of a permanganate ion in the reaction (1) was controlled by the measurement of optical absorbance of the solution using the photo colorimeter (Russian KFK 2) and light absorbance calibration curves.

13.2.2 PREPARATION OF SOLES OF AG NPS

The soles of silver NPs in a hexane were synthesized using the electro-chemical reactor consisting of a sacrificial silver anode (99.9% purity), and aluminum (99.9%) ring cathode which upon rotation crosses immiscible layers of an aqueous (0.05 M AgNO$_3$, doubled distilled water) and organic (Hexane, 1% oleic acid) solvents [17]. The experimental setup allows silver ions formed at the anode to discharge at the cathode surface poisoned by a surfactant (OA) which adsorbs at sites favorable for silver and atoms and inhibits the growth of silver nanoclusters. The latter being weakly adsorbed at the surface and strongly bounded to amphiphile OA molecules are easily washed out from the cathode upon rotation forming the stable soles of Ag-OA core-shell NPs in hexane. In the previous study, we demonstrated the ability to tune a particle size by variation in residence time τ_r, during which a metal cluster formed at a ring cathode in an aqueous electrolyte is allowed to adsorb amphiphile molecules of a surfactant from an organic phase [17]. In a present study, we carried out electro synthesis at the experimental conditions (cathode current density 7500 A cm^{-1}, ring cathode rotation rate 960 rev.min^{-1}, electrolyte temperature 20°C and $\tau_r = 36$ s), which leads to the formation of an Ag-OA sole containing 0.54 gl^{-1} NPs with an average particle size of 10–12 nm (Figure 13.2).

13.2.3   PREPARATION OF AG-MNO₂ HYBRIDS

For the preparation of a hybrid material 100 ml sole of Ag NPs in a hexane was mixed with 40 ml 0.2 M KMnO$_4$ aqueous solution under vigorous shaking during 1 h(until discoloring of permanganate solution) at ambient temperature. After the separation of organic and aqueous phases by centrifugation (0.5 h, 8,000 rev min^{-1}) a sole of spherical Ag-MnO$_2$ particles with an average size of 38 nm was obtained (Figure 13.3).

The chemical analysis of a sole containing a fine fraction of a nanocomposite separated by centrifugation shows the following composition: 29, 2 wt. % Ag 28, 16 wt. %, MnO$_2$, 4, 4–5 wt. % K and 38, 19% OA oxidation products.

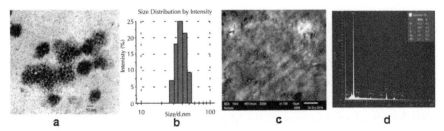

FIGURE 13.3   (a) SEM images of Ag-MnO$_2$ particles;(b) Histogram of particle size distribution;(c) TEM images of Ag-MnO$_2$ particles; and (d) EDX analysis of the sole obtained after centrifugation (8,000 rpm).

13.2.4 PREPARATION OF WATER FILTER AND EVALUATION OF BACTERICIDAL ACTIVITY

Woven textile filters (FHRC10EMB) were impregnated by treatment in an ultrasonic bath (100 Watt, frequency 40 KHz) containing 100 ml Ag-OA-MnO_2 sole during 1 hour. Bactericidal activity of the material was tested against gram-positive (streptococcus and staphylococcus) and gram-negative (*Escherichia coli*, *Pseudomonas aeruginosa*, Proteus) bacteria introduced into the test water of drinking quality. Microbial strains (obtained from G. Eliava Institute of Bacteriophages, Microbiology and Virusology) were tested according to standard method [20]. The kinetics of bacteria removal was evaluated from an analysis of bacteria concentration (lg cfu/100 mL) in a sample (10 mL) extracted from the filtration loop. Water filtration rate was 3 L/h.

13.2.5 PREPARATION OF CATALYSTS AND EVALUATION OF CATALYTIC ACTIVITY

Samples for catalytic tests were prepared by 1-himpregnation of a 1 g calcium aluminosilicate ($CaAl_2Si_2O_8$2.5 mm grain size) used as a supporting material in the ultrasonic bath (100 Watt, frequency 40 KHz) containing 5 ml Ag-OA sole. The catalytic activity toward CO oxidation reaction was studied by a standard test procedure at a 0.6 g catalyst load of the support and continuous flow of CO/CO_2 mixture at the rate of 30,000 h^{-1}. The CO conversion degree αwas calculated according to the equation:

$$\alpha = \frac{a-b}{a} \times 100\% \qquad (2)$$

where *a* and *b* are respectively the concentrations (%) of CO and CO_2, determined by gas chromatography (Gasochrom 3101).

13.3 RESULTS AND DISCUSSION

13.3.1 CHARACTERIZATION OF AG-OA AND AG-OA-MNO₂ NPS

13.3.1.1 Optical Absorption Spectra

UV-vis spectra of an OA in a hexane with and without addition of the $KMnO_4$ are shown at Figure 13.4. Soon after the permanganate addition

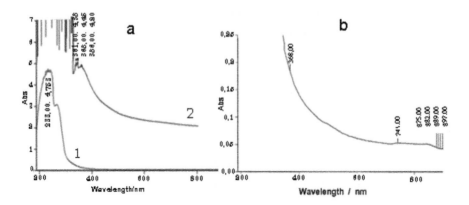

FIGURE 13.4 UV spectra of OA: (a1) In hexane. (b) In hexane, 5 min after KMnO$_4$ addition. (a2) 20 min after addition.

the characteristic for OA absorption peak at 233 nm (Figure 13.4a1) shifts to 368 nm and a number of new peaks in the red region characteristic of intermediate OA oxidation products are formed (Figure 13.4b). At the prolong action of permanganate ions the main peak splits to a series of peaks (386–361 nm) characteristic for the stable oleats and the manganese dioxide (360 nm)[21–24] are observed (Figure 13.4a2).

A UV-vis spectra of OA caped silver nanoparticles with and without addition of permanganate are shown at Figure 13.5. At the initial stage of a MnO$_{4-}$ – Ag-OA interaction depression and broadening of the characteristic for 10–12 nm size Ag NPs peak (404 nm, Figure 13.5a.1) caused by overlapping of the MnO$_2$ and Ag peaks and appearance of the peaks characteristic for oxidized organic products are observed (Figure 13.5b).

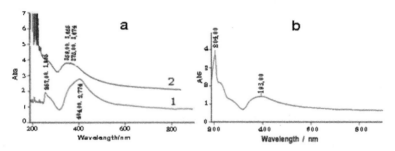

FIGURE 13.5 UV-vis spectra: (a1) Ag-OA nanosoles in hexane, (b) 5 min after permanganate addition. (a2) 20 min after permanganate addition.

After 20 min, contact of the reagents appearance of twin peaks at 359 and 378 nm (Figure 13.5a.2) indicates the formation of an emerged Ag-MnO$_2$ nanocomposite.

13.3.2   FT-IR SPECTROSCOPY

The chemical bonding of OA molecules to silver nanoclusters in toluene was studied in the previous chapter[17]. It was shown that the most remarkable effect of Ag-OA interaction is disappearance of the peak at 1712 cm^{-1}, assigned to C = O stretch bound and appearance of the two new bounds at 1635 and 1380 cm^{-1} which are characteristic of the asymmetric and symmetric carboxylate stretch. These effects were interpreted as the evidence of OA bonding to silver via two symmetrically coordinate oxygen atoms of the carboxylate head.

The FT-IR spectra of Ag-OA soles in a hexane with and without additions of permanganate ions are shown at Figure 13.6. It follows from these spectra that permanganate addition results in vanishing of the peak at 1712 cm^{-1}, and appearance of two new peaks at 1643, 13 and 1650 nm which only slightly change the position in time (Figure 13.6a–c). A new absorption peak appeared at 802, 28 cm^{-1}is assumed to be associated with Mn-O bending vibrations of [MnO$_6$] octahedral in MnO$_2$[22]. The disappearance of a peak assigned to a C = O stretch and appearance of two new bounds indicates the rearrangement of the stretching modes of a carboxylate owing to the interaction with both the silver and manganese oxide components of the composite. The formation of a peak at 1041, 42 cm^{-1} which disappears in time is in line with modification of UV-vis spectra in time, pointing to the formation of a short live intermediates accompanied OA oxidation.

13.3.3   XRD SPECTROMETRY

The structural properties of Ag NPs prior to and post oxidation of an OA shell as well as of a synthesized MnO$_2$ were characterized by the XRD spectra (Figure 13.7a–d). The XRD patterns of as prepared silver sole demonstrate the amorphous character of the material. The size of the silver particles calculated according to Debye Sheerer equation is about 18 nm (Figure 13.7a). After the calcinations at 400°C a fine dispersion of a crystalline phase with increased to 150 nm crystal size was observed (not shown).

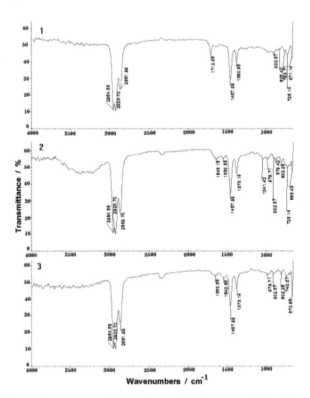

FIGURE 13.6 FT-IR spectra: (a) Ag-OA sole in hexane. (b) 5 min after addition of permanganate.(c) 20 min after permanganate addition.

The X-ray diffraction records of a MnO_2 prepared by the interaction of an aqueous potassium permanganate (40 ml 0.2 M) and a 1% solution of an OA in a 100 ml hexane illustrates an amorphous character of a colloidal MnO_2(a broad diffraction peak located around 20°C, Figure 13.7b). The diffraction maximum at 37°C is in agreement with the peak characteristic to Δ MnO_2 (JCPDS № 80–1098)[23, 24]. The X-ray diffraction patterns of as prepared Ag-MnO_2 shows that low crystallinity is characteristic also of the composite nanoparticles (18 nm size). A spectrum contains a peaks characteristic for both the individual MnO_2 [15, 23, 25] and silver NPs (Figure 13.7c) proving the formation of chemical bounds between the components of the composite. After calcinations at 800°C the X-ray patterns show clear crystalline structure of a composite with increased particle size up to 95 nm (Figure 13.7d). As expected, calcinations at elevated temperatures favor particles aggregation.

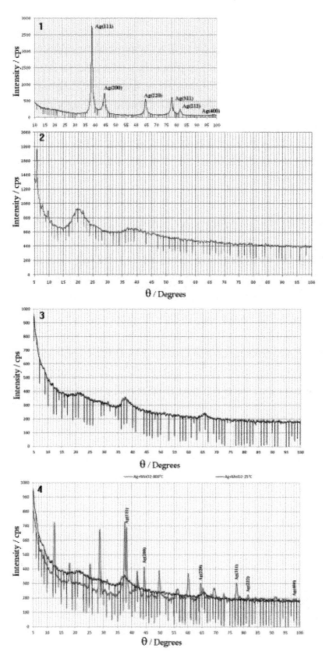

FIGURE 13.7 XRD patterns of: (a) As prepared Ag-OA NPs. (b) MnO$_2$ prepared via permanganate reduction by bulk OA. (c) As prepared Ag-MnO$_2$ NPs, (d) Ag-MnO$_2$ NPs after calcinations at 800°C during 2 hour.

13.3.4 THERMO GRAVIMETRIC ANALYSIS

The thermal gravimetric analysis (TGA) was performed to characterize the thermal stability of an Ag-MnO$_2$ nanocomposites and to compare to that of Ag-OA NPs (Figure 13.8). It follows from these data that both particles lose about 40% of their weight at about 500°C owing to evaporation of the oleic acid molecules and products of its oxidation. A shift of exothermic DTA peaks at 350°C and 450°C to the higher temperatures in case of Ag-MnO$_2$ indicates stronger bonding of organic molecules to the core center of the composite nanoparticles.

13.3.5 KINETICS OF PERMANGANATE REDUCTION

Kinetics of MnO$_2$ formation was studied by tracing decrease of the permanganate concentration in a reaction mixture containing Ag-OA sole in hexane and aqueous solution of potassium permanganate. To evaluate the effect of OA bonding to Ag core kinetics of-reduction of a permanganate ion by OA molecules dissolved in hexane was also studied. Time courses of both systems are presented at Figure 13.9. It follows from these data that reduction of permanganate ions is quite fast reaction and that both homogeneous (curve 1) and heterogeneous (curve 2) reactions follow the same

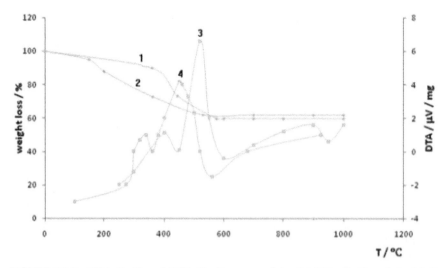

FIGURE 13.8 TGA (1, 2) and DTA (3, 4) curves of Ag-OA NPs (1, 3) and Ag-MnO$_2$ nanocomposites (2, 4).

kinetic regulations. The overall process involves three consecutive stage characterized by different rate constants. The silver core displays clear catalytic properties toward permanganate reduction reaction. The highest rate constant for heterogeneous formation of MnO_2 evaluated from curve 2 (Figure 13.9) under assumption on the first order reduction kinetics, k_{het} = 1.10, exceeds almost 1.6 times rate constant for homogeneous reduction reaction.

It is interesting to compare the kinetic data obtained in our study with studies performed in similar systems. The majority results relate to identification of the intermediates of OA oxidation. While there is consensus on composition of the final products (manganese dioxide, pelargonic and azelaic acids) views on the nature and composition of the intermediates diverge [18, 26–28]. The detailed study of oxidation kinetics of a monolayer of oleic acid spread over aqueous permanganate solution performed by the use of radiotracers reveals the complex character of OA oxidation reaction which involves rapid formation of the intermediate *cis-epoxy* acid in the OA monolayer (rate constant – k_1) and its farther fission yielding *azelaic* and *nonionic* acids (rate constant – k_2) instantaneously dissolved into an aqueous sub phase [28]. It is assumed that the overall rate of oxidation reaction is determined by the rate (k_3) of a parallel step of desorption of intermediate acid into an underlying sub phase. Though above conclusions relates to conditions differed from that of the present study an assumption on a slow transfer of intermediates through the organic-aqueous interface seems applicable to the system under investigation.

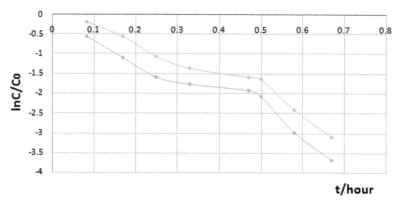

FIGURE 13.9 Time courses of potassium permanganate concentration in OA solution in hexane (1) and Ag–OA sole in hexane (2) at 25°C.

13.3.6 KINETICS OF BACTERIA REMOVAL

Figure 13.10 represents time courses of concentration of gram-positive (streptococcus and staphylococcus) and gram-negatives (*Escherichia coli*, *Pseudomonas aeruginosa*,Proteus) bacteria in a water passed though the woven filter impregnated by hybrid nanoparticles. The results indicate that Ag-MnO$_2$ hybrid material displays strong antibacterial activity resulting in almost 100% removal of all tested pathogens from the water.

13.3.7 CATALYSIS OF CO CONVERSION

The results of comparative assessment of the catalytic activity of Ag NPs and Ag-MnO$_2$ composites toward CO oxidation reaction are shown at Figure 13.11. Both oxidation curves follow almost similar patterns indicating the similarity of reactor kinetics. The composite catalyst displays higher activity: CO oxidation starts below the ambient temperature and the full conversion degree is achieved at lower temperatures. The beneficial action of the manganese oxide additives to the noble metal catalysts is commonly attributed to the ability of the MnOx to supply mobile oxygen for reoxidation of the core catalyst [29]. In the same way an improvement of catalytic activity

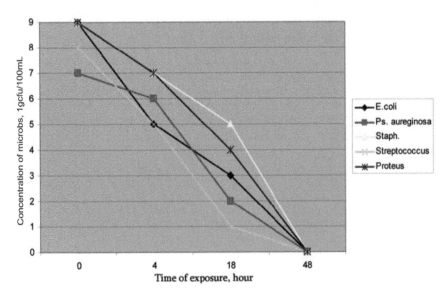

FIGURE 13.10 Antibacterial activity of the Ag-MnO$_2$ hybrid NPs.

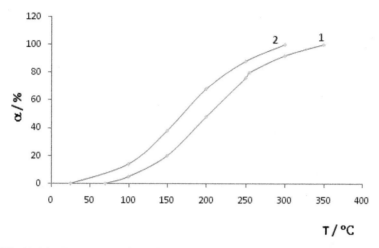

FIGURE 13.11 Temperature dependence of CO conversion catalyzed by as prepared catalyst: (1) Ag NPs. (b) Ag-MnO₂ NPs.

of silver nanoparticles by introduction into the composite of MnO_2 component can be reasonable explained byfacilitation of oxygen transfer from the manganese oxide shell to the silver core.

13.4   CONCLUSION

Ag-MnO₂ hybrid nanoparticles were synthesized following the novel bottom up two-step strategy which involves the electrochemical formation of oleic acid caped silver nanoparticles and partial substitution of the oleic acid shell by the MnO_2 via interfacial chemical reduction of a precursor, the potassium permanganate, by the oleic acid shell. At elevated temperatures amorphous Ag-MnO₂ nanocomposite loses organic components and acquires crystallinity.The kinetic study reveals the three step reduction process catalyzed by the silver core. The bactericidic and catalytic properties of the Ag-MnO₂ hybrids were also tested. The results revealed that filters composed from the hybrid materials removes almost 100% gram-negative and gram-positive bacteria from the contaminated water and converts CO to CO_2 with improved activity owing to the ability of MnO_2 to provide rapid supply of oxygen for reoxidation of the silver core. The proposed strategy can be successfully applied for the preparation $M_A M_B O_x$ nanocomposites involving metal oxides, which can be prepared via oxide precursor – nanoparticle shell interactions.

KEYWORDS

- antibacterial activity
- electrochemical-chemical synthesis
- heterogeneous catalysis
- hybrid nanoparticles
- nanoparticle characterization
- silver-manganese oxide

REFERENCES

1. Buck, M. R., Buck, R., Bondi, J. F. & Schaak, R. E. (2012). A total-synthesis framework for the construction of high-order colloidal hybrid nanoparticles. *Nature Chem., 4*, 37–44.
2. Bhanu P. S. Chauhan (Ed.). (2011). Hybrid Nanomaterials: Synthesis, Characterization and Applications, Willey, ISBN: 978-0-470-48760-0.
3. Goesmann, H. & Feldmann, C. (2010). Nanoparticulate functional materials. *Angewandte Chemie International Edition, 49*(8), 1362–1395.
4. Didier Astruc (Ed.). 2008. Nanoparticles and Catalysis. Willey.
5. Casavola, M., Buonsanti, R., Caputo, G., & Cozzoli, P. D. (2008). Colloidal strategies for preparing oxide-based hybrid nanocrystals. *European Journal of Inorganic Chem. 6*, 837–854.
6. *RebuttiniV., (2014).Dissertation Dr. Rer. Nat. Functional iron oxide-based hybrid nanostructure, Berlin*: Humboldt-Universität zu *Berlin*.
7. Mphenyana-Monyatsi, L., Mthombeni, N. H., Onyango, M. S. & Momba, M. N. B. (2012). Cost-effective filter materials coated with silver nanoparticles for the removal of pathogenic bacteria in groundwater. *Int. J. Environ. Res. Public Health, 9*, 244–271.
8. Franci, G., Falanga, A., Galdiero, S., Palomba, L., Rai, M., Morelli, G., et al. (2015). Silver nanoparticles as potential antibacterial agents. *J. Molecules, 20*, 8856–8874.
9. Ashton Acton Q.General Editor (2013). Benzene Derivatives – Advances in Research and Application, 281.
10. Zhen Ping Qu, Yibin Bu, Yuan Qin, & Qiang Fu. (2013). The improved reactivity of manganese catalysts by Ag in catalytic oxidation of toluene.*J. Appl. Catalysis B: Environmental*, 132-133, 353–362.
11. Wang Shouyan, XieJining, Zhang Tierui, Varadan Vijai K. (2009). Silver decorated γ manganese dioxide nanorods for alkaline battery cathode. *J. Power Source*, 186, 532–539.
12. Xiaodi Liu, Changzhong Chen, Yiyang Zhao, & Bin Jia. (2013). A review on the synthesis of manganese oxide nanomaterials and their applications on lithium-ion batteries. *J. of Nanomaterials*, Article ID 736375, 2013, 1–7.

13. Jie Cao, Qinghe Mao, Liang Shib & Yitai Qian. (2011). Fabrication of γ-MnO_2/α-MnO_2 hollow core/shell structures and their application to water treatment. *Journal of Materials Chemistry, 21*, 16210–16215.

14. Xing Zhang, Zheng Xing, Yang Yu, Qianwen Li, Kaibin Tang, Tao Huang, et al. (2012). Synthesis of Mn_3O_4 nanowires and their transformation to $LiMn_2O_4$ polyhedrons, application of $LiMn_2O_4$ as a cathode in a lithium-ion battery. *Cryst. Eng. Comm., 14*, 1485–1489.

15. Mahmoudiana, M. R., Aliasa, Y., Basiruna, W. J., Woia, P. M., & Sookhakian, M. (2014). Facile preparation of MnO_2 nanotubes/reduced graphene oxide nanocomposite for electrochemical sensing of hydrogen peroxide. *Sensor and Actuators B, 201*, 526–534.

16. Adil, S. F., Assal, M. E., Khan, M., Al-Warthan, A., Rafiq, M., & Siddiqui, H. (2013). Nano silver-doped manganese oxide as catalyst for oxidaion of benzyl alcohol and its derivatives: synthesis, characterization, thermal study and evaluation of catalytic properties. *J. Oxidations Communications, 36*(3), 778–791.

17. Agladze, T., Donadze, M., Gabrichidze, M., Toidze, P., Shengelia, J., Boshkov, N., et al. (2013). Synthesis and size tuning of metal nanoparticles. *Z. Phys. Chem., 227*, 1187–1198.

18. Garti N., & Avni E. (1981). Permanganate oxidation of oleic acid using emulsion technology. *JAOCS, 58*(8), 840–841.

19. ASTM E465–11, Standard Test Methods for Determination of Manganese(IV) in Manganese Ores by Redox Titrimetry.

20. American Public Health Association (APHA). (1998). Standard Methods for Examination of Water and Wastewater, 20th ed., APHA: Washington, DC, USA.

21. Jaganyi, D., Altaf, M., & Wekesa, I. (2013). Synthesis and characterization of whisker-shaped MnO_2 nanostructure at room temperature. *J. Appl Nanosci, 3*, **329–333.**

22. Allabad, **S.,** Syed**,** F. A., Assal, M. E., Khan, M., Alwart, A. & Siddqui, M. R. H. (2013). *J. Oxidation Communications, 3*,778–791.

23. Abulikemu Abulizi, Guo Hai Yang, Kenji Okitsu, & Jun-Jie Zhu. (2014). Synthesis of MnO_2 nanoparticles from sonochemical reductionof MnO_4 in water under different pH conditions. *J. Ultrasonics Sonochemistry, 21*(5), 1629–1634.

24. Avanish Kumar Srivastava (Ed.). (2013). Oxide Nanostructures: Growth, Microstructures, and Properties.

25. Moradkhani, D., Malekzadeh, M., & Ahmadi, E. (2013). Nanostructured MnO_2 synthesized via methane gas reduction of manganese ore and hydrothermal precipitation methods. *Trans. Nonferrous Met. Soc., 23*, 134–139.

26. Davies, J. T. & Rideal, E. K. (1961). Interfacial Phenomena. Academic Press: New York, *134* (3490), 1611–1612.

27. Doaa Muayad, Saadon Abdulla Aowda &Abbas A-Ali Drea. (2013). simulation study of oxidation for oleic acid by $KMnO_4$ using theoretical calculations. *J. Appl. Chemistry. 2*(1), 42–49.

28. Makio Iwahashi, Kunihiko Toyoki, Toshiyuki Watanabe,& Mitsuo Muramatsu. (1981). Radiotracer study on oxidation of oleic acid monolayer on aqueous permanganate solution. *J. Colloid and Interface Science, 79*(1), 21–32.

29. Xia, G. G., Yin, Y. G., Willis, W. S., Wang, J. Y., &Suib, S. L. (1999). Efficient stable catalysts for low temperature carbon monoxide oxidation. *J. Catalysis, 185*, 91–105.

CHAPTER 14

NEW COMPOSITE POLYMER ELECTROLYTE BASED ON PVdF WITH THE ADDITION OF TIO$_2$

G. S. ASKAROVA, K. A. ABLAYEVA, A. P. KURBATOV, and Ye. S. SIVOKHINA

Center of Physical and Chemical Methods of Research and Analysis, Tole bi Str. 96a, 050012, Almaty, Kazakhstan, E-mail: gertie858@gmail.com

CONTENTS

ABSTRACT

Nowadays, lithium ion batteries become one of the most important energy storage technologies in all fields of development in the contemporary world. The most promising direction in the optimization of lithium batteries is the synthesis of a solid polymer electrolyte (SPE). This type of electrolyte can exclude the majority of the safety and ecological hazards, due to its construction. The main disadvantage of all types of the SPE is a low ionic conductivity at ambient temperature. One of the ways to improve its conductivity is the incorporation of inorganic fillers into SPE to obtain a novel composite polymer electrolyte (CPE).

Membranes were prepared by casting of an organic polymer solution (solvent casting technique), followed by drying to evaporate the solvent. CPE based on PVdF were obtained to enhance their electrochemical properties. Micro- and nano-particles of titanium oxide (IV) was used as a filler, lithium per chlorate as a doping salt.

The effect of the filler was investigated by SEM, TG-DSC, Tensile strength and Galvanostatic pulse method. The CPE showed good results in ionic conductivity (1.01 mS/cm), thermal stability (range from 25–165°C) and mechanical strength (4.5% at 6.9 MPa).

14.1   INTRODUCTION

With the increasing safety demands on the lithium-ion battery solid polymer electrolytes are widely regarded as promising electrolytes in a new technical era. In comparison with traditional toxic liquid electrolytes they have superior performance in terms of no-leakage, high flexibility to cell geometry and high physical and chemical stability [1]. Composite polymer electrolytes have received a special attention as one of perspective types of SPE. These heterogeneous materials are composed of polymer matrix having distributed lithium-conducting solid electrolyte in the form of a separate phase [2]. The stability at the electrode/electrolyte interface, mechanical strength, safety features of lithium batteries and even ionic conductivity might be improved due to CPE construction. The main problem of SPE is still low ionic conductivity at room temperature, due to difficulties of migration of lithium cation in the volume of the polymer membrane. However, the incorporation of inert filler into the matrix is greatly being enhanced the amorphous region of the electrolyte medium which is helped for easy ionic motion [3].

Semi crystalline polyvinylidene fluoride (PVdF) is considered to be a perfect polymer matrix having high mechanical and thermal strength as well as sufficient electrochemical inertness. Furthermore, its high dielectric nature ($\varepsilon = 8.4$) helps for dissolving more ionic species in the polymer matrix. Nevertheless, membranes on pristine PVdF have low ion conductivity within 10^{-9}–10^{-10} S/cm at ambient temperature [3].

Among the several fillers in polymer electrolyte membranes, nano-sized titanium oxide (TiO_2) supports the ionic mobility due to its substrate characteristics, such as shape and surface nature, which effectively disturbs the order packing tendency of the host polymer chains [4–5]. Recently, the particle characteristics of the fillers are found to have a tremendous influence on the electrochemical properties of polymer electrolyte membranes. Indeed, in lithium battery studies, nano-sized fillers have exhibited higher ionic conductivity than the micro-sized particles [6].

In the present study, we have aimed to improve the ambient temperature ionic conductivity of the polymer electrolyte and show influence of fillers properties on it. This has been achieved by adding TiO_2 particles in different size and formation method. Many researches on the polymer electrolyte containing various fillers have been reported for a long time, however, there has been limited work where the comparison of electrochemical effect on polymer electrolytes of variously obtained nanoparticles of filler is performed.

14.2   EXPERIMENTAL PART

14.2.1   MATERIALS

Polyvinylidene fluoride (PVdF) (M_W: ~534,000), titanium dioxide (TiO_2) and acetone were purchased from Sigma Aldrich (USA) and used without further purification. Titanium tetrachloride ($TiCl_4$), dimethylformamide (DMF), lithium per chlorate ($LiClO_4$) under purification by recrystallization and vacuum drying, 10 M sodium hydroxide, 37% ammonia solution were also used in this work.

14.2.2   PREPARATION OF FILLER PARTICLES

Two methods of preparation of titanium dioxide nanoparticles were used for composites synthesis.

14.2.2.1   First Method of Synthesis of Nanoparticles of the Filler

Nanoparticles of titanium dioxide in the form of aqueous suspension were prepared by the following procedure. To do this, 1 ml of titanium tetrachloride was diluted in 25 mL of distilled water with constant vigorous stirring. Volatile titanium salt is hydrolyzed in water to form a titanium oxide by the following equation:

$$TiCl_4 + H_2O \rightarrow TiO_2 + 4H^+ + 4Cl^- \tag{1}$$

Stirring was continued for about four hours, resulting in the hydrolysis reaction solution turned colloidal titanium oxide particles. In order to wash slurry, concentrated ammonia solution was added to the system, and further stirred for another 4 h. The resulting slurry with large aggregates was filtered from the larger particles and filtrate was used for the further work. The prepared aqueous slurry was further ultrasonically dispersed. For the synthesis of a new series of composites procedure is repeated completely.

14.2.2.2   The Second Method of Synthesis of Nanoparticles of the Filler

Nanoparticles of titanium oxide in the form of a dry powder were prepared by a hydrothermal method. In this case, 3 g of titanium oxide TiO_2 powder was added to an aqueous solution of 10 M NaOH, with further vigorous stirring for 20 minutes. Subsequently, the solution was dispersed with ultrasonic waves for 30 minutes, and reaction was carried out hydrothermal treatment in a sealed Teflon autoclave for 48 hours at 150°C. The process was divided into stages for a week for ease of synthesis. The reaction products were washed with an aqueous solution 0.1 M HCl to pH = 7, to wash off the substance from a concentrated alkali. Titanates dried for 24 hours at 100°C and

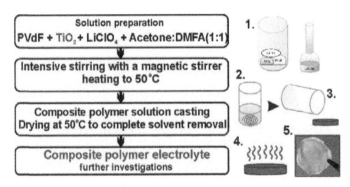

SCHEME 1   Preparation of CPE.

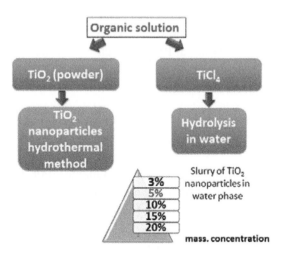

SCHEME 2   Preparation of organic solution.

treated to thermal annealing in air for 4 hours at 300–400°C. The result was a white powder.

14.2.3   PREPARATION OF COMPOSITE MEMBRANES BASED ON PVDF

Two series of samples synthesis were performed. According to the first one membranes were prepared by casting a solution of an organic polymer in advance by dissolving PVdF in a solvent mixture of N,N-dimethylformamide:Acetone = 1:1, with a measured amount of the prepared inorganic filler in the solid phase, followed by drying at 50°C to complete removing of the solvent. Alternative method of forming membranes has performed more complicated step of preparing a polymer solution. The inclusion of filler in the liquid phase in a form of a highly dispersed suspension is accompanied by the implementation of complete homogenization of the heterogeneous agent in the bulk of membranes. In addition, a lithium salt is added as a dopant.

14.2.4   OBJECTS OF STUDY

Composite polymer electrolyte based on polyvinylidene fluoride with various ways of filler addition as well as physic-chemical properties of the filler particles were the objects for investigation of electrochemical properties in dependence on concentration and filler type. There was obtained a composite polymer electrolyte based on PVdF, which was carried functionalization by distribution of high dispersion titanium dioxide and dissolution of lithium per chlorate in the resulting system.

14.2.5 CHARACTERIZATIONS AND MEASUREMENTS

The scanning electron microscope (SEM, Quanta 200i 3D) was used to observe the morphologies of polymer membrane and surface structure. The pictures were taken on the basis of the National Nanotechnology Laboratory of the open type.

Galvanostatic pulse method investigated the effect of different concentrations of the filler on the ionic conductivity of the CPE. Analysis of the results was carried out by pulse galvanostatic curves obtained from AUTOLAB PGSTAT-30.

Thermo gravimetric (TG) measurement and DSC analysis were performed on simultaneous thermogravimetry – differential scanning calorimetric-STA "NETZSCH STA 449F3 Jupiter" at 10°C min^{-1} under N_2 atmosphere in the temperature ranged 25–300°C under laboratory of inorganic chemistry of al-Farabi Kazakh National University.

The mechanical properties of PVdF and PVdF nanocomposite membranes were measured using a dynamometer TTM-50, allowing the sample to obtain stress-strain diagrams, the interpretation of which can receive a certain value of mechanical strength.

14.3 RESULTS AND DISCUSSION

14.3.1 CHARACTERIZATION OF COMPOSITES

Nanoparticles of titanium dioxide by the first method were introduced as an aqueous suspension in an organic polymer solution, and it was also subjected to intense agitation. However, the addition of titanium dioxide nanoparticles in the form of an aqueous slurry difficult process of complete dissolution of the polymer, so initially failed to receive only two samples with low filler content: 5% and 10% mass content. However, in the optimization method of producing nanoparticles, namely introduction sonication slurry, this problem has been eliminated. Due to the high intensity of the small volume of slurry dispersion quickly heated up to the evaporation of water, whereby the air foam formed product with some water remains, which is not a hindrance to the dissolving of the polymer in solution in the presence of titanium oxide nanoparticles.

Nanoparticles of titanium dioxide using the second method have been made in the form of a dry powder, which accordingly does not cause any problems and the need for revision techniques. Figure 14.1 shows the nanocomposite drying step in an oven, one can observe the characteristic opalescent hue as for nanoparticles solution.

14.3.2 MORPHOLOGIES OF COMPOSITES

The morphologies of the pure PVdF and PVDF-TiO$_2$ films were observed using SEM, as shown in Figure 14.2. The pure PVdF film (Figure 14.2a) showed a porous surface and homogeneous phase, whereas the addition of TiO$_2$ caused changes in the morphologies of the PVdF films, indicating structural changes in the composites. The porosity of PVdF-TiO$_2$ films that contained TiO$_2$ decreased with the increase of filler content. Moreover, surplus titanium dioxide particles deposited on the surface of the membrane and block the path of the lithium ion transport (Figure 14.2d). In the picture (d) you can see that a large excess of titanium oxide microparticles cause the destruction of the membrane. In samples with filler content fewer than 10%

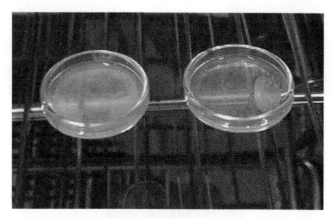

FIGURE 14.1 The composite polymer electrolyte of nanoparticles obtained using the second method, in step solvent evaporation.

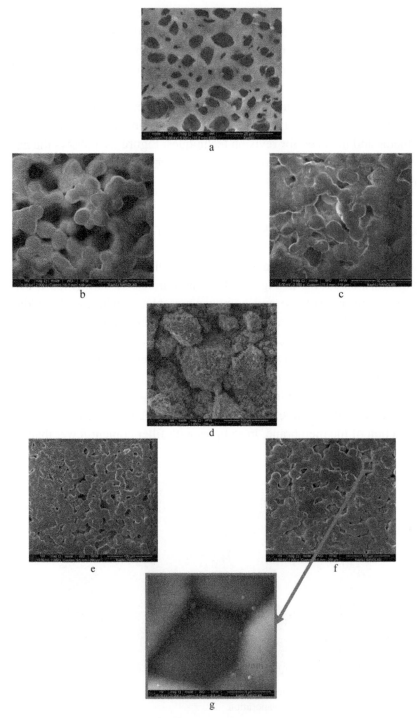

FIGURE 14.2 (Continued)

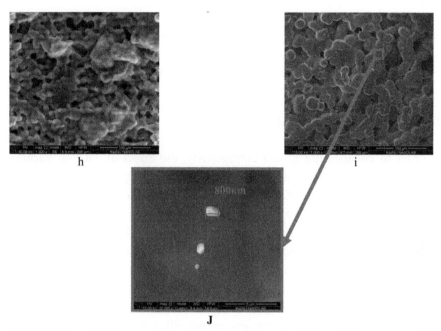

FIGURE 14.2 SEM images of the PVdF-TiO$_2$ films with different TiO$_2$ contents; (a) 0 wt.%, (b) 5 wt.% micro-TiO$_2$, (c) 10 wt.% micro-TiO$_2$, (d) 50 wt.% micro-TiO$_2$, (e) 5 wt.% nano-TiO$_2$, (f, g) 10 wt.% nano-TiO$_2$, (h) 5 wt.% nano-TiO$_2$, (i, j) 10 wt.% nano-TiO$_2$.

nanoparticles distributed in volume of polymer, the surface is detected only occasionally traced small fragments (Figure 14.2e–k).

When comparing the effect of the shape of titanium oxide particles incorporated in the form of a dry powder (Figure 14.2b), in the form of a water-air mass (Figure 14.2e), in dry powder form of nanoparticles (Figure 14.2h), there is a tendency to form a more developed at the membrane surface membranes with the introduction of dry filler, which is an important factor. When assembling the finished cell from the cathode, anode and CPE is required close contact, which can be ensured developed surface of the electrolyte.

Using SEM micrographs was determined approximate particle size of the polymer membrane manufactured according to the methods described above. The sample (Figure 14.2g), particles which were obtained by hydrolysis of the filler dimensions were equal to 300 nm, and the sample (Figure 14.2k), using a hydrothermal treatment in an autoclave – 800 nm. However, the sizes were determined by acting on the surface of the fragments of the heterogeneous phase at a magnification of 20,000 times.

SEM micrographs at the apparent porosity is not a key factor in improving the transport of lithium cation in the solid polymer electrolyte, but one cannot deny that it has on the beneficial effect. However, the transport mechanism of the lithium cation becomes quite complex, since in this case, the additional contribution to the ionic conductivity of the porosity, pore filling filler particles forced in homogeneity in the bulk of the membrane.

14.3.2.1 Electrochemical Investigations

In order to enhance the Li ion conductivity, the effect of TiO_2 in various forms and concentrations as a filler on ion conducting behaviors of CPE systems was s investigated. Dependences of ionic conductivity of the CPE prepared with different TiO_2 contents are presented in Figure 14.3.

As can be seen from the resulting graph (Figure 14.3) membrane compositions with the highest ionic conductivities were composites, which contained PVdF and 5 wt.% of micro-particles of TiO_2 ($6.89 \cdot 10^{-5}$ Sm/cm), PVdF – 5 wt.% water-air nanoparticle TiO_2 ($6.40 \cdot 10^{-5}$ Sm/cm) PVdF – 5 wt.% of dry nanoparticles TiO_2 ($1.35 \cdot 10^{-4}$ Sm/cm). This dependence shows that the ion conductivity increases as the filler concentration increases to a certain point, and after reaching the threshold concentration sharply decreases and remains constant. This phenomenon can be explained as follows. With the introduction of small amounts of additional filler the lithium ion transport paths are formed, in relation with which the ion conductivity increases. After overcoming a threshold concentration of filler it blocks the

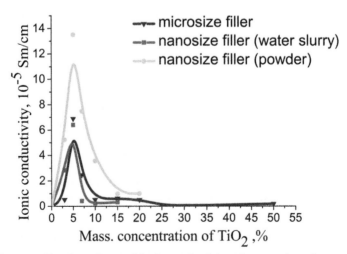

FIGURE 14.3 The dependence of ionic conductivity of composite polymer electrolyte based on the PVdF of content by mass of the filler particles at 25°C temperature.

active centers over the transfer of lithium, significantly reducing the ability of the membrane to carry out its ions. Comparing the impact of micro- and nanoparticles of titanium dioxide, it can be seen that the nanoparticles contribute to better coordination of the lithium ion in the polymer membrane than micro-particles. There is a possibility that during the synthesis of the polymer electrolyte composite with water and air particles, the aggregation of nanoparticles could occur, causing the deterioration of the electrochemical properties of the membrane. The ionic conductivity of pure PVdF is in 10^4 times less than CPE, so the real improvement was seen.

The titanium dioxide is a dielectric material, and it is not directly involved in the conduction process, but only facilitates the coordination of lithium ions inside the membrane, and facilitates the transfer of them, it was decided to increase the concentration of lithium ions of lithium per chlorate introduction into the composite polymer electrolyte. In this case, inert filler particles increase the amorphous polymer matrix, and it acts as a polymer-solvent for the lithium salt. On the basis of earlier work on this subject, it was decided to doping concentration of salt at 25%.

An interesting fact is noticed acquired softness membranes containing a lithium salt, compared with previous designs. This is probably due to the fact that for the formation of bonds between the polymer and salt, the first is more able to absorb the organic solvent or to absorb moisture from the air. Given seen especially by us, it was agreed that it is necessary to introduce an additional step in the method of preparation of the CPE. Therefore, drying in the convection-drying conditions has been added in a vacuum oven drying, to completely evaporate the solvent. These membranes were stored in a special container in an argon atmosphere prior to the measurement of the ionic conductivity in an organic electrolyte. Next composite polymer electrolytes with lithium perchlorate added were examined.

According to Figure 14.4, it can be seen that the filler concentration of 5% shows the greatest value of the ionic conductivity for all three types of membranes. This fact was confirmed in Figure 14.3, which shows the corresponding dependence without lithium salts. Thus, this concentration can be called optimal, however, in some theoretical considerations it is of interest. Nanoparticles, due to their higher specific surface according to theory should produce a different effect on the ionic conductivity at equal concentration of micro-particles. In other words, the smaller concentration of nanoparticles should produce the same effect as a large concentration of micro-particles. Furthermore, when comparing Figures 14.3 and 14.4 clearly

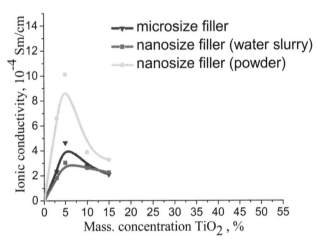

FIGURE 14.4 The dependence of ionic conductivity of composite polymer electrolyte based on the PVdF of content by mass of the filler particles with lithium salt at 25°C temperature.

drawn regular increase in ionic conductivity with the addition of CPE in their structure of the lithium salt, and the results are improved by one to two orders of magnitude.

Upon further examination of Figure 14.4, we have determined that the greatest ionic conductivity at room temperature showed a membrane containing 5% of nanoparticles of titanium oxide, 25% of lithium per chlorate (1.01 Sm/cm), which stands out the peak stresses.

Hence, from the above results of ion conductivity, it can be concluded that the measured value greatly depends on the qualitative and quantitative composition of the composite. The most successful have samples in the composition which contained as an inert phase filler – TiO_2 (5 wt.%), and an active phase – $LiClO_4$ (25 wt.%). In addition, to the filler particles are preferably introduced into the CPE in the form of a dry powder of nanoparticles.

To start the membrane in an industrial conveyor belt their electrochemical stability is very demanding. This will run the battery for a long time. In this work, we have studied the electrochemical stability on the basis of cycling polymer electrolyte with the best characteristics of ionic conductivity. The experiment was carried out in an organic solvent in a glove box under argon.

A more detailed review of the relaxation curves (Figure 14.5) and polarization (Figure 14.6) from galvanostatic curves shows that the curves

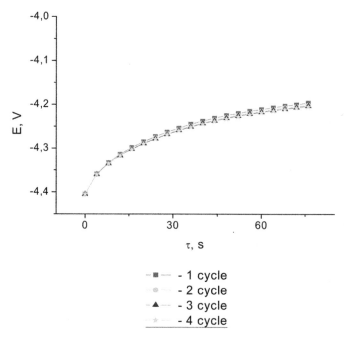

FIGURE 14.5 The relaxation curves of Faraday component in the polarization current 0.01 A.

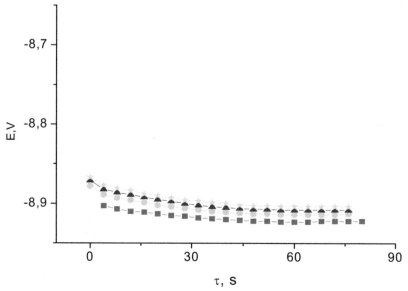

FIGURE 14.6 The relaxation curves resistive component in the polarization current 0.01 A.

of the relaxation process with each cycle are reversible, indicating that the unhindered passage of charged particles in the polymer membrane. The polarization curve after the first cycle of the polarization (Figure 14.6, curve 1) differs slightly from the polarization curves subsequent cycles. This may be due to processes occurring at the interface of the polymer membrane.

Thus, one can conclude that upon reaching thermodynamic equilibrium CPE electrochemical stable. Side processes in addition to the passage of the charge of the lithium ions are not observed.

14.3.2.2 Thermal Physical Investigations

An important factor for the Lithium batteries is the temperature stability of its components. This provides the desired degree of safety and battery performance. Therefore, we analyzed thermo gravimetric curves obtained membranes shown in Figures 14.7–14.9.

The comparison of the thermo gravimetric curves for pure PVdF (Figure 14.7) with the composite containing the filler nanoparticles (Figure 14.8) and a composite of the same composition, but with the addition of lithium per chlorate (Figure 14.9) was noticed an interesting pattern. The polymer membrane PVdF-nanoTiO$_2$ (5%) contained the least amount of moisture, so the weight loss was only 0.02% at the end of the test. This attests to the

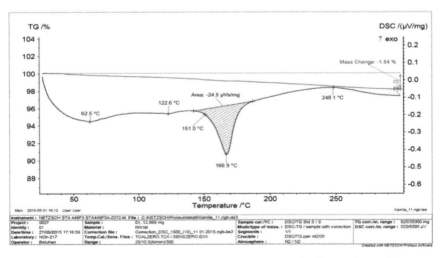

FIGURE 14.7   Thermo gravimetric curves TG-DSC for the pristine PVdF.

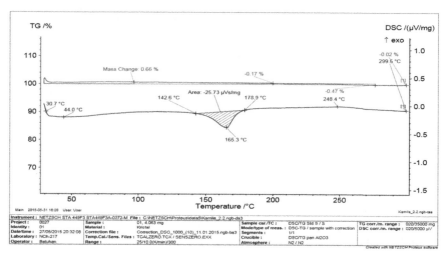

FIGURE 14.8 Thermo gravimetric curves TG-DSC for the polymer membrane PVdF-microTiO$_2$ (5%).

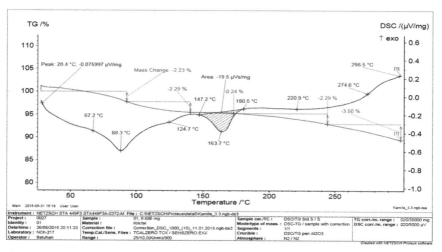

FIGURE 14.9 Thermo gravimetric curves TG-DSC for the polymer membrane PVdF-microTiO$_2$ (5%) – LiClO$_4$ (25%).

efficiency of drying in a vacuum oven, and this gives greater importance this synthesis stage. The pristine PVdF is not subjected to drying and stored in a normal atmosphere, so its weight loss was 1.84%. The greatest weight loss suffered sample PVdF- nanoTiO$_2$ (5%) – LiClO$_4$ (25%), which was dried out in a vacuum oven. As expected, the lithium salt influenced the ability of the polymer matrix to absorb moisture, the evaporation of which was about 10%

The shaded area corresponds to the specific heat of fusion at which indirectly can judge the crystalline of the polymer membrane. In the sample with pure PVdF shaded area larger than the other two samples, hence, one may suppose that crystalline of other two is less than in pristine sample.

From further analysis of the curves, one can see the characteristic groove is at temperature about 165°C for all three samples, but especially for the first two. This endothermic process is characteristic of the membranes in the composition containing PVdF and evidence of the immutability of the composition and the internal ordering of the polymer. In the third sample one can notice already two endothermic and one exothermic process at different temperatures, it is possible the formation process of a complex between the polymer chains and a lithium salt.

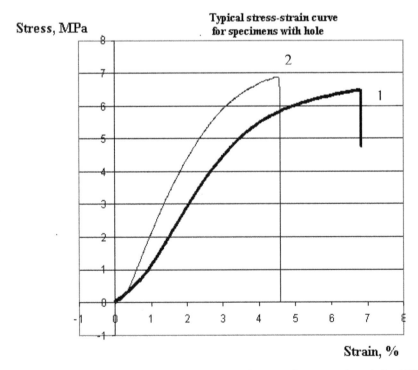

FIGURE 14.10 Diagram of stretching polymer membranes, where curve 1 – pristine PVdF film; curve 2 – PVdF-nanoTiO$_2$ (5%)-LiClO$_4$ (25%).

14.3.3 MECHANICAL TENSILE STRENGTH TESTS

The mechanical strength of the membrane can be measured by various methods, which should be considered when directional property of this parameter in the spatial directions in the membrane. A common method of measuring this parameter is to determine the tensile strength of membrane. Diagram of the experiment is shown in Figure 14.10. The maximum of the diagram corresponds to the rupture of the sample, the point of maximum strength. Comparing the maximum value of the supplied strain to the sample, one can say that CPE withstand more stress (6.9 MPa) compared to pure PVdF (6.5 Mpa), however, showed less significant stretch (4.5–7%). Hence, given values of the tensile strength of these membranes and comparing them with references, can be regarded as a synthetic CPE strong enough.

Comparing electrochemical studies and mechanical characteristics of the membrane, it can be concluded that good CPE should be mechanically strong, but at the same time sufficiently soft and flexible.

14.4 CONCLUSIONS

It was established that introducing the inorganic filler – TiO_2 in the polymer matrix creates additional coordination centers for lithium cation and suppresses polymer crystallization.

The research of membranes electrochemical properties showed a 10,000-fold increase of the ionic conductivity of the second series of membranes in comparison of first series composites at the room temperature. We found that the ionic conductivity of the polymer membrane is directly depending on particle shape and distribution of lithium-conductive agent in the matrix. The optimal ratio of composition accounted for 5% TiO_2 + 25% $LiClO_4$ in PVdF matrix. This CPE has high ionic conductivity (1.01 mSm/cm), thermally (range from 25–165°C) and electrochemically stable, mechanically strong (4.5% at 6.9 MPa) and safe.

ACKNOWLEDGMENTS

The financial support of the Science Committee of Ministry of Education and Science of the Republic of Kazakhstan Grant #4186/GF4 is gratefully acknowledged.

KEYWORDS

- composite polymer electrolytes
- lithium ion conductivity
- lithium polymer batteries
- solid polymer electrolytes
- TiO_2
- PVdF

REFERENCES

1. Cao, J., Wang, L., Fang, M., He, X., Li, J., & Gao, J. (2014). Structure and electrochemical properties of composite polymer electrolyte based on poly vinylidene fluoride hexa fluoro propylene/titaniaepoly(methyl methacrylate) for lithium-ion batteries. *Journal of Power Sources, 246*, 499–504.
2. Goodenough, J. B., & Kim, Y. (2010). Challenges for rechargeable Li batteries. *Chemistry of Materials, 22*(3), 587–603.
3. Rajendran, S., Kesavan, K., Nithya, R., & Ulaganathan, M. (2012). Transport, structural and thermal studies on nanocomposite polymer blend electrolytes for Li-ion battery applications. *Current Applied Physics, 12*, 789–793.
4. Appetecchi, G. B., Scaccia, S., & Passerini, S. (2000). Investigation on the stability of the Lithium-polymer electrolyte interface. *Journal of the Electrochemical Society, 147*(12), 4448–4452.
5. Caillon-Cavanier, M., Claude-Montigny, B., Lemordant, D., & Bosser, G. (2002). Absorption ability and kinetics of a liquid electrolyte in PVdF-HFP copolymer containing or not SiO_2. *Journal of Power Sources, 107*, 125–132.
6. Ki, S. K., & Soo-Jin, P. (2012). Influence of N-doped TiO_2 on lithium ion conductivity of porous polymeric electrolyte membrane containing $LiClO_4$. *Solid State Ionics, 212*, 18–25.

CHAPTER 15

DEXTRAN-POLYACRYLAMIDE AS NANOCARRIER FOR TARGETED DELIVERY OF ANTICANCER DRUGS INTO TUMOR CELLS

P. TELEGEEVA,[1] N. KUTSEVOL,[1] S. FILIPCHENKO,[1] and G. TELEGEEV[2]

[1]Taras Shevchenko National University, Faculty of Chemistry, 60 Volodymyrska Str., Kyiv, 0160, Ukraine

[2]Institute of Molecular Biology and Genetics of Natl. Acad. of Sci. of Ukraine, 150 Zabolotnogo Str. 150, Kyiv, 03680, Ukraine, E-mail: gtelegeev@ukr.net

CONTENTS

ABSTRACT

Drug targeting to specific organs and tissues is one of the crucial endeavors of modern pharmacotherapy. Controlled targeting at the site of action and reduced time of exposure of non-targeted tissues increase the efficacy of the treatment and reduce toxicity and side effects, improving compliance and convenience. Nanocarriers based on the branched copolymers dextran-graft-polyacrylamide were synthesized and characterized and were tested on phagocytic cells. It was shown that these nanoparticles are actively captured by phagocytic cells, and that they are not cytotoxic (90% ± 2% live cells at 0.01 mkg/ml).

The polymer nanoparticles loaded with cisplatinum at different concentrations from 0.01 to 0.1 mkd/ml yielded a percentage of living cells between 28% and 76%. Taking into account that our nanoparticles will act mainly on malignant phagocytic cells and do not affect healthy cells, they can thus potentially be used for the therapeutic treatment of tumor cells having phagocytic activity.

15.1 INTRODUCTION

Hybrid organic-inorganic materials, where molecular organic and inorganic fragments are combined, have been considered potentially attractive for the purpose of developing of new materials [1] with a broad spectrum of interesting properties. In comparison with organic and inorganic constituents and polymers separately, hybrid organic-inorganic materials have a lot of advantages [2–4].

15.2 EXPERIMENTAL PART

15.2.1 POLYMER NANOCARRIERS

As a nanocarrier we used a branched copolymer obtained by grafting polyacrylamide (PAA) chains onto dextran ($M_w = 7 \times 10^4$ g mol^{-1}) backbone [16] using a ceric-ion-reduce initiation method. This redox process initiates free radical sites exclusively on the polysaccharide backbone, thus preventing from the formation of homopolymer (PAA) [17, 18].

The detail of synthesis, identifications and analysis of internal polymer structure were described in Refs. [16, 17]. The theoretical number of grafting sites per polysaccharide backbone for the sample we used as polymer nanocarrier in the present work was equal to 5, and the related dextran-graft-polyacrylamide copolymer was referred as D70-g-PAA. The choice of this copolymer among the series of the branched samples synthesized based on our previous research. Namely this sample was the most efficient polymer matrices for Ag-sol in situ synthesis as well as for the nanoscale catalyst preparation [16, 19–21].

The D70-*g*-PAA copolymer was saponified by alkaline hydrolysis using NaOH to obtain branched polyelectrolyte, referred as D70-*g*-PAA(PE) throughout [16, 19]. The degree of saponification of carbamide groups to carboxylate ones onto PAA-granted chains determined by potentiometric titration was equal to 43% [19].

All polymer samples (the nascent and hydrolyzed ones) were precipitated into an excess of acetone, dissolved in bidistillated water, then freeze-dried and kept under vacuum for preventing them from additional hydrolysis. Potentiometric titration curves obtained for bidistilled water and for the non-ionic copolymer D70-*g*-PAA were the same. Thus, it can be concluded that the degree of the hydrolysis of the PAA moiety in the nascent copolymer was virtually zero.

15.2.2 SIZE-EXCLUSION CHROMATOGRAPHY

SEC analysis was carried out by using a multi-detection device consisting of a *LC-10 AD Shimadzu* pump (throughput 0.5 mL min^{-1}; Nakagyo-ku, Kyoto, Japan), an automatic injector *WISP 717+* from *Waters (Milford, MA, USA)*, three coupled *30-cm Shodex OH-pak* columns (803HQ, 804HQ, and 806HQ; Munich, Germany), a multi-angle light scattering detector *DAWN F* from *Wyatt Technology (*Dernbach, Germany*)*, and a differential refractometer *R410* from *Waters*. Distilled water containing 0.1 M NaNO$_3$ was used as fluent. Dilute polymer solutions ($c = 3$ g L$^{-1} < c^* = 1/[\eta]$) were prepared, allowing for neglect of intermolecular correlations in the analysis of light scattering measurement.

15.2.3 POTENTIOMETRIC TITRATION

Potentiometric titration of polyelectrolyte samples was performed using a pH meter pH-340 Economic Express, St. Petersburg, Russia). HCl (0.2 N) and NaOH (0.2 N) were used as titrants. Polymer concentration was 2 g L^{-1}. The polymer solution was titrated with HCl up to pH 2 and then with NaOH up to pH 12. Previously, a fine blank titration (titration of non-hydrolyzed polymer) was made. The absorption of OH$^-$ anions was calculated through the analysis of the titration curves and then the limits of these values were used to determine the conversion degree of carbamide groups into carboxylate ones. The measurements were performed at $T =$ 25°C under nitrogen.

15.2.4   FTIR SPECTROSCOPY

FTIR spectra were obtained on a Nicolet NEXUS-475 (USA) spectrophotometer in the range 4000–400 cm^{-1} using thin films (l = 6–9 μm). The films were cast from aqueous solutions of polymer without adding of cisplatin and with cisplatin.

15.2.5   THE BIOLOGICAL TEST FOR POLYMER-NANOCARRIERS EFFICIENCY

The cells culture of murine macrophage J774 and U-937 (hystocytic lymphoma) were used as the biological object for studying the efficiency of the polymer nanocarriers loaded with cisplatin as an anticancer agent. The cell culture was obtained from culture collection of Institute of Molecular Biology and Genetics of National Academy of Science of Ukraine.

The pharmaceutical cisplatin dosage form from "EBEVE" (Austria) was used for further encapsulation into polymer nonocarrier.

15.2.6   PHAGOCYTIC INDEX AND CELLS SORTING

The efficiency of absorption of the nanocarrier by a cell culture was determined by evaluating phagocytic index. Phagocytic index (PI) is the percentage of cells that have entered in the phagocytosis out of the total number.

The distribution of the phases of the cell cycle and apoptosis level was assessed using aFACS Calibur flow cytometer ("Becton Dickinson", USA) by the method described in [22].

15.2.7   TEST FOR CELL SURVIVAL

About 100 μl of cell was added to a culture plate in the nutrient medium. Cisplatin was added to nanocarriers at various concentrations. Cells were cultivated in an incubator with cisplatin as and also with nanocarriers loaded with cisplatin in a gaseous phase (5% CO_2) for one day. The number of live cells was analyzed using the inverted microscope Ulab XD 30 (China) in a Goryaev chamber, staining the cells with try pan blue following the standard procedure [23].

15.3 RESULTS AND DISCUSSION

The branched water-soluble copolymers dextran-graft-polyacrylamide in non ionic and anionic form was used as nanocarriers for targeted delivery of an anticancer drug into tumor cells.

The molecular parameters of the D70-g-PAA copolymer, as determined by size-exclusion chromatography (SEC), are reported in the Table 15.1.

SEC indicates that this sample possesses relatively low molecular weight polydispersity and the size of macromolecules relatively large. It is known that cells-phagocytes can trap the objects of 100–250 nmin size. Thus, the D70-g-PAA copolymer can be used for loading the therapeutically agents capable of killing the damaged cells-phagocytes. The peculiarities of the molecular structure of branched copolymers dextran-graft-polyacrylamide in nonionic and ionic forms are discussed in notes [16, 17, 19]. These copolymers are star-like, consisting of the compact dextran core and long polyacrylamide arms. As was reported in Refs. [16, 17], the average conformation of grafted PAA chains is controlled by the grafting ratio. For the D70-g-PAA copolymer used in the present research the PAA-grafted chains are highly extended, approaching their tethering point, and recover to a random conformation quite distant from this point.

The branched polyelectrolyte D70-g-PAA (PE) is characterized by an extremely expanded conformation of the grafted chains in solution. Alkaline hydrolysis of D70-g-PAA copolymers was not accompanied by irrelevant processes (the breaking or cross-linking of macromolecules), as was confirmed by SEC analysis of source and saponified samples [19].

The branched polymers, due to their more compact molecular structure, have a higher local concentration of functional groups in comparison to

TABLE 15.1 The Molecular Parameters of Polymer-Nanocarrier

Sample	$M_w \times 10^{-6}$	$R_{g, HM}$	M_w/M_n
D70-g-PAA	2.15	85	1.72

their linear analogs. These structure peculiarities of branched polymers are advantageous for application in nanotechnologies [16, 20, 21].

The D70-g-PAA and D70-g-PAA (PE) copolymers consist of biocompatible components – dextran and polyacrylamide, which are nontoxic, water soluble, their macromolecules are approximately 100 nm in size with ligands capable of coordinating multivalent metal ions [24], that is why our idea was to test them as nanocontainers-carriers for targeted delivery of the highly toxic therapeutic anticancer agent cisplatin.

It was shown that the D70-g-PAA nanocontainers which were absorbed by macrophages (Figure 15.1) and were not cytotoxic. The phagocytic index was equal to 84%. This fact permitted the testing of nanocontainers loaded with the cytotoxic chemotherapy drugs cisplatin on the cells.

Complexes of copolymer D70-g-PAA and an anionic derivative with cisplatin were synthesized. FTIR spectra analysis in Amide 1(1650–1660 cm^{-1}, C = O stretching) and Amide 2 (1615 cm^{-1}, N-H bending) region showed no strong interaction of the non-ionic copolymer with cisplatin (Figure 15.2). The spectra of the reaction product of cisplatin with D70-g-PAA (PE) revealed the drastic changes in the intensity of the absorption within the Amide 1 and Amide II bands, and for a band with a peak at 1570 cm^{-1} (COO-, stretching), respectively, indicating the complex formation of cisplatin with the copolymer carboxyl groups.

According to the FTIR data, the system D70-g-PAA(PE) + cisplatin were used in the experiments in vitro. Dose-dependent cytotoxicity of platinum complexes for J774 cells is shown in (Figure 15.3).

The nanoparticles loaded with cisplatin at different concentration in the range from 0.01 to 0.1 mkg/ml revealed a percentage of living cells between 28% and 76%.

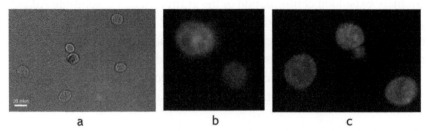

a   b   c

FIGURE 15.1   Nanocontainers D70-PAA5 tested on marine macrophages: nanocontainers D70-PAA5. (a) Murine macrophages microphotograph; (b) Murine macrophages under fluorescent microscope without nanocontainers. (c) Murine macrophages with nanocontainers in phagosomes under fluorescent microscope, stained with Acridine Orange.

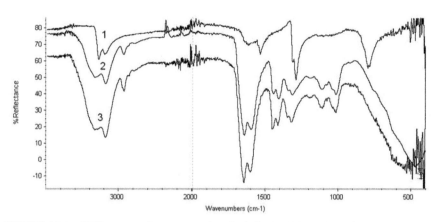

FIGURE 15.2 FTIR spectra for systems: 1 – cisplatin, 2 – D70-g-PAA, 3 – D70-g-PAA + cisplatin.

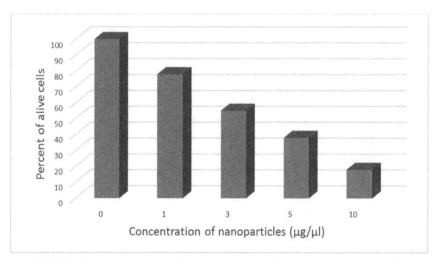

FIGURE 15.3 Cytotoxicity of nanocontainers with cisplatin for the J774 cells.

The distribution of the phases of the cell cycle and apoptosis level was assessed using a FACS Calibur flow cytometer. It has been shown that the effectiveness of growth inhibition on U937 culture (histolytic lymphoma) under the action of cisplatin in sub toxic concentration of 3 mol/l is the same as for the soluble form of the drug and for cisplatin encapsulated into polymer nanoparticles. Thus, cisplatin-containing nanoparticles does not lose its growth-inhibitory effect on these cells.

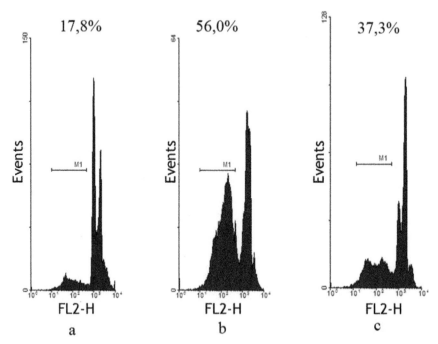

FIGURE 15.4 Histograms of distribution of U937 cells in content of DNA with the influence of cisplatin in solution or in conjunction with dextran containing nanoparticles. (a) control; (b) cisplatin solution of 3 μmol/l; (c) nanocontainers loaded with cytotoxic chemotherapy drugs cisplatin, 3 μmol/l.

The effects of cisplatin on the cell cycle were also identical to the soluble form of the drug, and the drug in nanoparticulate form (increasing fraction of cells in phases S and G2/M). However the level of induction of apoptosis in the preparation form of nanoparticles was somewhat inferior to that under the action of an equimolar concentration of cisplatin solution (Figure 15.4).

Combining these data with the kinetics of growth culture under treatment with different drugs platinum [25], it could be argued that the effect of cisplatin in the form of nanoparticles on leukemia cells leads to a change in the ratio between the growth-inhibitory and proapoptotic efficacy of platinum at an equimolar content of the active component. In addition, we must take into account that our nano particles will act mainly on phagocytic cells and will not affect healthy cells, thus realizing targeted orientation of nanoparticles with cisplatin.

15.4 CONCLUSIONS

The goal of cancer treatment is to kill as many cancer cells as possible without affecting healthy cells. The present research revealed the efficiency of nanodelivery anticancer systems based on the branched copolymer dextran-graft-polyacrylamide and cisplatin. These systems have shown several promising characteristics which can improve traditional chemotherapy. The main advantage of our approach is in target delivery of toxic drugs. It is assumed that it will be used for the treatment of tumor cells possessing phagocytic activity such as a leukemia M3-M5 (FAB classification) without damaging healthy tissue. Our experiment proved that nanocontainers the particles are non-toxic, which is important in the therapy session. Further studies are needed to turn concept of nanotechnology in into practical application (in vitro and in vivo) and to elucidate correct drug doses and optimal copolymer internal structure for ideal release therapeutically agent encapsulated in polymer molecule for the treatment of cancer cells.

KEYWORDS

- anticancer drug
- branched polymers
- cisplatin
- nanocarriers
- nanomedicine
- phagocytosis
- targeted delivery of drugs

REFERENCES

1. Martinho, N., Damge, C., & Reis, C. P. (2011). Recent advances in drug delivery systems. *Journal of Biomaterials and Nanobiotechnology, 2*, 510–526.
2. Mohanraj, V. J., & Chen, Y. (2006). Nanoparticles: A review. *Tropical Journal of Pharmaceutical Research, 5*, 561–573.
3. Calixto, G., Bernegossi, J., Fonseca-Santos, B., & Chorilli, M. (2014). Nanotechnology-based drug delivery systems for treatment of oral cancer: a review. *International Journal of Nanomedicine, 9*, 3719–3735.

4. Lattin, J. R., Belnap, D. M., & Pitt, W. G. (2011). Formation of eliposomes as a drug delivery vehicle. *Colloids and Surfaces B: Biointerfaces, 89,* 93–100.

5. Vlerken, L. E., Duan, Z., Little, S. R., Seiden, M. V., & Amiji M. M. (2008). Biodistribution and pharmacokinetic analysis of paclitaxel and ceramide administered in multifunctional polymer-blend nanoparticles in drug resistant breast cancer model. *Molecular Pharmaceutics, 5,* 516–526.

6. Xu, T., Zhang, N., Nichols, H. L., Shi, D., & Wen, X. Modification of nanostructured materials for biomedical applications. *Materials Science and Engineering, 27,* 579–594, 2007.

7. Liu, Y., Miyoshi, H., & Nakamura, M. (2007). Nanomedicine for drug delivery and imaging: A promising avenue for cancer therapy and diagnosis using targeted functional nanoparticles. *International Journal of Cancer, 120,* 527–2537.

8. Vlerken, L. E, & Amiji, M. M. (2006). Multifunctional polymeric nanoparticles for tumor-targeted drug delivery. *Expert Opinion on Drug Delivery, 3,* 205–216.

9. Vasir, J. K, & Labhasetwar, V. (2007). Biodegradable nanoparticles for cytosolic delivery of therapeutics. *Advanced Drug Delivery Reviews, 59,* 718–728.

10. Soppimath, K. S, Aminabhavi, T. M, Kulkarni, A. R., & Rudzinski W. E. (2001). Biodegradable polymeric nanoparticles as drug delivery devices. *Journal of Controlled Release, 70,* 1–20.

11. Ferrari, M. (2005). Cancer nanotechnology: Opportunities and challenges. *Nature Review Cancer, 5,* 161–171.

12. Grest, G. S., Fetters, L. J., & Huang, J. S. (1996). Star polymers: Experiment, theory, and simulation. *Advanced in Chemical Physics, 94,* 67–157.

13. Ballauff, M. (2007). Spherical polyelectrolyte brushes. *Polym. Sci, 32,* 1135–1151.

14. Widawski, G., Rawiso, M., & Francois, B. (1994). Self-organized honeycomb morphology of star-polymer polystyrene films. *Nature, 369,* 387–389.

15. Heinrich, M., Rawiso, M., Zilliox, J. G., Lesieur, P., & Simon, J. P. (2001). Small-angle X-ray scattering from salt-free solutions of star-branched polyelectrolytes. *Eur. Phys. J. Part E, 4,* 131–142.

16. Kutsevol, N., Bezugla, T., Bezuglyi, M., & Rawiso M. (2012). Branched Dextran-Graft-Polyacrylamide Copolymers as Perspective Materials for Nanotechnology. *Macro mol. Symp., 1,* 82–90.

17. Kutsevol, N., Guenet, J. M., Melnyk, N., Sarazin, D., & Rochas, C. (2006). Solution properties of dextran-polyacrylamide graft copolymers. *Polymer, 47,* ,2061–2068.

18. Owen, D. R., Shen, T. C., Harris, F. W., & Seymour, R. P. (1977). Structure Solubility Relationship in Polymers. Academic Press: New York.

19. Kutsevol, N., Bezuglyi, M., Rawiso, M., & Bezugla, T. (2014). Star-like destran-graft-(polyacrylamide-co-polyacrylic acid) copolymers. *Macro mol. Symp., 335,* 12–16.

20. Bezuglyi, M., Kutsevol, N., Rawiso, M., & Bezugla, T. (2012). Water-soluble branched copolymers dextran-polyacrylamide and their anionic derivatives as matrices for metal nanoparticles *in-situ* synthesis. *Chemik, 8(66),* 862–867.

21. Chumachenko, V., Kutsevol, N., Rawiso, M., Schmutz, M., & Blanck, C. (2014). *In situ* formation of silver nanoparticles in linear and branched polyelectrolyte matrices using various reducing agent. *Nanoscale Research Letters, 9,* 164.

22. Bhattacharjee, S., de Haan, L. H. J., & Evers, N. M. (2010). Role of surface charge and oxidative stress in cytotoxicity of organic monolayer-coated silicon nanoparticles towards macrophage NR8383 cells. *Particle and Fiber Toxicol., 7,* 25.

23. Strober W. (2001). Trypan blue exclusion test of cell viability. *Curr. Protoc. Immunol.,* May: Appendix: Appendix 3B.

24. Bezuglyi, M., Kutsevol, N., Bezugla, T., Rawiso, M., & Ischenko, M. (2012). Flocculation with Branched Copolymers in Ternary Component System: Kaolin/Polymer/Cu2+ Ions. Proceedings of the Eighth International Conference on the Establishment of Cooperation Between Companies and Institutions in the Nordic Countries, the Baltic Sea Region, and the World Conference on Natural Sciences and Environmental Technologies for Waste and Wastewater Treatment, Remediation, Emissions Related to Climate, Environmental and Economic Effects "Linnaeus ECO-TECH 2012," November 26–28, Kalmar, Sweden, 393–403.

25. Nicoletti, I., Migliorati, G., Pagliacci, M. C., Grignani, F., & Riccardi, C. (1991). A rapid and simple method for measuring thymocyte apoptosis by propidium iodide staining and flow cytometry. *J. Immunol Methods, 139,* 271–279.

CHAPTER 16

INVESTIGATION PHOTOTHERMAL AND PHOTOOPTICAL ENERGY CONVERSIONS IN SILVER AND GOLD NANOPARTICLES DOPED NANOCOMPOSITES FOR THE MODELING OF CANCER CELLS VISUALIZATION AND PHOTOTHERMAL CANCER THERAPY

K. CHUBINIDZE[1] and M. CHUBINIDZE[2]

[1]Tbilisi State University, 1 Ilia Chavchavadze Ave., Tbilisi, 0179, Georgia

[2]Tbilisi State Medical University, 7 Mikheil Asatiani St., Tbilisi, 0186, Georgia, E-mail: chubinidzeketino@yahoo.com

CONTENTS

ABSTRACT

We present two methods to calibrate light-to-heat and light-to-light conversions in the nanocomposites incorporated with silver and gold nanoparticles. First method relies on the optical properties of cholesteric liquid crystals, confined into microdroplets dispersed in the medium, and combines the advantages of high spatial resolution and good temperature accuracy with fast readout. The method can be used for any nanoparticles, disregarding their size or shape, for instance silver or gold based. In the second method, we demonstrate that the emission of visible light from the polymer matrix doped with luminescent dye and gold nanoparticles can be enhanced with the use of surface plasmon coupling. The visualization and control of optical and thermal energy conversions in nanostructures are key challenges in many fields of science with applications in areas as molecular sensing, detection and visualization cancer cells at the early stage of development, and photothermal cancer therapy.

16.1 INTRODUCTION

16.1.1 THERMOSENSITIVE LIQUID CRYSTAL MICRO THERMOMETERS FOR THE VISUALIZATION LIGHT-TO-HEAT CONVERSION

Systems of nanoparticles (NPs) dispersed in liquid crystals (LCs) have attracted attention to the possible development of novel materials based on the controlled assembly of the particles. An attractive way to satisfy this conformational freedom requirement would be to hosting the NPs within flexible, tunable and organic medium as LCs. LC phases are found in DNA, proteins, lipids and polysaccharides. Therefore, it is very important to know the properties of LC materials in order to better understand biological processes [1]. Due to this unique property of LC materials, the NP doped LCs are assumed to have many advanced optical characteristics, quite different from those of NP doped isotropic liquids, semiconductors and solid state systems [2–4]. The method of the visualization of light-to-heat conversions in the LC doped nanocomposites relies on the optical behavior of cholesteric liquid crystals (CLCs), confined into microdroplets dispersed in the medium, and combines the advantages of high spatial resolution and good temperature accuracy with fast readout (Figure 16.1).

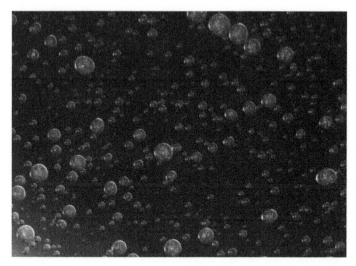

FIGURE 16.1   Emulsion of CLC microdroplets dispersed in water matrix.

In this work, we propose the idea to use thermotropic CLC microdroplets for the non-intrusive visualization and measurement of the temperature distribution at the micro scale. In particular, we focus on the visualization of the optical energy conversion to heat caused in metal NPs. Gold and silver NPs can efficiently release heat under optical excitation. When excited with a laser beam, the laser electric field strongly drives charge mobile carriers inside the NPs, and the energy gained by carrier turns into heat. Heat generation becomes especially strong in the regime of plasmon resonance [5]. The NPs temperature may rise significantly and the heat can propagate to the surrounding medium. In our method, the temperature surrounding the nanoparticles can be estimated monitoring the spectral shift of the selective reflection peak of a CLC. For this purpose, an emulsion of water, CLC and nanoparticles were investigated and the temperature in the environment surrounding silver NPs, as a function of the exposure time to laser radiation, was monitored (Figure 16.2).

16.2   EXPERIMENTAL PART I

16.2.1   MATERIALS

An emulsion of the cholesterol mixture in water and glycerol is prepared with the following percentage in weight: 95% (90% water + 10% glycerol)

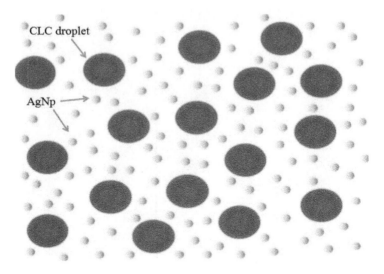

FIGURE 16.2 Schematic representation of CLC micro thermometer suspended in nanocomposite consisting of water doped silver NPs.

+ 5% CLC. Glycerol was added in order to reduce the evaporation rate of the emulsion. A cuvette with 1 mm gap was filled with the emulsion and stirred at 600 rpm, at room temperature for 25 min. As a result, a homogeneous distribution of CLC microdroplets suspended in an aqueous glycerol matrix was obtained. The diameter range of CLC microdroplet was 10–15 µm. Depending on the CLC concentration and stirring speed, CLC microspheres with different diameter sizes and packing density may be obtained. To visualize the temperature distribution inside the cuvette when the emulsion was heated up from room temperature, the cuvette was placed on a heating stage. Figure 16.3(a) shows the local temperature at different heights in the cuvette, also showing the vertical temperature gradient. Blue color at the bottom of the cuvette corresponds to the highest temperature, and it turns to green, yellow, red and brown as we move to the top of the cuvette, indicating a decrease in temperature. Figure 16.3(b, c) shows the same microscopic area of a liquid crystal cell at 25°C (b) and 42°C (c). The final nanocomposite emulsion (NCE) is obtained mixing cholesterol liquid crystal, glycerol and the NP aqueous solution in the following concentrations in weight: 95% (90% Silver NP suspension + 10% glycerol) + 5% CLC. As in the previous case, glycerol is used to reduce the evaporation rate of water during experiments. Finally, the NCE is stirred at

FIGURE 16.3 (a) Temperature distribution in a cuvette containing CLC microdroplets in a water and glycerol mixture. Optical microscope images of CLC microdroplets in a liquid crystal cell. Red color (b) is visible at room temperature, 25°C, while blue (c) is visible at 42°C.

600 rpm, at room temperature to produce homogeneously distributed CLC microdroplets.

To visualize the effect of energy harvesting from silver nanoparticles, the experimental setup sketched in Figure. 16.4 was used.

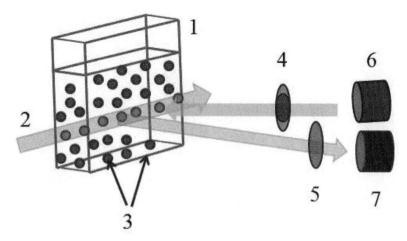

FIGURE 16.4 Experimental setup: cuvette (1), impinging laser beam @ 457 nm (2), CLC micro thermometers (3), the optical shutter (4), edge filter (5), light source (6), fiber-coupled spectrometer (7). The laser beam travels inside the cuvette, parallel to its larger faces.

The "heat trace" penetrates inside the cuvette, highlighting the profile of the temperature distribution. In Figure 16.5, the color distribution inside the cuvette (a) before and (b) immediately after the switching off of the laser beam is shown. The red arrow indicates the area exhibiting a larger temperature (green) than the surroundings (yellow-red). The highest temperature was obviously reached along the previous laser beam propagation path. An high-pass filter can be added in front of the camera to cut the impinging laser light, in order to follow the dynamics of the heat propagating inside the sample.

16.1.2 GOLD NANOPARTICLES STIMULATED LIGHT EMISSION ENHANCEMENT

In this section, we present gold nanoparticles (GNPs) formed and incorporated together with luminescent dye Nile blue (Nb) into a film of polyvinyl alcohol (PVA). Light enhancement of a luminescent dye strongly depends on the factors that can be manipulated to obtain a desired luminescence: the distance between the luminescent dye and the GNPs and the spectral overlap between the wavelength of pumping light source and GNPs plasmon resonance peak. GNPs are among the most extensively studied nanomaterials. They are known to be the most stable metal nanoparticles [6, 7]. Numerous

FIGURE 16.5 (a) Image of the laser beam propagating inside the cuvette, and (b) red arrow indicating the highest local temperature in correspondence of the previous laser beam propagation path immediately after switching off the laser beam.

studies have been reported on the synthesis, property and application development of gold clusters, colloids, and nanoparticles [8–11]. In addition, combining the properties of gold nanoparticles with those of known organic dyes has already led to many interesting applications including sensing of biologically relevant molecules [12, 13].

16.3 EXPERIMENTAL PART II

16.3.1 MATERIALS

As the initial components we utilized next materials: macro monomer PVA with a number average molecular weight (Mw) 85,000–124,000, 99% hydrolyzed. Colloidal mono dispersed GNPs with 40 nm in size, with concentration 7.15×10^{10} nanoparticles/ml and molecular weight 196.97 g/Mol dispersed in an aqueous buffer (0.02 mg/ml), and glycerol solution 86–89% (T), with density 1.252 g/ml at 25°C. All components were purchased from Sigma-Aldrich. As the luminescent dye we used Nile Blue (Nb) (Figure 16.6).

We have, therefore, chosen the PVA because of its ease of process ability, solubility, mechanical and thermal stability. Light enhancement of a luminescent dye strongly depends on the factors that can be manipulated to obtain a desired luminescence: the distance between the luminescent dye

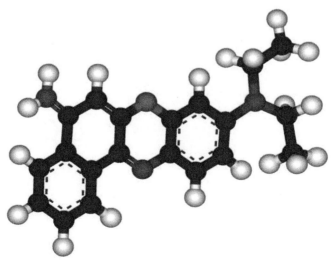

FIGURE 16.6   Structural formula of luminescent dye Nb.

and the GNPs and the spectral overlap between the wavelength of pump-ing light source and GNPs plasmon resonance peak. The Figure 16.7 shows single GNP, which is surrounded by Nb molecules.

We have prepared and investigated two nanocomposites. First one, PVA matrix doped with Nb and another one is the PVA matrix doped with Nb and GNPs. Absorbance and luminescence spectra of the samples were recorded by multi fiber optic spectrometer (Avaspec-2048, "Avantes"). Photoexitation of the nanocomposites were performed at $\lambda = 532$ nm, using MSL-III-532 diode laser, with 5 mW power. The optical ("LOMO"), and confocal scan-ning microscopes ("Leica"), were involved to investigate a polymer nano-composites at micro scale. In the description of the nature of the energy transfer from an organic luminescent dye to a GNP a crucial role plays the distance dependence between the luminescent dye and the surface of a GNP. The altered electromagnetic field around the metal nanoparticle changes the properties of a dye that is placed in the vicinity. It can cause two enhance-ment effects: the first is an increase in the quantum efficiency of the dye and the second is an increase in the excitation rate of the dye. The induced col-lective electron oscillations associated with the surface plasmon resonance, give rise to induce local electric fields near the nanoparticle surface. Energy

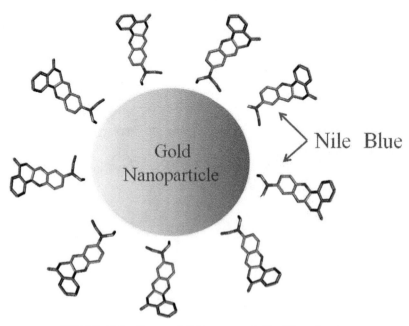

FIGURE 16.7   Image of GNPs surrounded by Nb molecules.

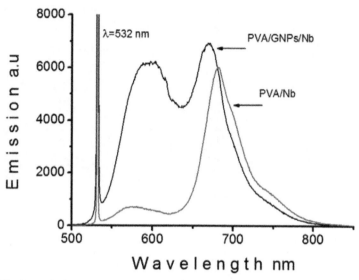

FIGURE 16.8 Induced light emission from PVA/Nb and PVA/Nb/GNPs polymer composites when pumped by laser with wavelength coincide to the absorption of GNPs.

transfer from luminescent organic dyes to GNPs is generally considered to be the major process leading to the excited-state activation/deactivation of the dyes. As the result we obtained a strong enhancement of the luminescence, when by the laser light was irradiated a nanocomposite PVA/GNPs/Nb. Figure 16.8, displays the luminescence spectra emitted from PVA/Nb and from PVA/GNPs/Nb. It is obvious that the light intensity emitted from the nanocomposite PVA/GNPs/Nb 690 is much stronger than the light intensity emitted from PVA/Nb composite.

Finally, based on the experimental results and calculated data from the equations, we found the distance between the dye molecules and GNPs, which are statistically distributed in the polymer medium, was equal 0.5 ± 0.15 nm. To visualize the distribution of GNPs/Nb pairs in PVA matrix we used polarized light microscope and confocal microscope. We found that some amount of GNPs/Nb 690 is prone to aggregate and form the clusters, which are the condensed quantity of GNPs and Nb 690 dyes. The vivid "halos" surrounding the clusters demonstrate the enhancement of light brightness, which confirms the strong energy transfer between GNPs and Nb (Figure 16.9).

At the end of our experiments we have demonstrated an advances of the proposed concept in biology, in particular for the cell labeling and tracking in biological tissues. For this reason we used pieces of lungfish tissues. One

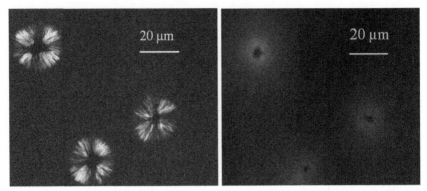

FIGURE 16.9 Images of GNPs/Nb clusters under (a) optical, and (b) confocal microscopes.

was labeled with Nb luminescent dye and other with Nb luminescent dye/ GNPs. As shown in Figure 16.10, a light emission from tissue labeled with Nb/GNPs, is much stronger, than from the tissue labeled just with Nb/luminescent dye.

16.4 CONCLUSIONS

We describe two novel methods that describe optical to thermal and optical to optical energy conversions in metal nanoparticles doped nanocomposites. First method enables to evaluate the temperature of the medium surrounding silver NPs as a function of the exposure time to light radiation. This method relies on the optical properties of cholesterol liquid

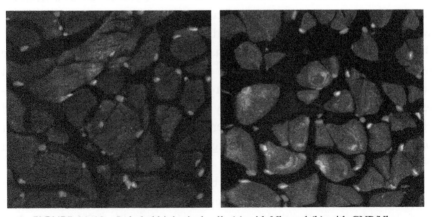

FIGURE 16.10 Labeled biological cells (a) with Nb, and (b) with GNP/Nb.

crystals, confined into microdroplets dispersed in the medium, and combines the advantages of high spatial resolution and good temperature accuracy with fast readout. The proposed method can be used for any NPs, disregarding their size or shape, for instance silver or gold based. The micron size droplets act as micro thermometers, providing a local visualization of the temperature, which is well suitable for applications in the biological field, in particular for the visualization and calibration of temperature distribution near the biological cells, for the application in plasmonic photo-thermal therapy. Photodynamic therapy can cause burns, swelling, pain, and scarring in nearby healthy tissue. CLC micro thermometers allow us to adjust and calibrate the irradiated light intensity and exposure time, which must be controlled carefully so that, to destroy the cancer cells and left alive, healthy cells. In the second method is described an enhancement in luminescence intensity, using GNPs and luminescent dye complex, which will lead to many applications in bimolecular labeling to produce novel optical contrast agents with high sensitivity and specificity. Obtained results may have a great importance in chemotherapy and cancer diagnosis.

ACKNOWLEDGMENTS

This work was supported by the Shota Rustaveli National Science Foundation under grant DO/171/8–314/14

KEYWORDS

- cancer cells
- cholesteric liquid crystals
- luminescent dye
- microdroplets
- polymer nanocomposite
- silver and gold nanoparticles
- surface plasmon
- visualization

REFERENCES

1. Rey, A. D. (2010). Liquid crystal models of biological materials and processes. *Soft Matter*, *6*, 3402–3429.
2. Sun, S. H., Murray, C. B., Weller, D., Folks, L., & Moser, A., (1989). Mono disperse FePt nanoparticles and ferromagnetic FePt nanocrystal super lattices. A 2000 *Science* 287.
3. Chen Hao Ming, Liu Ru-Shi, Asakura Kiyotaka, Lee Jyh-Fu, Jang Ling-Yun, & Hu Shu-Fen. (2006). Fabrication of Nano rattles with Passive Shell. *J. Phys. Chem. B 110*, 19162–19167.
4. Zhang Qingbo, Lee Jim Yang, Yang Jun, Boothroyd Chris, & Zhang Jixuan. (2007). Size and composition tunable Ag–Au alloy nanoparticles by replacement reactions. Nanotechnology 18, 245605.
5. Govorov, A. O., & Richardson, H. H. (2007). Generating heat with metal nanoparticles. *Nano Today 2*(1), 30–38.
6. Borriello, A., Agoretti, P., Cassinese, A., D'Angelo, P., Mohanraj, G. T., & Sanguigno, L., (2009). Electrical Bistability in Conductive Hybrid Composites of Doped Polyaniline Nanofibers-Gold Nanoparticles Capped with Dodecane Thiol, *Journal of Nanoscience and Nanotechnology, 9*(11), 6307–6314.
7. Hu, C. W., Huang, Y., & Tsiang, R. C. -C. (2009). Thermal and Spectroscopic Properties of Polystyrene/Gold Nanocomposite Containing Well-Dispersed Gold Nanoparticles, *Journal of Nanoscience and Nanotechnology. 9*(5), 84–3091.
8. Daniel, M. C., & Astruc, D., (2002). Gold Nanoparticles: Assembly, Supramolecular Chemistry, Quantum-Size-Related Properties, and Applications toward Biology, Catalysis, and Nanotechnology. *Chemical Reviews, 104*(1), 293–346.
9. Burda, C., Chen, X., Narayanan, R., & El-Sayed, M. A. (2005). Chemistry and Properties of Nanocrystals of Different Shapes. *Chemical Reviews, 105*(4), 1025–1102.
10. Katz, E., & Willner, I., (2004). Integrated Nanoparticle-Biomolecule Hybrid Systems: Synthesis, Properties, and Applications. *Angewandte Chemie International Edition, 43*(45), 6042–6108.
11. Kogan, M. J., Bastus, N. G., Amigo, R., Grillo-Bosch, D., Araya, E., Turiel, A., Labarta, A., Giralt, E., & Puntes, V. F. (2006). Nanoparticle-Mediated Local and Remote Manipulation of Protein Aggregation. *Nano Letters, 6*(1), 110–115.
12. Shang, L., Qin, C. J., Wang, T., Wang, M., Wang, L. X., & Dong, S. J. (2007). Fluorescent conjugated polymer-stabilized gold nanoparticles for sensitive and selective detection of cysteine. *J. Phys. Chem. C, 111,* 13414–13417.
13. Griffin, J., Singh, A. K., Senapati, D., Rhodes, P., Mitchell, K., Robinson, B., Yu, E., & Ray, P. C. (2009). Size and Distance Dependent NSET Ruler for Selective Sensing of Hepatitis C Virus RNA. *Chemistry A European Journal, 15,* 342–351.

CHAPTER 17

SYNTHESIS AND CHARACTERIZATION OF CONDUCTIVE POLYURETHANE COMPOSITES CONTAINING POLYANILINE-CUO NANOCOMPOSITES

Z. HESARI[1] and B. SHIRKAVAND HADAVAND[2]

[1]Department of Chemistry, Faculty of Science, Science and Research of Tehran Branch, Islamic Azad University, Tehran, Iran, E-mail: zeinab.hesari@yahoo.com

[2]Department of Resin and Additives, Institute for Color Science and Technology, Tehran, Iran

CONTENTS

ABSTRACT

In this chapter, we synthesized conductive urethane acrylate with added PANi-CuO nanocomposites into urethane acrylate matrix. PU-PANi-CuO composites were prepared and then curing of the composites were completed in the ultraviolet curing device. The morphological studies (SEM) showed uniform distribution PANi-CuO nanocomposites in polyurethane matrix. Glass transition temperature (Tg) value of the PU-PANi-CuO composites to higher value which indicates that PU-PANI-CuO composites are more stable than PANi-CuO. Electrical conductivity of urethane acrylate increased with increasing content of PANi-CuO nanocomposites.

17.1 INTRODUCTION

Conducting polymers have attracted significant attention in recent decades because of their potential applications in various fields such as electromagnetic interference (EMI) shielding, rechargeable battery, chemical sensor, corrosion devices and microwave absorption [1–5].

Polyurethane (PU) is one of the most important polymers and widely used in commercial applications, including construction, automotive, food packaging, storage, transportation, textiles, foot-wear and wound dressing materials. Polyurethanes have appropriate mechanical properties such as film-formation, so with the preparation of composites of polyaniline and polyurethane can improve the mechanical properties of polyaniline [6, 7]. PANi-CuO nanocomposite is a conductive nanocomposite. Therefore, with added PANi-CuO nanocomposites into other non-conductive polymers such as polyurethane created a conductive polymer nanocomposite.

In this research, urethane acrylate was synthesized and then PU-PANi-CuO composites were fabricated with different contents of PANi-CuO nanocomposites and then curing of the films was completed in the ultraviolet curing device. Morphological, electrical conductivity, thermal properties of the products were characterized by electron microscopy (SEM), thermo gram analysis (DSC) and four point probe resistivity measurements.

17.2 EXPERIMENTAL PART

17.2.1 SYNTHESIS OF PU-PANI-CUO COMPOSITES

PU-PANi-CuO composites was synthesized with PANi-CuO nanocomposites into PU matrix. First step, 4.44 gr sophorone diisocyanate (IPDI) and

1.18 gr 1,6-Hexanediol mixing and then added acetone (20 ml) as solvent and two drops of catalyst (DBTDL) at 45°C and 1 h stirred. In second step, 2.6 gr Hydroxy ethyl methacrylate (HEMA), 2 cc trimethylol propan tri-acrilat (TMPTA) and 0.05 gr of PANi-CuO nanocomposites was added to reaction mixture and 3 h stirred. Finally, 0.33 gr benzophenone and 0.33 gr tripropylamine as initiator poured into reaction mixture, after 15 minutes reaction completed and product obtained.

17.2.2 PREPARATION OF PU-PANI-CUO COMPOSITE FILMS

PU-PANi-CuO nanocomposite films were preparation by coating device on glass substrates to thick 12 microns. Afterward, films under UV light were irradiated in Ultraviolet curing device and curing processing of the films completed.

17.3 RESULTS AND DISCUSSIONS

17.3.1 SCANNING ELECTRON MICROSCOPY ANALYSIS

The morphologies of PU-PANi and PU-PANi-CuO (5 wt%) nanocomposites films are shown in Figure 17.1. The SEM images showed that PANi-CuO nanocomposites have been distributed in the urethane matrix and the films structure was modified by increasing nanocomposites.

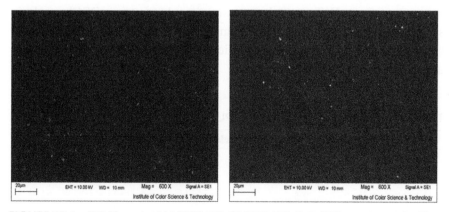

FIGURE 17.1   SEM images of (a) PU-PANi, (b) PU-PANi-CuO (5%) nanocomposite films.

17.3.1.1 Electrical Conductivity

Table 17.1 shows the variation of conductivity of pure PU, and PU-PANi-CuO (1–5wt%) hybrid nanocomposites films. The conductivity of the films increases with the increase PANi-CuO nanocomposites of in the PU matrix.

17.3.2 *THERMAL STABILITY*

Differentiating scanning calorimetric (DSC) analysis of PU-PANi and PU-PANi-CuO composites showed glass transition temperature (Tg) value. Glass transition temperature (Tg) value of the PU-PANi-CuO composites to higher value (95°C for PU-PANi-CuO nanocomposites, 87°C for PU-PANi) which indicates that PU-PANi-CuO composites are more stable than PU-PANi.

17.4 CONCLUSIONS

In this study, we synthesized conductive urethane acrylate with added PANi-CuO nanocomposites into urethane acrylate matrix. Urethane acrylate -PANi-CuO composites synthsized and then PU-PANi-CuO composites films were prepared and curing of the films is completed in the ultraviolet curing device. The morphological studies (SEM) showed uniform distribution of the PANi-CuO nanocomposites in the PU matrix. Electrical conductivity measurements of PU increase with increasing content of PANi-CuO nanocomposites. Differentiating scanning calorimetric (DSC) analysis showed that glass transition temperature (Tg) value of the PU-PANi-CuO composites to higher value which indicates that PU-PANi-CuO composites are more stable than PU-PANi.

TABLE 17.1 Conductivity of Pure PU, and PU-PANi-CuO (1–5 wt%) Composites Films

Sample	CuO nano particles (wt%)	$\sigma\ 10^{-3}$ (at 300 K) S cm^{-1}
Pure PU	0	1.5
PU-PANi-CuO	5	0.11
PU-PANi-CuO	2.5	0.9
PU-PANi-CuO	1	0.3

KEYWORDS

- curing
- electrical conductivity
- glass transition temperature
- morphological
- PANi-CuO nanocomposites
- urethane acrylate

REFERENCES

1. Fang, F. F., Choi, H. J., Ahn, W. S., (2009). *Compos. Sci. Technol., 69,* 2088–2093.
2. Geetha, S., Kumar, K. K. S., Meenakshi, S., Vijayan, M. T., Trivedi, D. C., (2010). *Compos. Sci. Technol., 70,* 1017–1022.
3. Kan, J. Q., Pan, X. H., Chen, C., (2004). *Compos. Sci., 19,* 1635–1640.
4. Zhang, W., Xiao, H., Fu, S. Y. (2012). *Compos. Sci. Technol., 72,* 1812–1817.
5. Rahaman, M., Chaki, T. K., Khastgir, D., (2012). *Eur. Polym. J., 48,* 1241–1248.
6. Liu, B-T., Syu, J-R., Wang, D.-H. J. (2013). *Colloid and Interface Science., 393,* 138–142.
7. Qu, R., Gao, J., Tang, B., i Ma, Q., Qu, B., Sun Ch. (2014). *Applied Surface Science., 294,* 81–88.

PART III

MATERIALS AND PROPERTIES

CHAPTER 18

FUNDAMENTALS AND CHALLENGES IN CURING PROCESS

MARC J. M. ABADIE

Institute Charles Gerhardt of Montpellier Aggregates, Interfaces and Materials for Energy (ICGM – AIME, UMR CNRS 5253), University Montpellier, Place Bataillon, 34095 Montpellier Cedex 5, France, E-mail: abadie@univ-montp2.fr, marc@ntu.edu.sg

CONTENTS

ABSTRACT

In structural materials and composites based on thermo set matrices, it is important to control reactions related to the cross-linking network. A fully cured system appears when a physical properties become non-variable and reach a plateau such as Tm, Tg, hardiness, etc.

The gel time is defined as the time when a liquid resin (open time) starts to form micro gels, before the gelatin (initial cure) and the formation of solid state (final cure). Above this time, processing cannot be properly done. Therefore, it is important, when processing a composites, to be able to measure and control the gel time.

We discuss on this chapter about the possibility to regulate the gel time according the domain concerned, that is, Marine and Aerospace applications.

18.1 INTRODUCTION

Matrices play a crucial role in the formation of a controlled and homogeneous 3D structure for assuring good properties to the materials [1, 2]. Most of the polymer networks belong to the imperfect network type; thus they are: inhomogeneous, contain cycles or loops, contain dangling ends, entanglements of molecular chains between cross-links (Figure 18.1). Nowadays, with the development of composites and nano-composites, the presence of reinforcing agents or nano-fillers may have influence in the development of the network of thermo sets and therefore affect the mechanical properties [3–5].

Unsaturated polyester (UPR) and Vinyl ester resins (VER) are widely used for fabricating small to large FRP structures (boat hulls, decks, and other marine applications) due to their cost and their facility to process at room temperature (SCRIMP™ technology) [6].

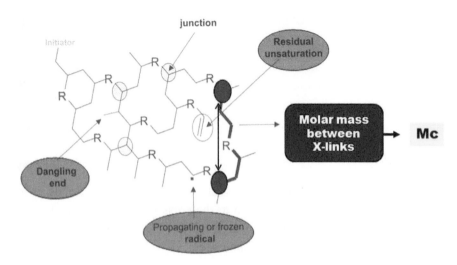

FIGURE 18.1   Insight into a 3 D.

18.2 EXPERIMENTAL PART

18.2.1 MATERIALS

1. ***For Marine applications,*** we have used two types of industrial resins, UPR and VER in solution of styrene, which are cross-linked by different type of peroxides.

   UPR – TOROLITHE® **4004 ISO T** [7] was provided by Reich hold and VER – DERAKANE® **411** [8, 9] from Dow Chemicals. These products are formulated for curing with methyl ethyl kentone peroxide (MEKP) initiator (improperly called "catalyzor") and may be used for fabrication of general FRC [10].

   The resins are dissolved in styrene which react as co-reactive solvent, respectively 42% wt. and 45% wt. Different peroxides such as methyl ethyl ketone peroxide (MEKP), liquid cyclo hexanone peroxide (LCHP), liquid ketone peroxide (LKP) or liquid acetylacetone peroxide (LAAP) have been used, all of them in solution in dimethylphthalate @ 63%. The cobalt octoate in solution @ 6% in phthalate is used as accelerator.

2. ***For Aerospace applications*** we have studied some flexible resins such as commercial sealants based on Polysulfides modified urethanes epoxies – qualified to Boeing BMS 5–63 for their MRO processes and cross-link by hardener (Table 18.1).

18.2.2 CHARACTERIZATION

Kinetics is made of viscosity or temperature *vs.* time curves for Marine applications and by hardness (Vickers) *vs.* time curves for Aerospace applications. In both cases, the gel time define the time beyond which the matrix cannot be processed anymore.

TABLE 18.1   Chemical Class and Name of Materials of Certified Aviation Seal Ants

Materials Name	Chemical Class
DAPCO 18–4F (BMS 5–63)	Polysulfide (100 parts)
Hardener B4	Polysiloxane (7 parts)
Pro-Seal 870 (BMS 5–95)	Polysulfide (100 parts)
Hardener	Manganese dioxide (7 parts)

Characterization of cure kinetics of thermosetting resins is obtained by TROMBOMAT (TROMBOTECH) actually RHEOTECH[TR] proposed by Material Engineering (France) [7, 8, 9] (Figure 18.2a). This equipment allows to determine the cure kinetics reliably and without damaging the measuring system, the characteristic points of a cross-linking resin as the gel time (viscosity), conventional and thermal reactivity, the exothermic peak, and some other job characteristics such as the reactivity of the system considered (Figure 18.2b).

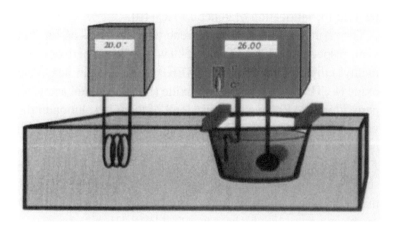

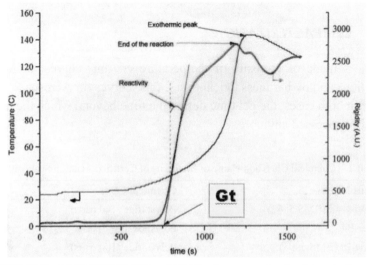

FIGURE 18.2   (a) TROMBOMAT, and (b) Kinetics curves for the determination of Gt.

18.2.3 OBJECTS OF STUDY

The gel time is defined as the time when a liquid resin (open time) starts to form microgels, before the gelatin (initial cure) and the formation of solid state (final cure). Above this time processing cannot be properly done. Therefore, it is important, when processing a composites, to be able to measure and control the gel time. The objects for the investigations were concerned by two domains: Marine where we need to increase gel time to be able to process the composite; and Aerospace where the gel time has to be reduced to assure fast and effective curing of resins and composites.

18.2.4 METHODS OF MEASUREMENTS OF ELECTRIC CONDUCTING CHARACTERISTICS OF MATERIALS

For Marine application, gel time was determined by TROMBOMAT® and for Aerospace application, the completion of the cross-linking reaction was measured by hardness (Vickers) vs. time.

18.3 RESULTS AND DISCUSSION

18.3.1 MARINE APPLICATION

Different parameters have been investigated to increase the gel time in optimizing the system resin/accelerator/initiator. The unsaturated polyester resins react by their double bond along the backbone chain due to the use of maleic anhydride during their synthesis.

The initiator system is symmetric peroxide containing hydro peroxides (MEKP, LCHP) or peroxides (LAAP) groups in their structure for most of them. The active centers are created by a red/ox reaction of the hydro peroxide (Scheme 1).

RO° radicals can react on styrene/UPR but not ROO° engaged in chain hydro peroxidation reactions. In the case of symmetric or non-symmetric peroxide, the reaction is slower and need to be initiated by hydro peroxides (Scheme 2).

We have investigated the influence of different parameters on the gel time, such as, the effect of the concentration of hydro peroxide (MEKP) (Figure 18.2a), the effect of the nature of peroxide (Figure 18.2b), the effect

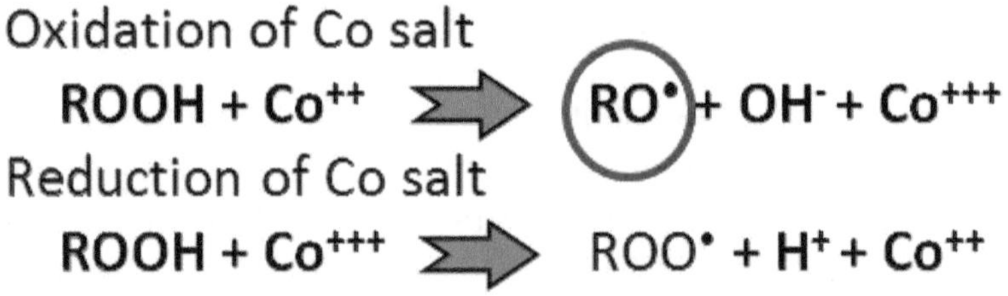

Oxidation of Co salt

$$ROOH + Co^{++} \Rightarrow \boxed{RO^{\bullet}} + OH^- + Co^{+++}$$

Reduction of Co salt

$$ROOH + Co^{+++} \Rightarrow ROO^{\bullet} + H^+ + Co^{++}$$

SCHEME 1   Decomposition of hydro peroxide.

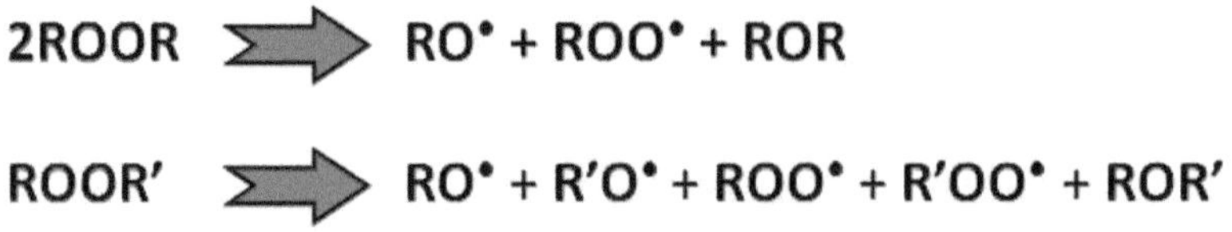

$$2ROOR \Rightarrow RO^{\bullet} + ROO^{\bullet} + ROR$$

$$ROOR' \Rightarrow RO^{\bullet} + R'O^{\bullet} + ROO^{\bullet} + R'OO^{\bullet} + ROR'$$

SCHEME 2   Decomposition of peroxide.

of temperature on the gel time (Figure 18.2c) and the effect of [MEKP] on the gel time at different temperatures (Figure 18.2d).

Therefore, the kinetic results show that:

- when the temperature $T°C$ and the accelerator [CoNaphth] are fixed, the gel time Gt increases when peroxide [MEKP] decreases;
- when the temperature $T°C$ and the peroxide [MEKP] are fixed, the gel time Gt increases when the accelerator [CoNaphth] decreases;
- when the peroxide [MEKP] and the accelerator [CoNaphth] are fixed, the gel time Gt increases when the temperature $T°C$ decreases.

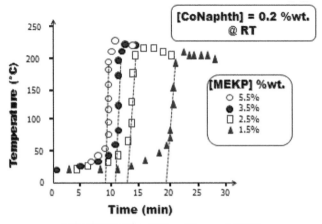

FIGURE 18.2A $T°C$ *vs*. Time w. MEKP.

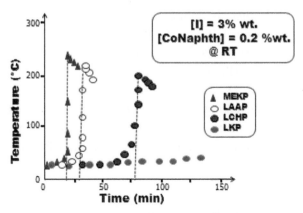

FIGURE 18.2B $T°C$ *vs.* Time for different peroxides.

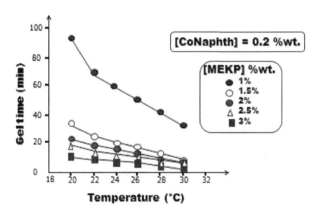

FIGURE 18.2C Gel time *vs.* $T°C$ w. MEKP.

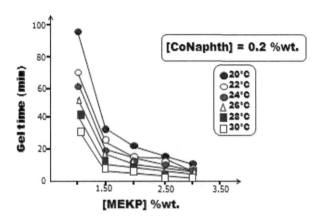

FIGURE 18.2D Gel time *vs.* [MEKP] at different °C.

We have also studied the influence of the double bonds on the kinetics by comparing two resins UPR and VER of similar molar mass, respectively 3,200 Da and 3,300 Da. We do observed that the gel time is much higher for the VER than the UPR (Figures 18.3a and 18.3b).

The data in Figure 18.3a do not permit to define with precision the gel time, the one given by taking the slope of the curves is not the beginning of the crosslink reaction which appears since the formation of micro gels (see Figure 18.2b). However, we clearly observe that the UPR cross-link faster than the VER.

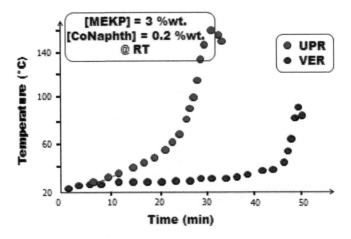

FIGURE 18.3A $T°C$ vs. time.

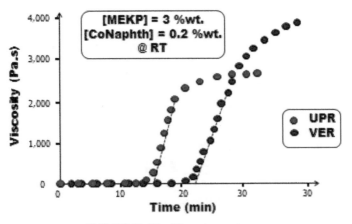

FIGURE 18.3B   Viscosity vs. time.

In Figure 18.3b, where we plot the "poiseuille" viscosity in Pa.s *vs.* time (determination of the gel time much accurate), the respective curves confirm that the gel time is lower for the UPR. These results are related to the number of double bonds, respectively ≈ 5.84 (along the backbone chain) and 2 (at each end of the chain).

The degree of cross-linking Xc (*) for the UPR is higher than the VER, respectively 3.75×10^{-3} and 12.1×10^{-3}, due to the difference of the number of double bonds DB (Table 18.2).

X_c is calculated according the following equation:

$$(*) \ X_c = 1/M_c$$

with $M_c = M_0/(f_0-2)$; M_0 and f_0 being respectively the average molar mass and the functionality of the resin considered.

We also have investigated not only the effect of complexing agent but also the effect of different solvents mixed with styrene or used alone. All experiments have been conducted on a VER, the Derakane® 411–45 with styrene 45% wt.

- **Influence of complexing agent**

  We have tested different complexing agents such as Pedersen crown ether compounds, Kryptates and retarding agent as well (tertio butyl catechol). Penta-2,4-dione and Hepta-2,6-dione have been considered. The penta-compound plays the role of pince of the Co salt in its cis conformation (Figure 18.4a), whereas the hexa one cannot completely complex the cobalt salt due to the size of the pince (Figures 18.4a and 18.4b).

  This difference in the size of the pince explains that the effect of penta-2,4-dione on the cobalt salt is higher than the hepta-2,6-dione.

TABLE 18.2 Comparison Between UPR and VER Properties

	UPR	VER
MW (Da)	3,200	3,300
Gel time* (min)	15.5	25.0
Nbr DB	5.84	2
M_0	3,200	3,300
f_0	11.68	4
Xc	3.75×10^{-3}	1.21×10^{-3}

*Gel time according Figure 18.3b.

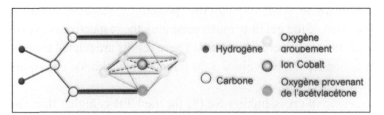

FIGURE 18.4A   Cis-conformer of Penta-2,4-dione.

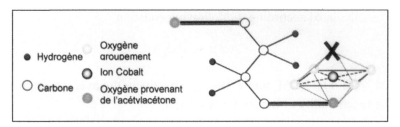

FIGURE 18.4B   Trans-conformer of Hepta-2,6-dione.

Therefore, the gel time is increased two times for 0.4 ‰wt. Going from 30 min for hepta-2,4-dione to 60 min for the penta-2,6-dione.

- **Influence of different solvents**
  Different solvents- methyl methacrylate MMA, dicyclopentadiene DCPD, diaalyl phthalate DAP or vinyl toluene VT – have been tested on a VER Derakane® 411–45 with 45%wt. styrene St, 2%wt. of MEKP and 0.2%wt of CoNaphth @ RT.

  The addition of a small amount of co-solvent 10%wt. and 20%wt.—MMA, DCPD, DAP or VT to styrene leads to the decrease of the gel time – Table 18.3.

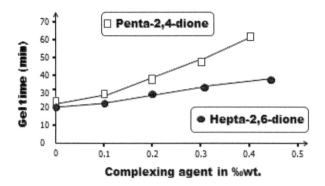

FIGURE 18.5   Gel time *vs.* Complexing agent.

TABLE 18.3 Effect of Co-Solvent Mixed to Styrene on Gel Time Gt

Co-solvent	Gel time (min) Styrene 90 wt.%	Gel time (min) Styrene 80 wt.%
MMA	49.8	66.2
DCPD	57.7	68.9
DAP	65.8	74.7
VT	**66.7**	**85.9**

However, when the co-solvent is pure, we see from Figure 18.6 that the VT has a low gel time compared to St. The highest value of the Gt is given with a mixture MMA/VT 50%wt. as co-solvent (Figure 18.6; Table 18.4).

- **Aerospace Application**

  If in Marine application it is interesting to increase the gel time for processing, before the apportion of micro-gels, in Aerospace application, it is interesting to minimize the gel time. Therefore, we need in Maintenance, Repairing and Overall (MRO) to:

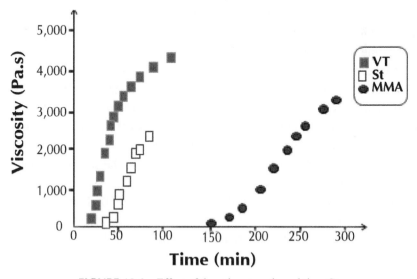

FIGURE 18.6 Effect of the solvent on the gel time Gt.

TABLE 18.4   Optimization of the Gel Time Gt

Co-solvent (wt.%)	Gel time (min)
Vinyl toluene (100%)	12
Styrene (100%)	41
MMA/VT 0.50 wt.%	63

- find solutions for repairing structural damages of composite;
- assure fast and effective curing of resins and composites;
- shorten delivery time, as aircrafts are grounded and off of service until structural damages are not repaired;
- new technologies and processes has to be developed.

The curing behavior of two certified aviation sealants: DAPCO 18–4F silicone firewall (qualified to Boeing BMS 5–63) and Pro-Seal 870 (qualified to Boeing BMS 5–95) with respective hardeners have been investigated.

The objective was to reduce the curing time for these sealants to less than 2 h without its chemical modification of the formulation.

We have exposed the two sealant formulations to a 250 Watts Xe lamp doped with Ytrium. Wide infrared emission spectrum – similar to the rapid curing device, 28Vdc Thermo reactor made by SunAero, which perfectly matches the infrared absorption spectrum of Aircraft Sealants and Paints – Optimum energy transfer.

Hardness *vs.* cure time results for DAPCO 18–4F and for Pro-Seal 870 sealants cured @ room temperature are presented Figure 18.7.

DAPCO 18–4F sealant cured at 60°C, cross-links at a shorter time, going up to 7 days @ RT to 10h @ 65°C (Figure 18.8).

With the 250 W Xe/Yt lamp the time to crosslink the sealant is reduced to 90 min (Figure 18.9).

Similar results are observed with sealant BMS 5–95 B2. Plateau obtained after 14 days (Hardness = 50 S) without IR compared to plateau reached after 2 h (Hardness = 76 S) with the used of condensed IR.

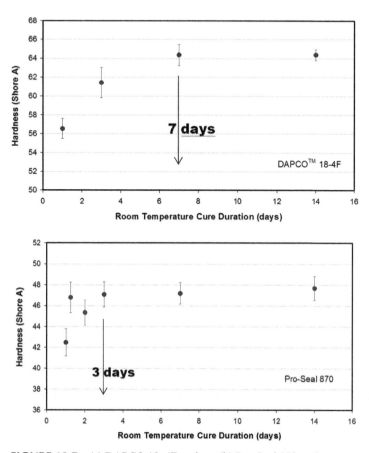

FIGURE 18.7   (a) DAPCO 18–4F sealant; (b) Pro-Seal 870 sealant.

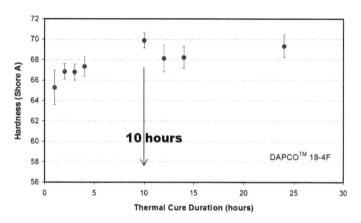

FIGURE 18.8   Hardness vs. Time for DAPCO 18-4F

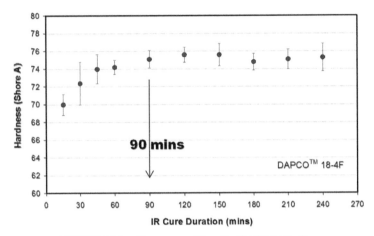

FIGURE 18.9   Harness vs. Time for DAPCO 18-4F

18.4   CONCLUSIONS

The gel time Gt is an important control parameter for the optimization of the 3D structure for the matrix of a composite. This parameter can be tuned either for Marine or Aerospace applications.

In Marine application, the gel time Gt can be increased either by decreasing the concentration of catalysor [MEKP] and/or the concentration of activator [CoNaphth], using complexing agent or by adding an adequate co-solvent (MMA, VT) to the styrene.

However, in Aerospace application, the gel time has to be reduced for repairing structural damages of composite and therefore shorten the delivery time when grounded for the new generation of civil air planes with around 50% of the primary structure – including the fuselage and wings, are made of composite materials (Airbus 350 XWB and Boeing 787 Dream Liner).

ACKNOWLEDGMENTS

Dr. Demosthene Sakkas for Marine application and Dr. Wanda Yu. Abadie-Voytekunas for Aerospace application are gratefully acknowledged for their participation to these results.

KEYWORDS

- catalytic system
- crosslinking reactions
- cure modeling
- gel time
- kinetics
- polyester
- silicone
- vinyl ester

REFERENCES

1. Hay, J. N., O'Gara, P., (2006). Recent developments in thermo set curing methods. Proceedings of the Institution of Mechanical. Part G, *Journal of Aerospace Engineering, 220*(G3), 187–195.
2. Diamanti, K., Soutis, C., (2010). Structural health monitoring techniques for aircraft composite structures. *Progress in Aerospace Sciences, 46*(8), 340–352.
3. Braga, D. F. O., Tavares, S. M. O., da Silva, L. F. M., Moreira, P. M. G. P., de Castro, P. M. S. T. (2014). Advanced design for light weight structures: Review and prospects. *Progress in Aerospace Sciences, 69,* 29–39.
4. Vargas, M. A., Vazquez, H., Guthausen, G., (2015). Non-isothermal curing kinetics and physical properties of MMT-reinforced unsaturated polyester (UP) resins. *Thermochimica Acta, 611*, 10–19.
5. Karger-Kocsis, J., Mahmood, H., Pegoretti, A., (2015). Recent advances in fiber-matrix interphase engineering for polymer composites. *Progress in Materials Science, 73*, 1–43.
6. Aktas A., Krishnan L., Kandola B., Boyd S.W., Shenoi R.A., A cure modeling of an unsaturated polyester resin system for the simulation of curing of fiber-reinforced composites during the vacuum. *Journal of Composite Materials*, volume 49 (20): 2529-2540 - http://dx.doi.org/10.1177/0021998314549820
7. Abadie, M. J. M., Sakkas, D., (1992). Kinetics of crosslinking reactions of intarurated isophthalic polyester by DSC: Etude cinetique de la reticulation du polyester isophthalique insature par DSC; Isothermal study. *European Polymer Journal, 28*(8), 873–879.
8. Abadie, M. J. M., Mekhissi, K., Burchill, P. J. (2002). Effect of processing conditions on the curing of a vinyl ester resin. *Journal of Applied Polymer Science, 84*(6), 1146–1154.
9. Abadie, M. J. M. (2002). Handbook of Polymer Blends and Composites, *Kinetic of Curing,* Chapter 9, 375–402.
10. Muckbaniani, O., Abadie, M. J. M., Tatrishvili, T., (2015). High-Performance Polymers for Engineering-Based Composites, Edited by CRC Press, June 2015.

CHAPTER 19

NEW POLYMER METAL COMPLEXES BASED β-DIKETONES AND LANTHANIDES FOR OLEDS

I. SAVCHENKO,[1] A. BEREZHNYTSKA,[2] and Ya. FEDOROV[2]

[1]*Kyiv National Taras Shevchenko University, Department of Chemistry, Kyiv, Ukraine*

[2]*V. I. Vernadsky Institute of General and Inorganic Chemistry NASU, Kyiv, Ukraine, E-mail: iras@univ.kiev.ua*

CONTENTS

ABSTRACT

The complexes of Nd, Er, Eu, Gd, Yb, Sm, Pr, Lu, Ho with 2-methyl-5-phenylpenten-1-3,5-dion (mphpd) and allyl-3-oxo-butanoate (allyl) were synthesized. The polycomplexes on their basis and copolymers with styrene and N-vinylcarbazole in ratio 5:95 were obtained by free-radical polymerization and the kinetics of polymerization was studied by dilatometric method at the first time. The results of above studies have shown that the configuration of the chelate unit is unchanged during the polymerization. The similarity of monomers electronic absorption spectra with polymers spectra confirms the identical coordinative environment of lanthanide ions in both cases. The method of dynamic light scattering and the results of electronic microscopy showed that the obtained polymer systems are nanoscale. The luminescent spectra of obtained metal complexes in solid state and solutions are investigated and analyzed. The solubilization of europium β-diketonate complexes with phenanthroline was shown to change luminescence intensity in such complexes.

19.1 INTRODUCTION

Efficiency is crucial for organic light emitting diodes (OLEDs) to be energy-saving and to have a long lifetime for display and solid state lighting applications. Numerous approaches have been proposed to attain high efficiency OLEDs through the synthesis of novel organic materials, the design of light extraction structures and the design of efficiency-effective device architectures.

There has been increasing recent interest in light-emitting devices based on thin organic films of electroluminescent polymers or small molecules. Such systems offer several potential advantages over the more traditional inorganic light-emitting devices, including relative ease of production and processing [1–3]. While many different systems have been examined, all are essentially variations on a theme. Typically, these devices consist of one or more organic layers situated between a low work function metal cathode such as calcium and a higher work function anode, often transparent indium-tin oxide (ITO). When a sufficient potential bias is applied across the electrodes, electrons are injected from the cathode into the conduction band of the luminescent layer, and holes are injected from the anode into the valence band. Under the influence of the applied potential, electrons and holes migrate to a plane within the organic film, where they meet. In an electroluminescent film, there is a reasonably high probability of photon emission due to electron hole recombination. When this occurs, photons of an energy determined by the band gap (or HOMO/LUMO gap, as the case may be) are produced.

The number of charge carriers injected from the two electrodes is generally not equal. In this case, the current passed through the film will be determined by the majority carriers, while the emission of light will be limited by the minority carriers. Optimum emission efficiency is thus obtained when the rates of hole and electron injection are matched. The exact mechanism whereby charge injection occurs is a matter of some controversy and, indeed, may vary for different materials and device construction motifs. That point notwithstanding, there is at least an empirical relationship between the rates of charge carrier injection and the proximity of the Fermi level energies of the respective electrode to the energy of the band (or orbital) into which charge is injected. In other words, the Fermi energy of the anode should match the valence band energy, while the Fermi energy of the cathode should match the conduction band energy. The simplest class of OLEDs is those composed of a single polymer layer between the two electrodes.

The trivalent lanthanide ions are well known for their unique optical properties such as line-like emission band and high quantum efficiency owing to the special 4f electron configuration [4–6]. Nevertheless, the f–f electron transition belongs to the forbidden transition, which results in the relatively low-absorption coefficient. Thereby, researchers construct the complexes of lanthanide ions with organic ligands, which can not only strongly absorb the energy and transfer the energy to central lanthanide ions but also expel water molecule from the first coordination sphere and protect lanthanide ions from vibrational coupling through "antenna effect" [7, 8] (Scheme 1).

Due to their unique photo physical properties that aid in shielding 4f electrons from interactions with their surroundings by the filled $5s^2$ and

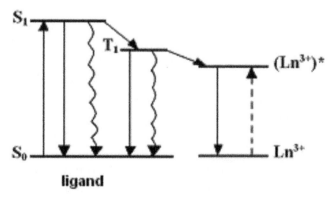

SCHEME 1   The processes of energy transfer in lanthanides complexes ("antenna effect").

$5p^6$ orbital, lanthanide ions have been well known as important components in phosphors, lasers and optical amplifiers [9–11]. However, the direct Ln^{3+} photo excitation is not very efficient, with the low molar absorption coefficients limiting the light output. Some organic legends such as aromatic carboxylic acids etc. are well known to be efficient sensitizers for the luminescence of lanthanide ions, whose organic chromophores typically present effective absorption and a much broader spectral range than the corresponding Ln^{3+} ions can absorb energy to be transferred to nearby Ln^{3+} ions by an effective intramolecular energy transfer process. These chelates possessing the effective emission in the near-UV, visible and NIR spectral regions are of great interest for a wide range of optical applications, such as tunable lasers, amplifiers for optical communications, components of the emitter layers in multilayer organic light-emitting diodes, light concentrators for photovoltaic devices and so on [12–20].

Using the monomer complex has a number of disadvantages connected with aggregation or crystallization of the film. Therefore, there is a necessity of the polymeric materials synthesis. It is well known that metal polymers are mainly produced by intercalation of metals in the polymer ligand matrix. This method has a lot of disadvantages such as partial degradation of the polymer chain and low yield of the synthesized polymers as well as low coordination level which results in composition heterogeneity. All these shortcomings have an influence upon physical characteristics of obtained compounds.

The aim of this work were synthesis of Nd, Er, Eu, Gd, Yb, Sm, Pr, Lu, and Ho with 2-methyl-5-phenylpenten-1-3,5-dion (mphpd) and allyl-3-oxo-butanoate (allyl) and phenanthroline as well as copolymers based on them with styrene and investigations of optical properties of metal-containing polymeric systems depending on influence of phenanthroline configured-in the complex coordination sphere on the luminescence properties and concentration of rare earth elements complexes in polymeric materials. With aim of the simplification of the layers incorporation in the electroluminescent cell can be superposed the emission layer (rare earth elements complex), hole conduction layer (such as N-vinylcarbazol) and electronic conduction layer (1,10-phenanthroline) in one macromolecule. We report our findings on the synthesis and photo physical properties of N-vinylcarbazolecopolymers with Pr, Nd, Er, and Eu with 1,10-phenanthroline complexes.

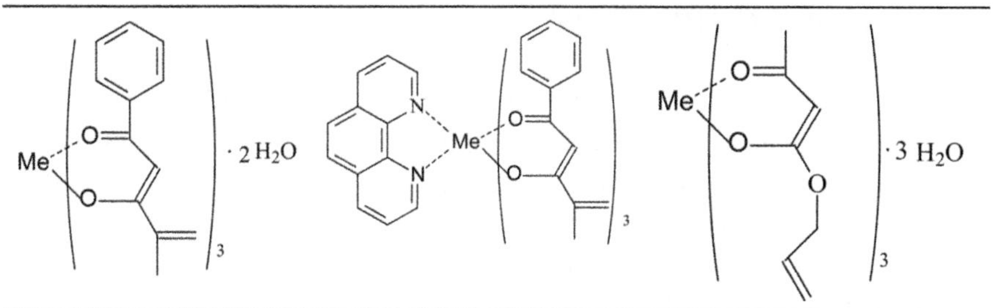

| Ln(mphpd)$_3$ | Ln(mphpd)$_3$. phen | Ln(allyl)$_3$ Me – Nd, Er, Eu, Gd, Yb, Sm, Pr, Lu, Ho |

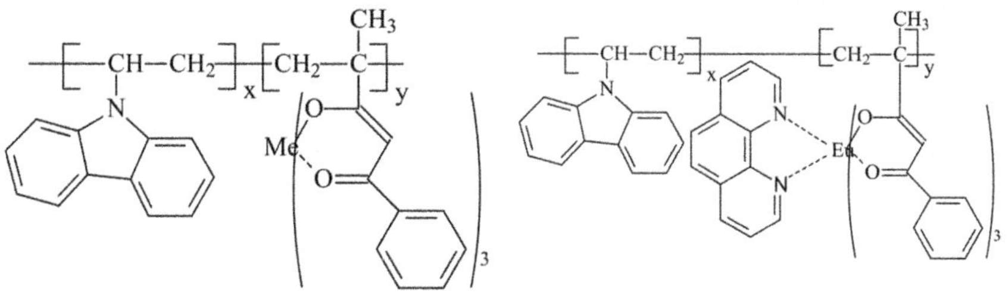

Me – Er, Pr, Nd

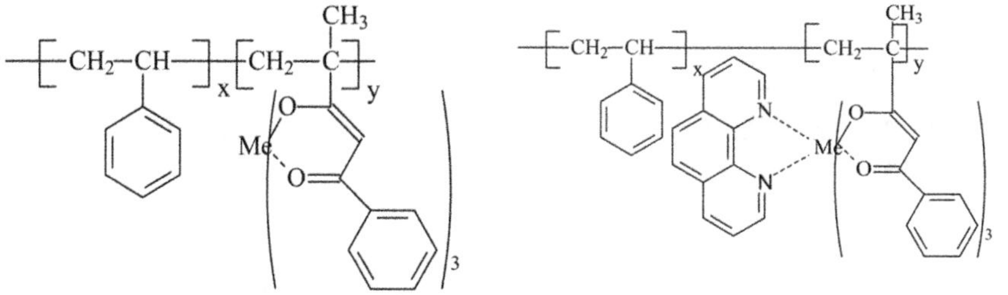

Me – Nd, Er, Yb, Sm, Pr, Lu

SCHEME 2   Chemical structures of obtained compounds.

19.2   EXPERIMENTAL PART

19.2.1   MATERIALS

We synthesized two kinds of monomeric ligands: allyl-3-oxo-butanoate and 2-methyl-5-phenylpentene-1-3,5-dione. NMR (mphpd): ^{1}H (D$_2$O) δ (ppm): 3.27 (singlet, 3H, CH$_3$); 3.47(singlet, 1H, = CH-); 5.27 (singlet, 1H, = CH$_2$); 5.58 (singlet, 1H, = CH$_2$); 7.15–760 (multiplet, 5H, Ph).

19.2.2   CHARACTERIZATION

Complexes obtained by an exchange reaction between equimolar amounts of lanthanide acetate and sodium 2-methyl-5-phenylpentene-1–3,5-dione-salt or allyl-3-oxo-butanoatein a water-alcohol solution at pH9–95 with a slight excess of the ligand.

The polymerization was carried out in 10 wt.% DMF solution of monomers with 2,2′-azobisisobutyronitrileAIBN as free radical initiator (1 wt.% with respect of monomers mass) at 80°C for more than 8 hours in thermostat. The polymerization mixture was poured into methanol. The solid precipitate was filtered, dissolved in DMF, and reprecipitated into methanol and then dried at 20°C overnight.

19.2.3   OBJECTS OF STUDY

Complexes on the basis of some diketones and lanthanides were the objects for the investigations of some optical properties in dependence of additional ligand configured-in the complex coordination sphere and concentration of rare earth elements complexes in polymeric materials.

19.2.4   METHODS OF MEASURES OF CHARACTERISTICS OF MATERIALS

The synthesized compounds have been studied by NMR, IR-, and electronic absorption, diffuse reflectance spectroscopy and thermal analysis. The infrared spectra were recorded in KBr tablets at a range of 4000–400 cm^{-1} with Spectrum BX II FT – IR manufactured by Perkin Elmer, Nicolet Nexus 670 FTIR spectrometer. Thermo grams were recorded on a TA instruments Q-1500 D apparatus by system of Setsys evolution-1750 at a heating rate of 5°C/min from room temperature up to 500°C in platinum capsule in presence carrier Al_2O_3(anhydrous).

The electronic absorption spectra were recorded using spectrophotometer Shimadzu "UV-VIS-NIR Shimadzu UV-3600" and the diffuse reflectance spectra were obtained using the Specord M-40 spectrophotometer in the range of 30,000–12,000 cm^{-}1. The excitation and luminescence spectra of solid samples were recorded °n a spectrofluorometer "Fluorolog FL 3–22," "Horiba JobinYvon" (Xe-lamp 450W) with the filterOS11. The

InGaAs photo resistor (DSS-IGA020L, Electro-Optical Systems, Inc., USA) cooled to the temperature of liquid nitrogen was used as a radiation detector for infrared region. The excitation and luminescence spectra were adjusted to a distribution of a xenon lamp reflection and the photomultiplier sensitivity. The particle size studying was performed at 25°Causing the equipment from "Zeta Sizer Nano by Malvern." Photomicrographs were obtained by scanning electron microscope "Hitachi H-800" (SEM).

19.3 RESULTS AND DISCUSSION

The monomeric complexes of lanthanides were prepared in an aqueous alcohol solution at pH 8–10.

$$Ln(NO_3)_3 + 3Na(mphpd) \rightarrow Ln(mphpd)_3 + 3NaNO_3$$

The monomeric and polymeric complexes of Nd, Er, Eu, Gd, Yb, Sm, Pr, Lu, and Ho with 2-methyl-5-phenylpenten-1–3,5-dion, allyl-3-oxo-butanoate(allyl) were synthesized at the first time. The metallopolymers on their basis, copolymers-2-methyl-5-phenylpentene-1-dione-3,5with N-vinylcarbazole in ratio 5:95, which containing Pr, Nd, Er, and Eu with 1,10-phenanthroline and copolymers of 2-methyl-5-phenylpentene-1-dione-3,5 with styrene in ratio 5:95, which containing Eu, Yb and Eu, Yb with 1,10-phenanthrolinewere obtained by free-radical polymerization with the initiator AIBN.

The elemental analysis was carried out to establish the composition of the complex (e.g., Eu(mphpd)$_3$·2H$_2$O – 20.18% Eu, (theor (20, 29)); Eu(mphpd)$_3$·Phen – 16.91% Eu, (theor (17, 2))).The elemental analysisformetal ionswas carried outwith ICPE Shimadzu 9000 atomicemission spectrometer.

The kinetics of polymerization of obtained monomeric complexes was studied by dilatometric method and kinetic parameters of radical polymerization of complexes were calculated (Table 19.1).

The kinetic parameters of radical polymerization have very good agreement with literature data the rate constant is somewhat overstated [21]. Since no literature data for lanthanide complexes polymerization, the obtained results were compared with similar complexes for 3d metals. Probably, the growth rate constant due to the high resistance of polymeric lanthanide compared to metallopolymers of 3d metals. The kinetic parameters of

TABLE 19.1   The Kinetic Parameters of Radical Polymerization of Metal Complexes

	$Er(mphd)_3 \cdot 2H_2O$	$Pr(mphd)_3 \cdot 2H_2O$	$Ho(mphd)_3 \cdot 2H_2O$	$Eu(mphd)_3 \cdot 2H_2O$	$Nd(allyl)_2 \cdot 3H_2O$	$Ho(allyl)_3 \cdot H_2O$
$V_{pol} \times 10^4$, mol/(l·s)	0.28	1.18	0.22	0.27	1.52	0.54
$V_{red} \times 10^4$, s^{-1}	2.10	8.80	3.4	5.43	9.36	4.88
$K_{\Sigma} \times 10^3$, l$^{1/2}$/(mol$^{1/2}$·s)	2.70	11.27	4.35	7.77	12.03	6.84

polymerization depend on electronic and steric properties of β-diketonates. The influence of nature of organic ligand on polymerization rate are deduced, in case *allyl* the polymerization rate above *mphpd* are observed owing to steric restrictions under polymerization of complexes.

With the aim of the identification of ligand functional groups coordination method to metal ions were studied the IR spectra of synthesized compounds (Table 19.2).

In the IR spectra of the synthesized complexes and polymetallo complexes in 1500–1600 cm^{-1}, there are bands corresponding to stretching vibrations of the ν (C-O) and ν (C-C), which confirms the bidentate cyclic coordination of the ligand to the metal ions (Table 19.2). At the same time a higher frequency band should be attributed to the stretching vibrations ν (CC) and a lower frequency to the stretching vibrations of the ν (CO). The region of 1680–1710 cm^{-1} contains stretching vibrations ν (C = C). In the case of polycomplexes this band disappears or is significantly reduced in intensity, indicating the presence of only the terminal unsaturated groups. Fairly broadband of coordinated water molecules is observed in the region 3300–3500 cm^{-1}.

In IR spectra of the polycomplexes comparatively with monomeric complexes the location of main absorption bands are shifted in short-wave spectrum region, their intensity is lower significantly, the intensity is decreased especially which corresponds to vibration of the double bond. Presented results are confirmed of the polymer complex formation.

The differential thermal analysis (DTA) of obtained compounds for the identification of the complexes hydrated composition and their temperature behavior was performed.

TABLE 19.2 Some Distinctive Absorption Bands of Metallic Complexes and Metallopolymers and Copolymers with N-Vinylcarbazole (VC)

Complex	v(M-O)	v_{as}(C-O)	v_{as}(C-C)	v_s(C-O)	v_s(C = C)
Nd(mphpd)$_3$·3H$_2$O	420	1466	1559	1580	1671
[Nd(mphpd)$_3$]$_n$	430	1465	1557	1557	1654
Er(mphpd)$_3$·H$_2$O	420	1465	1557	1594	1676
[Er(mphpd)$_3$]$_n$	425	1470	1575	1575	—
Eu(mphpd)$_3$·3H$_2$O	420	1467	1554	1594	1666
Eu(mphpd)$_3$·(phen)	409	1419	1556	1593	1660
[Eu(mphpd)$_3$·(phen)]$_n$	418	1421	1557	1594	—
[Eu(mphpd)$_3$]$_n$	430	1470	1560	1560	1667
Pr(mphpd)$_3$·2H$_2$O	420	1418	1554	1592	1679
[Pr(mphpd)$_3$]$_n$	413	1467	1552	1600	1666
Lu(mphpd)$_3$·2H$_2$O	406	1470	1555	1590	1671
[Lu(mphpd)$_3$]$_n$	412	1459	1560	1590	1679
Lu(mphpd)$_3$·(phen)	418	1422	1560	1595	1660
[Lu(mphpd)$_3$·(phen)]$_n$	435	1426	1566	—	—
Ho(mphpd)$_3$·2H$_2$O	416	1421	1556	1596	1676
[Ho(mphpd)$_3$]$_n$	419	1461	1558	1570	1661
Er(OH)(allyl)$_2$·H$_2$O	416	1420	1518	1640	1720
[Er(OH)(allyl)$_2$]$_n$	418	1412	1518	1638	—
Pr(allyl)$_3$·2H$_2$O	415	1411	1516	1638	1717
[Pr(allyl)$_3$]$_n$	412	1410	1516	1638	—
[Nd(ally)$_3$]·3H$_2$O	416	1420	1518	1640	1720
[Nd(allyl)$_3$]$_n$	418	1412	1518	1638	—
Ho(allyl)$_3$ · H$_2$O	417	1414	1520	1624	1636
[Ho(allyl)$_3$]$_n$	419	1416	1520	1630	—
Er(mphd)$_3$co-VC	420	1454	1557	1595	—
Eu(mphpd)$_3$(phen)-co-VC	420	1452	1560	1598	—
Nd(mphpd)$_3$co-VC	420	1420	1552	1594	—
Pr(mphd)$_3$co-VC	420	1418	1554	1596	—

In the case of praseodymium complex with endothermic and weight loss 5.1% ($\Delta m_{theor.}$ = 4.9%), corresponds to two water molecules, respectively. Weak endothermic at 215°! caused by melting complex mass loss observed in this case, a small (2.5%). The process of decomposition of the complex starts with the cleavage of one molecule of ligand corresponding exothermic at 257°C and 26% weight loss ($\Delta m_{theor.}$ = 25.3). The following heating is attended by exoeffects at temperature 337, 384, 419, 460°C, 484°C, and results in the total complex decomposition Δm = 29%. The total weight is 72% loss for praseodymium complex.

Based on the mass loss curve and DTA for complex Pr(allyl)$_3$·2H$_2$O at 106°C detached one uncoordinated water molecule corresponding endo-effect and weight loss while 3% approximately, $\Delta m_{theor.}$ = 2.91%. Further heating to 126°C endothermic and accompanied by mass loss 6.7%, which corresponds to the two molecules of water ($\Delta m_{theor.}$ = 6.0%). Decomposition of complexes begins with cleavage of two allyl (C$_2$H$_3$O) substituents at the temperature range 180–270°C exothermic and mass loss 14.7% ($\Delta m_{theor.}$ = 14.3%). Further heating in the range 267–847°C accompanied by a set of exothermic 333, 390, 425, 650, 780°C and endothermic 567, 720°C. This is due to the destruction of organic molecules. Weight loss is 23.5% in this temperature range.

The presented results allow to assume that the complexes composition corresponds to the formula, Pr(mphpd)$_3$·2H$_2$O, Pr(allyl)$_3$·2H$_2$O.

The mass loss is 5.3% (Δm_{theor} = 4.7%) for the erbium complex Er(mphpd)$_3$·2H$_2$O at 125°C, which corresponds to the decoupling of two coordinative water molecules accordingly. The small exoeffect in the region 190°C can be conditioned by the polymerization process but the complex melting. The process of the complex decomposition begins with the removal of one ligand molecule, which corresponds of the exoeffect at 248°C and the mass loss 28% (Δm_{theor} = 24.5). The following heating is attended by exoeffects at temperature 337, 384, 419, 460°C, 484°C and results in the total complex decomposition Δm = 15%. The total weight loss is 66% for the erbium complex. The presented results allow to assume that the complexes composition corresponds to the formula, Er(mphpd)$_3$·2H$_2$O, Er(OH) (allyl)$_2$·H$_2$O.

Based on the results of thermal analysis, the dehydration of unsaturated complexes occurs at higher temperatures 120–150°C than in the case of ace-tyl acetonates. The neodymium complex dehydration occurs at the tempera-ture 130°C and is attended by the small endoeffect with the mass loss 7%

($\Delta m_{\text{reop.}}$ = 7.13%), which corresponds to the decoupling of three coordinative water molecules. The small endoeffect at the temperature 210°! probably is conditioned of the complex melting temperature, the mass loss is insignificant (2%). The further temperature increase is attended by the small exoeffect at the 225°!, which corresponds to the beginning of the complex polymerization. The significant exoeffect and the mass loss 25.5% ($\Delta m_{\text{reop.}}$ = 25%) at 285°! correspond to the removal of one ligand molecule. The following heating is attended by exoeffects at temperature 340, 360, 385, 425, 457°C and results in the total complex decomposition Δm = 18%. The total weight loss is 76% in investigated temperature interval.

Electronic absorption spectra (EAS) of metallopolymers are analogous to EAS of monomeric complexes (Figures 19.1 and 19.2). But, main absorption bands of polycomplexes undergo the long-wavelength shift 10–110 cm^{-1} that is indicated about the decay of the metal bond with ligand in the polymer. The decrease of the intensity of all absorption bands for metallopolymers and their bathochromic shift in comparison with monomeric spectra indicates about polymer structure formation. The similarity of monomers EAS with polymers spectra confirms of the identical coordinative environment of lanthanide ions in both cases.

Electronic absorption spectra of Er(OH)(allyl)$_2$ demonstrate (Figure 19.1) transitions are shifted at the short-wave region of spectrum relatively with aquaion which are indicated about the complex formation with less coordination number of central atom and enhancement of ionic component in bond erbium oxygen.

Thus, coordination number is 8 for Er(mphpd)$_3$·2H$_2$O and [Er(mphpd)$_3$]$_n$, coordination number is 7 for Er(OH)(allyl)$_2$·H$_2$O and [Er(OH)(allyl)$_2$]$_n$.

The shift of the main absorption bands in the long-wavelength region, compared with the spectrum of aqua-ion, and increase their intensity indicates the formation of a complex.

The electronic spectra of Nd (III) complexes (Figure 19.2a) it corresponds to the ion Nd^{3+}, and have a set of transitions from the $^4I_{9/2}$ main (quantum) state. The shift of the main absorption bands in the case of monomeric complex in the long wavelength region, compared with the spectra of aquaions, and increase in intensity indicates the formation of metal complexes, as well as indirectly confirms increasing in the covalence of the metal–ligand bond (Table 19.3).

The electronic spectra of Pr (III) complexes (Figure 19.1b) it corresponds to the ion Pr^{3+}, and composed of four absorption bands, which is founded

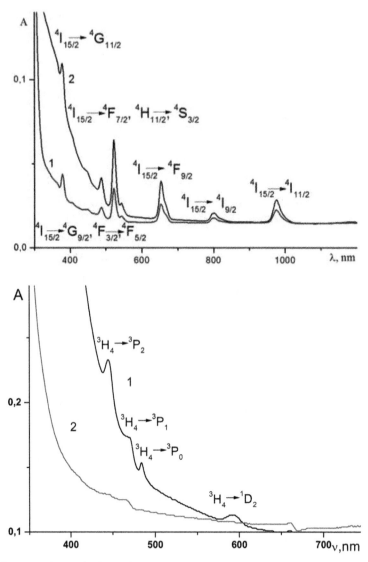

FIGURE 19.1 Electronic absorption spectra (a) 1 – Er(OH)(allyl)$_2$,2 – [Er(OH)(allyl)$_2$]$_n$;(b) 1 – Pr(mphpd)$_3$, 2 – [Pr(mphpd)$_3$]$_n$.

in visible spectrum part. The absorption bands of monomer metal complex ($^3H_4{\rightarrow}^3P_0$) compared with aqua-ions shifted to the longer wavelengths, which indicates a formation of compounds with greater coordination number than Pr inorganic salt.

Electronic spectra of the monomer as well as metal polymeric complexes have a set of bands corresponding to europium ion (Table 19.4). Shift of

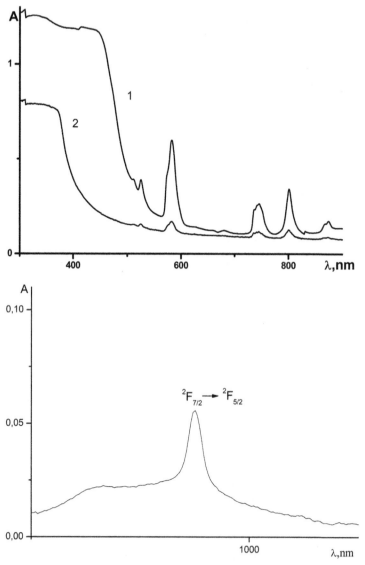

FIGURE 19.2 Electronic absorption spectra of (a) 1 – Nd(mphpd)$_3$·2H$_2$O and 2 – [Nd(mphpd)$_3$]$_n$; (b) Yb(mphpd)$_3$·2H$_2$O.

the main absorption bands in the long wavelength region in comparison with the spectra of aqua-ions, and their increase in intensity indicates the formation of metal complexes. Electronic spectra of the Yb(mphpd)$_3$·2H$_2$O have a singular transition band $^2F_{7/2} \rightarrow ^2F_{5/2}$ corresponding to ytterbium ion. A slight shift of the maximum which is observed in absorption spectra of the

TABLE 19.3 Energy Transition in Nd Electronic Absorption Spectrum

Transition	Nd^{3+}, cm^{-1}	Nd(mphpd)$_3$, cm^{-1}	Δ, cm^{-1}
$^4I_{9/2} \rightarrow ^2P_{1/2}$	23,064	23,255$_{arm}$	200
$^4I_{9/2} \rightarrow ^4G_{9/2}$	19,560	19,520	40
$^4I_{9/2} \rightarrow ^4G_{7/2}$	19,160	19,050	90
$^4I_{9/2} \rightarrow ^4G_{5/2}$	17,360	17,152	200
$^4I_{9/2} \rightarrow ^4F_{9/2}$	14,720	14,700	20
$^4I_{9/2} \rightarrow ^4F_{7/2}$	13,480	13,400	80
$^4I_{9/2} \rightarrow ^4H_{9/2}$	12,560	12,470	90
$^4I_{9/2} \rightarrow ^4F_{3/2}$	11,560	11,430	130

TABLE 19.4 Energy Transition in Er Electronic Absorption Spectrum

Transition	Er^{3+}, cm^{-1}	Er(mphpd)$_3$, cm^{-1}	[Er(mphpd)$_3$]$_n$, cm^{-1}	Er(OH)(allyl)$_2$, cm^{-1}	[Er(OH)(allyl)$_3$], cm^{-1}
$^4I_{15/2} \rightarrow ^4D_{7/2}$	39,215	—	—	38,834	38,460
$^4I_{15/2} \rightarrow ^4G_{11/}$	26,370	26,350	26,350	26,420	26,490
$^4I_{15/2} \rightarrow ^4F_{7/2}$	20,480	20,523	20,450	20,480	20,480
$^4I_{15/2} \rightarrow ^4H_{11/2}$	19,160	19,120	19,120	19,160	19,160
$^4I_{15/2} \rightarrow ^4S_{3/2}$	18,400	18,382	18,420	18,420	18,420
$^4I_{15/2} \rightarrow ^4F_{9/2}$	15,320	15,270	15,320	15,320	15,320
$^4I_{15/2} \rightarrow ^4I_{9/2}$	12,480	12,450	12,500	12,500	12,500

complex Yb (mphpd)$_3$·2H$_2$O in comparison with the spectra of aqua-ions indicates the formation of metal complex and a weakening of the metal–ligand bond (Figure 19.2b).

Similarity of electronic absorption and diffuse reflectance spectroscopy (Figure 19.3) show a similar structure of the complexes in solution and poly-crystalline state. A slight shift of the maximum which is observed in absorption spectra of the complex Eu(mphpd)$_3$Phen indicates a replacement of water molecules in the nearest coordination environment without significant changes in the coordination polyhedron geometry.

For the europium complex in the visible spectrum it is set of bands at corresponding to the transition $^7F_0$ main (quantum) state. Absorption bands for complexes of erbium and europium are shifted in the long wavelength region, compared with the spectra of corresponding metal salts on account of the complexation (Table 19.5).

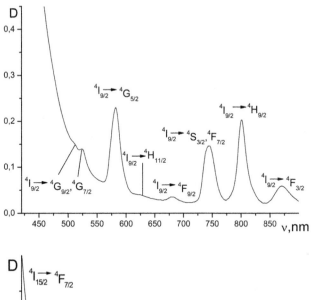

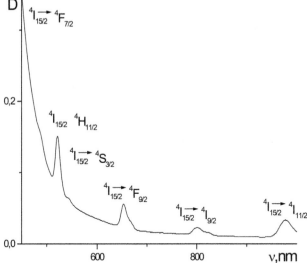

FIGURE 19.3 Diffuse reflection spectra: a) Nd(mphpd)3 b) Er(mphpd)3.

The luminescence spectra of the synthesized compounds were recorded in the solid state and in a solution.

The praseodymium complex Pr(mphpd)$_3$·2H$_2$O emit the visible luminescence (Figure 19.4a) from two existed states ^{3}P$_0$ and ^{1}D$_2$ in solid state at room temperature, like β-diketonate and carbochelates [22]. A band pattern of the Pr^{3+} ion transition with the ^{3}H$_4$ main (quantum) state is observed.

TABLE 19.5 Energy Transition in Eu Luminescence Spectrum

Transition	Eu(mphpd)$_3$,cm^{-1}	[Eu(mphpd)$_3$]$_n$,cm^{-1}	Eu(mphpd)$_3$(Phen)], cm^{-1}	[Eu(mphpd)$_3$(Phen)]$_n$, cm^{-1}
$^5D_0 \rightarrow {}^7F_0$	17,331	17,301	17,331	17,331
$^5D_0 \rightarrow {}^7F_1$	16,978	16,920	16,977	16949
	16,849		16921	16844
$^5D_0 \rightarrow {}^7F_2$	16,287	16340	16340	16313
	16234	–	16233	–
$^5D_0 \rightarrow {}^7F_3$	15,408	15,385	15,384	15,385
$^5D_0 \rightarrow {}^7F_4$	14,468	14535	14490	14556
	14347	14327	14347	14327

The luminescent spectra of the erbium complex Er(mphpd)$_3\cdot$2H$_2$O and metallo polymer are analogous, the band, which corresponds $^4I^{13/2}$ $^4I^{15/2}$ transition, is observed (Table 19.2; Figure 19.4b).

19.3.1 THE INFLUENCE OF ADDITIONAL LIGAND AND AGGREGATE STATE ON LUMINESCENT PROPERTIES OF NEW COMPLEXES

The luminescence spectra of europium β-diketonatescomplexes in all samples are similar to each other and approving the structure similarity of coordination polyhedrons which are distorted anti prism. All the samples have an equal number of the magnetic and electric dipole transitions as well as the forbidden transitions.

The transition and^5 $D_0 \rightarrow {}^7F0$in luminescence spectra of Eu compounds appears as a symmetrical single line andindicates a presence of one luminescence center. High-intensity lines caused by electric dipole transition $^5D_0 \rightarrow {}^7F2$compared with relatively low intensity magnetic dipole transition$^5D_0 \rightarrow {}^7F_1$suggests not center-symmetric nature of the environment for all investigated compounds. The luminescence intensity of complexes based on Eu(III) and Yb(III)-phenanthroline is greater in comparison with complex without additional ligand (Figure 19.5). The phenanthroline displaces the coordinate water out of the coordinate sphere which is quenching agent.

The Sm^{3+} ion with 4f^{5}configuration has complicated energy levels and different conceivable transitions inter f levels. It can be seen that the red

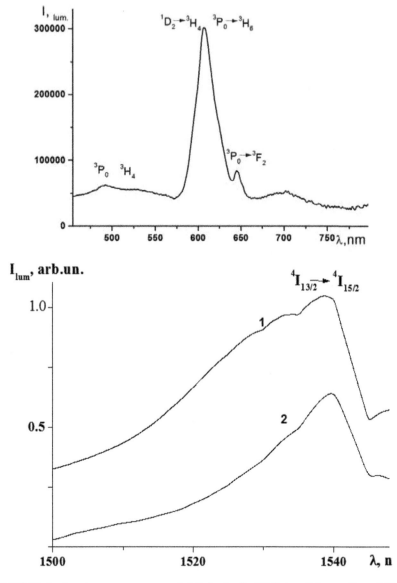

FIGURE 19.4 Luminescence spectra of a) Pr(mphpd)$_3 \cdot 2H_2O$ in solid state (λ_{ex} = 364 nm, 298K) b) Er(mphpd)$_3 \cdot 2H_2O$ (1) and [Er(mphpd)$_3$]$_n$; (2) in solid state (λ_{ex} = 362 nm, 298 K).

light of Sm^{3+} ion consists of three emission peaks in the visible region at 562, 599, and 645 nm, which are ascribed to the intra-4f-shell transition from the excited level $^4G^{5/2}$ to the ground levels $^6H^{5/2}, ^6H^{7/2}$, and $^6H^{9/2}$, respectively (Figure 19.6b).

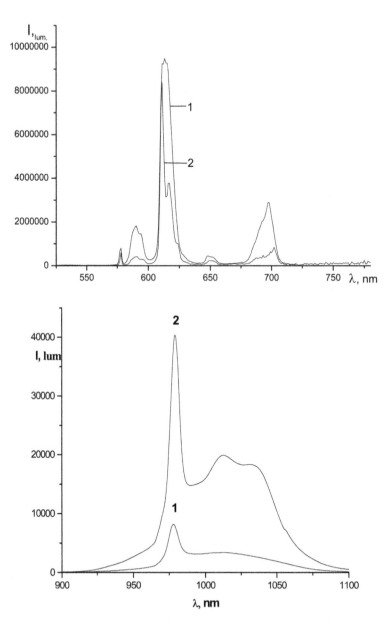

FIGURE 19.5 Luminescence spectra of (a) 1 – Eu(mphpd)$_3$·Phen; 2 – Eu(mphpd)$_3$, in solid state, T = 298 K, λ = 362 nm; (b) 1 – Yb(mphpd)$_3$, 2 – Yb(mphpd)$_3$Phen.

The excitation spectrum of Nd(mphpd)$_3$ in the solid state consists of a single band with maximum at 363 nm. In the excitation spectrum of the solution is observed long-wavelength shift of the band (Δλ = 15 nm) is

observed (Figure 19.6a). In the excitation at the maximum of this band a
4f-luminescence of samples is observed in the solid state and in the solution.
The luminescence spectrum consists of the three bands corresponding to the
transitions from the excited level of ion Nd(III) $^4F_{3/2}$ into the multiples of the

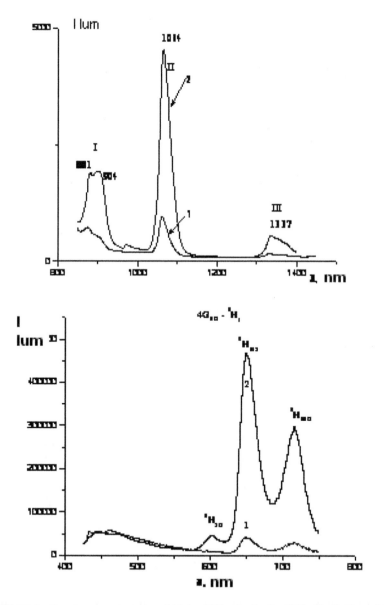

FIGURE 19.6 Luminescence spectra of (a) 1 – Nd(mphpd)$_3$, 2 – Nd(mphpd)$_3$Phen; (b) 1 –
Sm(mphpd)$_3$, 2 – Sm(mphpd)$_3$Phen.

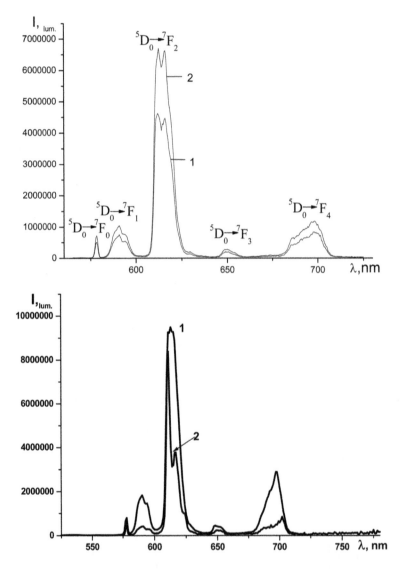

FIGURE 19.7 Luminescence spectra of (a) 1 – [Eu(mphpd)$_3$]$_n$, 2 – [Eu(mphpd)$_3$·Phen]$_n$ in solution CHCl$_3$, T = 77 K, λ = 362 nm; (b) 1 – Eu(mphpd)$_3$, 2 – Eu(mphpd)$_3$·Phen in solid state, T = 298 K, λ = 362 nm.

main (quantum) level ^{4}I$_j$, j = 9/2 (I, 875 and 888 nm), 11/2 (II, 1061 nm) and 13/2 (III, 1332 nm).

The excitation spectrum of the sample Nd(mphpd)$_n$ is similar to the above spectra. In the excitation at 362 nm for the solid samples a 4f-luminescence

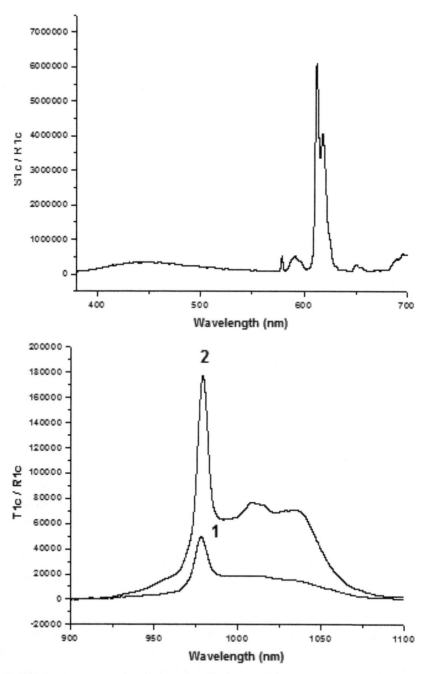

FIGURE 19.8 Luminescence spectra of (a) Eu(mphpd)$_3$·Phen-co-styrene in solid state, T = 298 K, λ = 358 nm; (b) 1 – Yb(mphpd)$_3$·3H$_2$O-co-styrene; 2 – Yb(mphpd)$_3$·Phen-co-styrene in solid state, T = 298 K, λ = 336 nm.

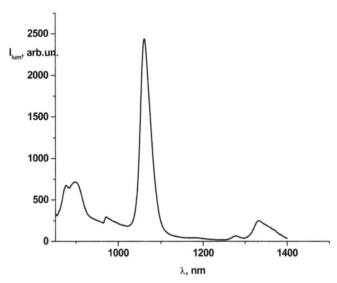

FIGURE 19.9 Luminescence spectrum of Nd(mphpd)$_3$·2H$_2$O-co-VK in solid state (λ_{ex} = 350 nm, 298K).

of Nd (III) is observed, with almost the same intensity compared to the solid sample Nd(mphpd)$_3$.

The bands maxima corresponding to the transitions $^4F_{3/2} \rightarrow ^4I_{11/2}$, и$^4F_{3/2} \rightarrow ^4I_{13/2}$, undergo a long-wavelength shift of 2 nm in comparison with the spectra of the solid state monomers.

The intensive red luminescence is observed for europium complex and metallopolymer. The most intensive band corresponds to transition $^5D_0 \rightarrow ^7F_2$ in region 610–630 nm. The band with considerably less intensity are in spectrum region 575–580, 585–600, 650–670 and 680–705 nm which correspond to transitions $^5D_0 \rightarrow ^7F_j$, j = 0, 1, 3, and 4 accordingly. The band corresponding to transition $^5D_0 \rightarrow ^7F_0$ show in the image of the symmetric single line (Figure 19.7) that allows to assume of the one-luminescence center.

As forphenantroline complexes, obviously phenantroline is a part of the complex and forms an adduct but not a mixed-complex due to spaciousness of the diketonate fragment. Based on the number of the splitting components we can assume a significant rhombic distortion.

The near-term environment influences weakly on the level position of 4f^n-electronic coordination of lanthanides ions. However, these levels split under the influence of crystal field that allow to determine the environment symmetry of the luminescence center, the splitting value was studied. Symmetry type can be estimated qualitatively in concordance with the shtarks splitting of the europium ion transition in luminescence spectra. The qualitative appearance of transition splitting of energy levels (Shtarks effect) for presented compounds indicates about appreciable hexagonal deformation inside of the crystal field, that corresponds to presence of two components $^5D_0 \rightarrow ^7F_1$-transition and two components $^5D_0 \rightarrow ^7F_2$-transition in the spectrum. The symmetry type without measurement of polarized characteristics no established definitely.

The luminescence intensity of metallopolymer with Eu is lesser in solution than in frozen solution or solid state because of the water molecules in solvent as OH-oscillator quenchers (Figure 19.7). Europium luminescence spectra at 77 K allow to establish the short-range coordination environment symmetry.

The luminescence intensity of complexes based on Eu(III) Sm(III) Nd(III) Yb(III)-phenanthroline is greater in comparison with complex without additional ligand (Figures 19.6 and 19.7). The phenanthroline displaces the coordinate water out of the coordinate sphere which is quenching agent.

The solubilization of Eu(III), Sm(II), Yb(III) β-diketonate complexes with phenanthroline, was shown to change luminescence intensity in these complexes. The luminescence intensity of these complexes is greater in comparison with complex without additional ligand.

19.3.2 THE INFLUENCE OF CONCENTRATION OF EMISSION COMPONENTS ON LUMINESCENT PROPERTIES OF NEW COMPLEXES

The luminescence intensity of copolymer of styrene with Eu(III)-phenanthroline complex (Figure 19.8a) is identical practically with homopolymer of Eu(mphpd)$_3$ Phen [6].

The luminescence intensity of copolymer of styrene with Yb(III)-phenanthroline complex is greater in several times in comparison with copolymer with Yb(mphpd)$_3$·3H$_2$O (Figure 19.8b) and respective monomeric complexes.

The luminescence intensity of copolymer VK with Nd(mphpd)$_3$ complex is greater in comparison with respective monomeric and polymeric complexes (Figure 19.9).

Thus, obtained copolymers which containing 5% emission components (lanthanide complex) in polymer chain only are comparably with suitable

TABLE 19.6   The Luminescence Quantum Yield of Monomeric and Polymeric Metallocomplexes

Complex		The luminescence quantum yield, $\varphi \times 10^3$
Eu(mphpd)$_3$ · H$_2$O	Solution	32
[Eu(mphpd)$_3$]$_n$	Solution	38
Eu(mphpd)$_3$(Phen)	Solution	34
[Eu(mphpd)$_3$(Phen)]$_n$	Solution	24
[Nd(mphpd)$_3$]$_n$	Solid	1.72
	Solution	0.86
		1.21 (CHCl$_3$–d_6)
Nd(mphpd)$_3$ · H$_2$O	Solid	0.31
	Solution	0.12
		0.26 (CHCl$_3$–d_6)
Yb(mphpd)$_3$ · 2H$_2$O	Solid	0.22
	Solution	0.13
Yb(mphpd)$_3$ · Phen	Solid	0.32
	Solution	0.15

gomopolymers concerning luminescence properties and can be perspective for optical application.

Luminescence quantum yield of monomeric and polymeric metallo-complexes were calculated (Table 19.5), without erbium. The 4f-luminescence of erbium is weak relatively and is detected in the range 1450–1560 nm, what corresponds to the bandwidth (the maximum half-height) for Er(mphpd)$_3$·2H$_2$O – 18 FWHM, and Er(mphpd)$_n$ – 19 FWHM, therefore the luminescence quantum yield of its complexes no established definitely. The luminescence quantum yield of polycomplex based on Eu(III) is maximum (Table 19.6), which out numbers the values for known low-molecular Eu complexes on several exponents.

Decreasing of φ for [Eu(mphpd)$_3$(Phen)]$_n$ takes place probably due to the fact that the phenantroline in the polymer matrix rather shields emission centers of europium ions and then reduces the number of non-emission losses and displaces the solvent molecules from the metal coordination sphere.

Investigations of a particle size were performed at 25°C on the Zeta Sizer Malvern instrument. The results showed (Figures 19.10–19.13) that the systems obtained are polydisperse with a predominance of particles 6–12 nm for Er(mphpd)$_3$·2H$_2$O, 10–170 nm for Er(OH)(allyl)$_2$ and 10–50 nm for

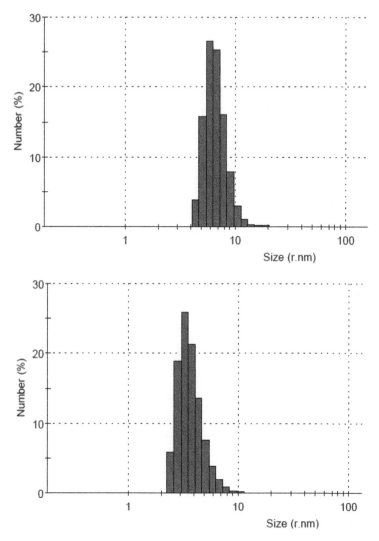

FIGURE 19.10 The polydisperse assignment of particles in system: (a) Pr(mphpd)$_3$, (b) [Pr(mphpd)$_3$]$_n$.

metallopolymer [Er(mphpd)$_3$]$_n$, 20–200 nm for [Er(OH)(allyl)$_2$]$_n$, 6–10 nm for Pr(mphpd)$_3\cdot$2H$_2$O, and 4–9 nm for metal complex [Pr(mphpd)$_3$]$_n$.

Monomeric complex Er(OH)(allyl)$_2$ and metallopolymer based of them were fixed in chloroform what the large size of particles of these compounds was explained.

Since the primary purpose of these studies is associated with obtaining of new nanomaterials which can be used as layers in the organic light

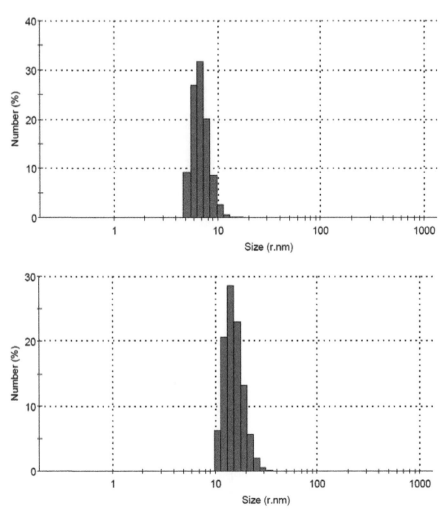

FIGURE 19.11   The polydisperse assignment of particles in system: (a) Er(mphpd)$_3$, (b) [Er(mphpd)$_3$]$_n$.

emitting diodes, it is necessary to choose a solvent which promotes good spreading, and therefore high adhesion of the complex to the substrate surface, allowing to obtain homogeneous films by deposition from solutions. Studies of particle size of themonomeric and polymeric complexes (25°C) in the dimethylformamide, chloroform and ethyl alcohol were conducted. The results showed that the smallest particle size observed in the chloroform solution.

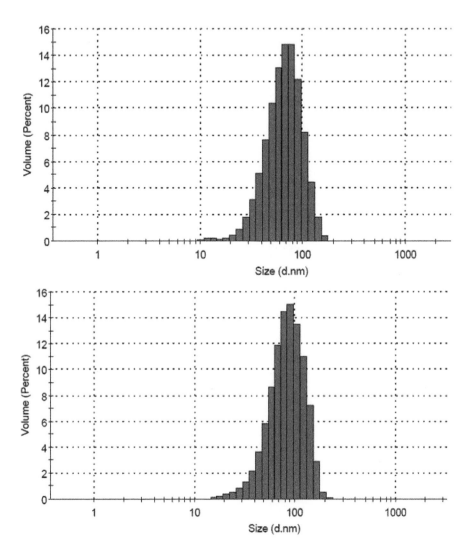

FIGURE 19.12   The polydisperse assignment of particles in system: (a) Er(OH)(allyl)$_2$ (b) [Er(OH)(allyl)$_2$]$_n$.

The system is homogeneous in terms of uniform distribution of metal throughout the polymer matrix as seen from the micrographs of powdered samples (Figures 19.14 and 19.15).

The method of dynamic light scattering and the results of electronic microscopy showed that the obtained polymer systems are nanoscale (the copolymer excepted).

Present electron diffraction (Figure 19.16) confirms the absence of a short-range and long-range order so the metal polymer is X-ray amorphous.

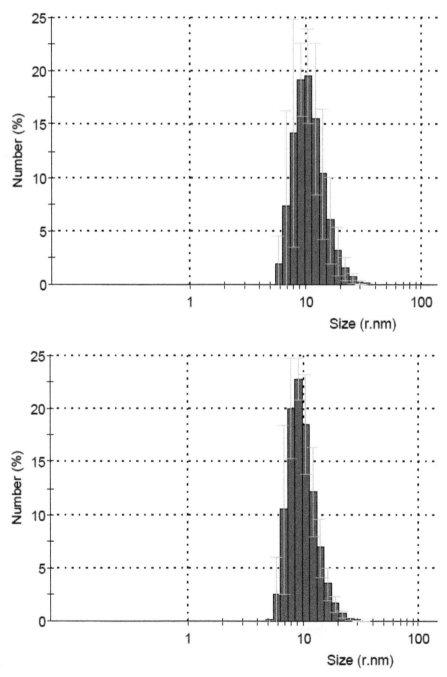

FIGURE 19.13 The polydisperse assignment of particles in system [Eu(mphpd)$_3$] and [Eu(mphpd)$_3$]$_n$.

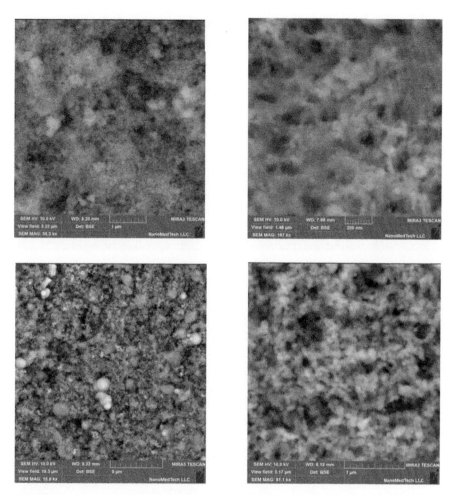

FIGURE 19.14 SEM micro photo of metallopolymers powders: (a) Er(mphpd)$_3$, (b) [Er(mphpd)$_3$]$_n$, (c) Er(OH)(allyl)$_2$, (d) [Er(OH)(allyl)$_2$]$_n$.

19.3.3 LIGHT-EMITTING ORGANIC PLANAR HETERO STRUCTURES

Multilayer semiconductor structures with their thick several orders smaller than their linear dimensions were fabricated on the basis of the synthesized metal polymers [Sm(mphpd)$_3$]$_n$ and [Eu(mphpd)$_3$]$_n$.

This device consists of two organic layers situated between a low work function metal cathode such as calcium and a higher work function anode, often-transparent indium-tin oxide (ITO). The emission layer

FIGURE 19.15   SEM micro photo of powders: (a) Pr(mphpd)$_3$ 2H$_2$O, (b) [Pr(mphpd)$_3$]$_n$,(c) Pr(mphd)$_3$-co-styrene,(d) scale 500 nm (a, b, c), 1 μm.

– polycomplexes[Eu(mphpd)$_3$]$_n$, [Sm(mphpd)$_3$]$_n$, hole conduction layer – the mixture PEDOT:PSS were used.

Emission layer ([Eu(mphpd)$_3$]$_n$, [Sm(mphpd)$_3$]$_n$) was applied to the surface (PEDOT:PSS) by spin-coating in nitrogen filled sealed box from chloroform solutions (with concentration 10^{-2} –10^{-3} M). Cathode (Al) was applied by magnetron metal spraying. All devices have the structure with two organic layers ITO/ PEDOT:PSS/metal complexes/Al. The current–voltage characteristics of samples are exponential dependences which characterized

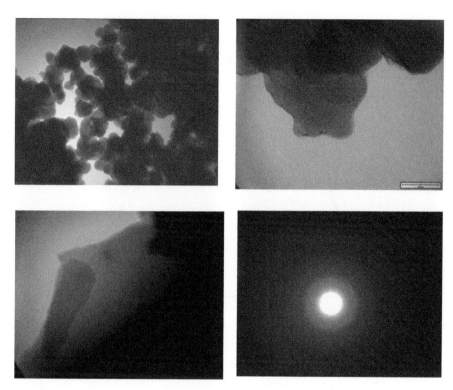

FIGURE 19.16 TEM micro photo of metallopolymer[Eu(mphpd)$_3$]$_n$: powder (a, b), film (c) and electron diffraction, (d) U = 150 кV, cursor – 50 nm.

for *p-n* transition. The weak electroluminescence (the voltage 21–22 V) of obtained devices was observed.

19.4 CONCLUSIONS

In conclusion, a nowel structures – complexes of Nd, Er, Eu, Gd, Yb, Sm, Pr, Lu, and Ho with 2-methyl-5-phenylpenten-1–3,5-dionand allyl-3-oxo-butanoate, polycomplexes on their basis and copolymers with styrene and N-vinylcarbazole in ratio 5:95 has been proposed. The investigations conducted in the present work-allowed to determine the composition, structure and symmetry of coordination polyhedron of the monomeric complexes and polycomplexes on their basis obtained for the first time. The results of the above study showed that the configuration of the chelate unit is unchanged during the polymerization.

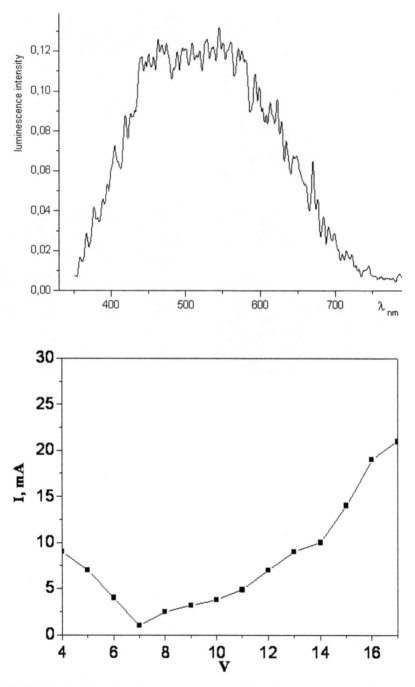

FIGURE 19.17 Electroluminescence spectrum (a) and voltammetry of the ITO/
PEDOT:PSS/[Sm(mphpd)$_3$]$_n$/Al: Rotational speed 1000 rpm (b).

The similarity of monomers electronic absorption spectra with polymers spectra confirms of identical coordinative environment of lanthanide ions in both cases.

The results of the chemical analysis correspond to the calculated composition, all the functional groups of a macro ligand are bound to the metal ion.

It was shown that all synthesized compounds are nano systems.

The luminescence quantum yield of polycomplex based on Eu(III) is maximum.

In this way, synthesized polycomplexes of lanthanides meet the requirements of the ligands in complexes for design OLED:

1. the triplet level of ligand is applicable by emission level of metal;
2. ligands have hole transporting and electron transporting properties for injection charge and exit on generation in complex;
3. complexes are deposited slightly on substrate without aggregates and crystals and are combined with other components;
4. the probability of radiation less deactivation of resonance level is little in comparison with the probability of radiation transition.

The present investigation suggests that obtained polycomplexes were positional candidates as materials for organic light-emitting devices and other optical applications.

KEYWORDS

- luminescence
- organic light emitting diodes
- polymer complexes
- spectra
- structure
- β-diketonate complex

REFERENCES

1. Komoda, T., Ide, N., Varutt, K., Yamae, K., Tsuji, H., Matsuhisa, Y., (2011). High-performance white OLEDs with high color-rendering index for next-generation solid-state lighting. *J. Soc. Inform. Display 19,* 838–846.

2. Jou, J., Kumar, S., Agrawal, A., Li, T., Sahoo, S., (2015). Approaches for fabricating high efficiency organic light emitting diodes. *J. Mater. Chem. C, 3,* 2974–3002.
3. Savchenko, I., Bereznitskaya, A., Fedorov, Ya. (2014). Structural organization of polymer metal complexes with water or phenanthroline and their influence on luminescence properties. *Chem. & Chem. Technology. 7,* N 4, 423–428.
4. Yamae, K., Tsuji, H., Kittichungchit, V., Ide, N., Komoda, T., (2013). Highly efficient white organic light-emitting diodes with over 100 lm/W for next-generation solid-state lighting. *J. Soc. Inform. Display, 21,* 529–540.
5. Amegadzea Paul, S. K., Noha, Y., (2014). Development of high-performance printed polymerfield-effect transistors for flexible display. *J. Information Display, 15*(4), 213–229.
6. Savchenko, I., Bereznitskaya, A., Smola, S., FedorovYa, Ivakha, N., (2012). Novel polymer metal complexes as precursors for electroluminescent materials. *Functional Mater., 19, N4,* 541–547.
7. Cho, H., Jin Ch, Kim, E., Yoo, S., (2014). Polarizer-free, high-contrast-ratio organic light-emitting diodes utilizing micro cavity structures and neutral-density filters. *Journal of Information Display, 15*(4), 195–199.
8. Ding, L., Sun Ya, Chen, H., Zu, F., Wang Zh, Liao, L., (2014). A novel intermediate connector with improved charge generation and separation for large-area tandem white organic lighting devices. *J. Mater. Chem. C, 2,* 10403–10408.
9. Savchenko, I. A., Berezhnytska, A. S., Fedorov, Ya. V., Trunova, E. K. (2014). Copolymers of rare earth elements complexes with unsaturated β-diketones and n-vinylcarbazole for OLEDs. *Mol. Cryst. Liq. Cryst., 590,* 66–72.
10. LaiCh, Lee Yu, Tsai, T., ChangCh, Wu, M., (2014). Highly Reliable and Low-Cost Fabrication of Warm-White LEDs Using Composite Silica Photonic Crystals.Internat. *J. Photo Energy.* 1–6.
11. Savchenko, I. A., Berezhnytska, O. S., Trunova, E. K. (2015). New nanosized systems based lanthanide diketonate complexes for OLEDs. Ink: Nanocomposites, Nanophotonics, Nanobiotechnology and Applications Springer International Publishing Switzerland ISBN: 978–3–319–06610–3. *Proc. in Phys. 156,* Chapter 6, 85–94.
12. Mishchenko, A., Berezhnytska, O., Savchenko, I., (2014). Novelytterbium (III) β-diketonatesas a precursors of nir-emitting materials. *Mol. Cryst. Liq. Cryst., 590,* 49–57.
13. Jiu, H., Liu, G., Zhang, Z., et al. (2011). Fluorescence enhancement of Tb(III) complex with a new β-diketone ligand by 1, 10-phenanthroline. *J. Rare Eart., 29,* 741–745.
14. Kuo, Y., Chi-Chou, L., (2013). A light emitting device made from thin zirconium-doped hafnium oxide high-k dielectric film with or without an embedded nanocrystal layer. *Appl. Phys. Lett., 102,* 31117(1–7).
15. Eliseeva, S., Bunzli, J. C. G. (2010). Lanthanide luminescence for functional materials and bio-sciences. *Chem. Soc. Rev., 39,* 189–227.
16. Mech, A., Monguzzi, A., Meinardi, F., (2010). Sensitized NIRerbium(III) emission in confined geometries: A new strategy for light emitters in telecom applications. *J. Am. Chem. Soc., 132,* 4574–4576.
17. Chen, Z., Ding, F., Hao, F., Guan, M., et al. (2010). Synthesis and electroluminescent property of novel europium complexes with oxadiazole substituted 1,10-phenanthroline and 2,2-bipyridine ligands. *New J., Chem., 34,* 487–494.
18. Peng, Zh, Qing, Zh. (2014). Sensitization of europium (III) luminescence in water with β-diketone-poly(ethylene glycol) macro ligand. *Sci. China Chem., 57*(2), 243–247.

19. Yu-Guang, L., Zhong-Ping, G., Hong-Bing, G., et al. (2015). Synthesis and luminescent properties of ternary complex Eu(UVA)$_3$Phen in nano-TiO$_2$. *Optoelectronics Let., 11*(1), 41–44.
20. Youyi Sun, Binghua Yang, GuizhenGuo, Yaqing Liu, Guizhe Zhao. (2011). Controlled Formation of Sm(III) Doping Polymer Thin Films Based on a New Macroligand with b-Diketonate. *J. Inorg. Organomet. Polym., 21,* 395–399.
21. Zub, V. Ya., Berezhnitskaya, A. S., Savchenko, I. S., Voloshanovskii, I. S., Gudich, I. N., Mazurenko, E. A., Shevchenko, O. V. (2004). Synthesis and Polymerization in Unsaturated Co β-Diketonates.Rus. *J. Coordinat. Chem., 30*(10), 709–712.
22. Voloshyn, A., Shavaleev, N., Kazakov, V., (2001). Luminescence of praseodymium (III) chelates from two existed states ($^3P_0$ and $^1D_2$) and its dependence on ligand triplet state energy. *J. Lumin., 93,* 199–204.

CHAPTER 20

INTERGEL SYSTEMS: HIGHLY EFFECTIVE INSTRUMENTS FOR RARE EARTH ELEMENTS EXTRACTION FROM INDUSTRIAL SOLUTIONS

T. K. JUMADILOV and R. G. KONDAUROV

JSC "Institute of Chemical Sciences after A. B. Bekturov,"
Sh. Valikhanov st. 106, Almaty, Republic of Kazakhstan,
E-mail: jumadilov@mail.ru

CONTENTS

ABSTRACT

Intergel system polymethacrylic acid hydrogel (gPMAA) and poly-2-methyl-5-vinylpyridine hydrogel (gP2M5VP) behavior in aqueous medium and solution of 10-water lanthanum sulfate in dependence from time was studied. During sorption of La^{3+} ions electrochemical properties of solutions and volume-gravimetric properties of hydrogels change significantly. Under affect of mutual activation sorption capacity of initial hydrogels in intergel system increases.

20.1 INTRODUCTION

Production of rare metals, rare earth metals and their compounds in Republic of Kazakhstan can be characterized as unstable, not corresponding to its potential. In some enterprises production of these metals decreased sharply and suspended. However, the world demand for rare earth products increases every day. Due to this fact production of clean rare and rare earth metals and their compounds is highly profitable. Consequently, production of pure rare and rare earth metals and their compounds may be priority direction in future for Republic of Kazakhstan.

Previous studies showed that in result of remote interaction hydrogels, which are in intergel systems, have significant changes in volume-gravimetric, sorption properties [1–7]. In this regard, the goal of the work is to compare electrochemical and volume-gravimetric properties of intergel system polymethacrylic acid hydrogel (gPMAA) – poly-2-methyl-5-vinylpyridine hydrogel (gP2M5VP) in an aqueous medium and in lanthanum sulfate solution and their change during lanthanum sorption.

20.2 EXPERIMENTAL PART

20.2.1 MATERIALS

Hydrogels of polymethacrylic acid and poly-4-vinylpyridine are initial hydrogels for intergel system. Distilled water and 0.005 M 10-water lanthanum sulfate are medium of investigation of electrochemical and conformational properties.

20.2.2 CHARACTERIZATION

Hydrogels of polymethacrylic acid were synthesized in the presence of cross-linking agent N,N-metilen-bis-acrylamid and redox system $K_2S_2O_8$,

$Na_2S_2O_3$. The hydrogel of poly-2-methyl-5-vinylpyridine (gP2M5VP) was synthesized from a linear polymer in dimethylformamide in presence of epichlorohydrin at 60°C.

Hydrogels synthesized in an aqueous medium is an intergel pair polymethacrylic acid gel – poly-2-methyl-5-vinylpyridine gel (gPMAA-gP2M5VP). Swelling coefficients of hydrogels are $K_{sw(gPMAA)}$ = 10.1 g/g, $K_{sw(gP2M5VP)}$ = 0.46 g/g.

20.2.3   OBJECTS OF STUDY

Intergel system on the basis of polymethacrylic acid and poly-2-methyl-5-vinylpyridine in water medium and in 10 lanthanum sulfate solution were the objects for investigations of electrochemical and conformational properties in dependence of hydrogels molar ratio in time.

20.2.4   METHODS

For measurement of electro conductivity conductometer "MARK 603" (Russia) was used, pH of solutions was measured on pH meter "Seven Easy" (METTLER TOLEDO, China). Swelling coefficient K_s was defined by weighting of hydrogel swollen samples on electronic scales "SHIMADZU AY220" (Japan).

Experiments were carried out at room temperature. Study of intergel systems was made by this way: each hydrogel was located in separated glass weighing filter pores of which are permeable for low-molecular ions and molecules, but it is not permeable for hydrogels dispersion.

Then weighing filter with hydrogels were located in glasses with distilled water and lanthanum sulfate solutions. Electric conductivity and pH of over gel liquid were measured in absence of weighing filters with hydrogels in the glass. Swelling coefficient was calculated as the difference of weights of weighing bottle with hydrogel and empty weighing filter according to this equation:

$$K_{sw} = m_2 - m_1/m_1$$

where m_1 – weight of dry hydrogel, m_2 – weight of swollen hydrogel.

χ, μS/cm

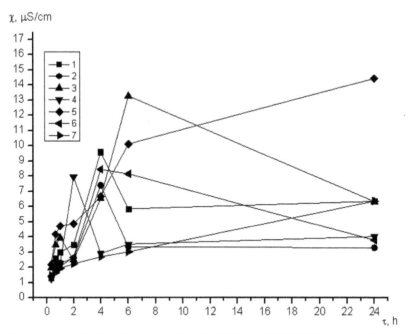

Curves' description (molar ratio of gPMAA:gP2M5VP, mol:mol):
1–6:0; 2–5:1; 3–4:2; 4–3:3; 5–2:4; 6–1:5; 7–0:6

FIGURE 20.1 Dependence of water solutions specific electric conductivity in presence of intergel system gPMAA-gP2M5VP from time.

20.3   RESULTS AND DISCUSSION

20.3.1   STUDY OF ELECTROCHEMICAL AND VOLUME-GRAVIMETRIC PROPERTIES CHANGE OF POLYMETHACRYLIC ACID HYDROGEL (GPMAA) AND POLY-2-METHYL-5-VINYLPYRIDINE HYDROGEL (GP2M5VP) INTERGEL SYSTEM IN AN AQUEOUS MEDIUM

Obtained dependencies of specific electric conductivity show (Figure 20.1) that electric conductivity of solutions increases with time for all ratios of gPMAA:gP2M5VP. However, character of electric conductivity change is different for various ratios. With increase of contact time with water areas of maximum and minimum electric conductivity appear. At gPMAA:gP2M5VP ratios 4:2 and 2:4 areas of maximum conductivity are found. Minimum electric conductivity is due to binding of cleaved proton from carboxyl group by

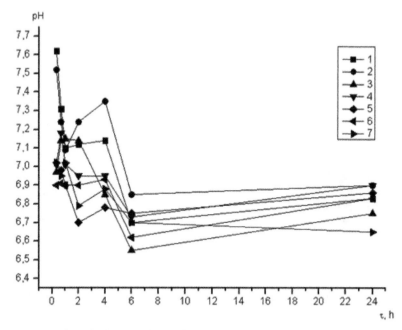

Curves' description (molar ratio of gPMAA:gP2M5VP, mol:mol):
1–6:0; 2–5:1; 3–4:2; 4–3:3; 5–2:4; 6–1:5; 7–0:6

FIGURE 20.2 Dependence of water solutions pH in presence of intergel system gPMAA-gP2M5VP from time.

nitrogen heteroatom of vinyl pyridine. High conductivity values point that at certain ratios of hydrogels in intergel system carboxyl group dissociation prevails over proton association process by nitrogen heteroatoms.

Based on the results of hydrogen ions concentration change (Figure 20.2) in aqueous medium in presence of intergel systems gPMAA:gP2M5VP it can be concluded that cleaving rate of H^+ during polyacid dissociation is higher than H^+ binding rate by polybasis. At ratio 5:1 pH of medium increases, and this phenomenon is consequence of the fact that with decrease of swelling rate there is decrease of –COOH groups dissociation rate. With polybasis share increase solutions pH decreases gradually, minimum point of pH is seen at gPMAA:gP2M5VP ratio 4:2 (at 6 hours of interaction). Low concentrations of hydrogen ions are also observed at ratio 5:1, which almost corresponds with results of electric conductivity.

In result of polymer hydrogels long-range effect polyacid swelling coefficient (K_{sw}) change occurs (Figure 20.3). Significant increase of swelling

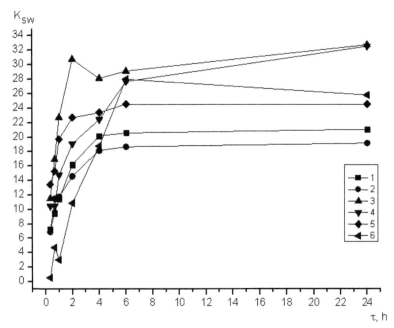

Curves' description (molar ratio of gPMAA:gP2M5VP, mol:mol):
1–6:0; 2–5:1; 3–4:2; 4–3:3; 5–2:4; 6–1:5

FIGURE 20.3 Dependence of polymethacrylic acid hydrogel swelling coefficient in presence of poly-2-methyl-5-vinylpyridine hydrogel in an aqueous medium from time.

is observed at gPMAA:gP2M5VP = 4:2 ratio. Also swelling coefficient increase is observed at ratio 2:4. This phenomena's point that conformational state of internode links is determined predominantly by ionization, not by concentration of charged groups, which were formed during functional groups dissociation.

Measurement of polybasis swelling coefficient in an aqueous medium point to its increase with polyacid share (Figure 20.4), what occurs due to additional swelling of gP2M5VP. As it is known, poly-2-methyl-5-vinylpyridine swells weakly as weak polybasis. Increase of gP2M5VP K_{sw} point to additional activation of polybasis links. It can be only due to cleaved proton binding by nitrogen atoms of polybasis.

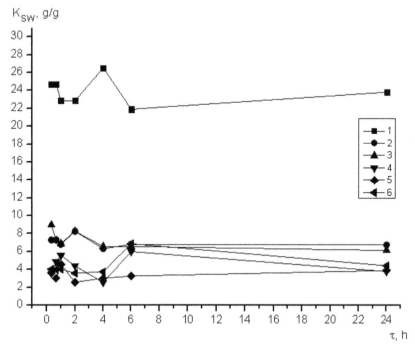

FIGURE 20.4 Dependence of poly-2-methyl-5-vinylpyridine hydrogel swelling coefficient in presence of polymethacrylic acid hydrogel in an aqueous medium from time.

20.3.2 STUDY OF ELECTROCHEMICAL AND VOLUME-GRAVIMETRIC PROPERTIES CHANGE OF POLYMETHACRYLIC ACID HYDROGEL (GPMAA) AND POLY-2-METHYL-5-VINYLPYRIDINE HYDROGEL (GP2M5VP) INTERGEL SYSTEM AT LANTHANUM IONS SORPTION FROM LANTHANUM SULPHATE 10-WATER

Situation during lanthanum sorption is similar to intergel system behavior in an aqueous medium – specific electric conductivity increases with time (Figure 20.5). Areas of maximum and minimum electric conductivity appear with time. After 1 hour of interaction minimum point at gPMAA:gP2M5VP ratio 3:3 becomes noticeable, also minimum is observed in presence of only polyacid (ratio 6:0). At 3 hours of remote interaction maximums appear at ratio 4:2, and at 6 hours there is predominance of maximum conductivity noticeable when there is a complete domination of basis hydrogel (ratio 0:6); while the minimums are observed in both cases in the presence of

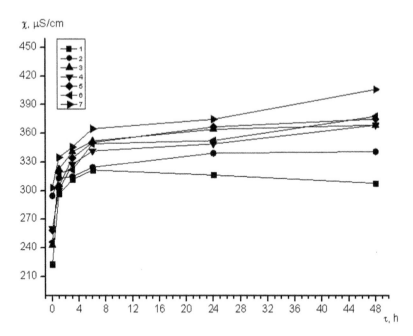

Curves' description (molar ratio of gPMAA:gP2M5VP, mol mol):
1–6:0; 2–5:1; 3–4:2; 4–3:3; 5–2:4; 6–1:5; 7–0:6

FIGURE 20.5 Dependence of 10-water lanthanum sulfate solutions specific electric conductivity in presence of intergel system gPMAA-gP2M5VP from time.

polymethacrylic acid individual hydrogel (ratio 6:0) and for hydrogels ratio 5:1, respectively. After 24 hours there is conductivity increase at ratios 4:2 and 2:4. After that further increase of electric conductivity with pronounced maximum can be observed in presence of poly-2-methyl-5-vinylpyridine hydrogel (ratio 0:6).

For detailed explanation of low and high values of electric conductivity it is necessary to study processes of hydrogels ionization and dissociation. During carboxyl group dissociation a proton is formed, further binding of what is implemented by vinyl pyridines nitrogen heteroatom. This phenomenon is main reason of low conductivity. Hydrogels swelling in lanthanum sulfate solutions occurs due to their interaction with water molecules. At first there is an ionization of carboxyl groups, after that – dissociation to carboxyl at anions –COO– and hydrogen ions H^+. At carboxyl group and water molecules dissociation H^+ and OH- ions are formed. Proton binding causes ionization of poly-2-methyl-5-vinylpyridine cationic hydrogel.

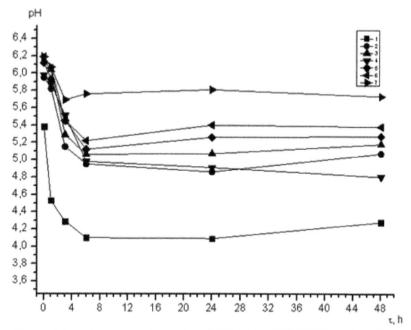

Curves' description (molar ratio of gPMAA:gP2M5VP, mol:mol):
1–6:0; 2–5:1; 3–4:2; 4–3:3; 5–2:4; 6–1:5; 7–0:6

FIGURE 20.6 Dependence of 10 water lanthanum sulfate solutions pH in presence of intergel system gPMAA-gP2M5VP from time.

Ionized carboxyl group of polymethacrylic acid binds lanthanum ions. All these interactions lead to decrease of total concentration of positive charges in solution.

At hydrogels certain ratios in intergel system there is a predominance of carboxyl groups dissociation over proton binding by nitrogen heteroatoms, what, in turn, may be the main reason of high conductivity values. Rate of polybasis proton binding decreases due to conformational changes in poly-2-methyl-5-vinylpyridine hydrogel internode links. Consequently, formation of intermolecular cross-links like $\geq N\ldots H^+\ldots N\equiv$ at charged NH^+ groups causes to polymer coils folding, and, as a result, proton binding degree decreases. Also it should be noted that there is an increase of conductivity due to formed SO_4^{2-} ions in lanthanum sulfate dissociation.

In Figure 20.6, the dependence of lanthanum sulfate solutions pH in presence of intergel system gPMAA – gP2M5VP from time is shown. With predominance of polyacid there is an increase of hydrogen ions concentration with time. With polybasis share increase solutions pH gradually increases and reaches maximum at hydrogels ratio 0:6. There is a pH decrease with time, what may point to additional proton release. Comparing hydrogen ions concentration low values and high conductivity values at gPMAA: gP2M5VP ratios 1:5 and 0:6 it can be concluded that conductivity increase occurs due to OH⁻ and SO₄²⁻ anions concentration increase.

For description of obtained results it is necessary to analyze ionic equilibrium in solutions. In lanthanum sulfate solutions the following processes occur:

1. Dissociation of internode links –COOH groups:

$$-COOH \rightarrow COO^-...H^+ \rightarrow -COO^- + H^+$$

It should be taken to attention that at first there is ionization with ionic pairs formation, after that ionic pairs partly dissociate to separate ions.

2. Nitrogen atom in pyridine ring undergoes ionization and dissociates partially:

$$\equiv N + H_2O \rightarrow \equiv NH^+ ... OH^- \rightarrow \equiv NH^+ + OH^-$$

3. Further nitrogen atom interacts with proton, which was cleaved from carboxyl group:

$$\equiv N + H^+ \rightarrow \equiv NH^+$$

4. H⁺ and OH⁻ ions, formed as a result of functional groups interaction with water molecules, form water molecules:

$$H^+ + OH^- \rightarrow H_2O$$

5. La³⁺ and SO₄²⁻ ions form during lanthanum sulfate dissociation:

$$La_2(SO_4)_3 \leftrightarrow 2La^{3+} + 3SO_4^{2-}$$

6. Formed La³⁺ ions are binded by –COO- groups:

$$La^{3+} + 3-COO- \rightarrow (-COO)_{3La}$$

In result of these interactions there is a state, where counter ions of some part of hydrogels charged functional groups are absent. Concentration of ionized groups without counter ions is in dependence of initial molar ratios of polymer networks and other factors.

Appearance of hydroxyl ions high concentration is possible in case of vinyl pyridine links interaction with water molecules.

Ions concentration in solutions is in direct dependence of swelling rate and hydrogels concentration in salt solutions. Swelling and deprotonization rate are in dependence of hydrogels nature, crosslink degree, dispersion and the longitudinal size of polymer hydrogels.

Appearance of H^+ ions excess is due to high rate of swelling, COOH dissociation and low swelling rate of polybasic functional groups and their low concentration. OH^- ions concentration increase occurs due to low swelling rate and low concentration of COOH groups, and high swelling rate and interaction of basic functional groups with H^+ ions. This is possible in case of occurring of 2nd reaction, where hydroxyl anions are released in solution.

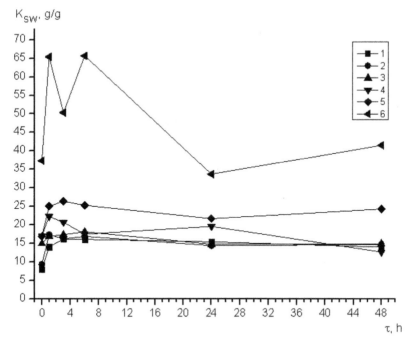

Curves' description (molar ratio of gPMAA:gP2M5VP, mol:mol): 1–6:0; 2–5:1; 3–4:2; 4–3:3; 5–2:4; 6–1:5

FIGURE 20.7 Dependence of polymethacrylic acid hydrogel swelling coefficient in presence of poly-2-methyl-5-vinylpyridine hydrogel in 10 water lanthanum sulfate solution from time.

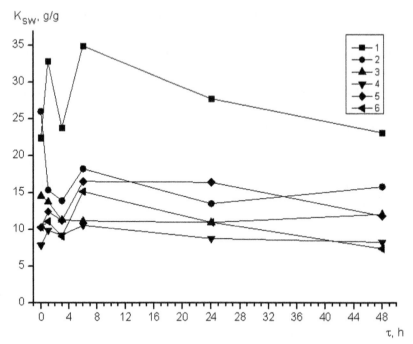

Curves' description (molar ratio of gPMAA:gP2M5VP, mol:mol):
1–6:0; 2–5:1; 3–4:2; 4–3:3; 5–2:4; 6–1:5

FIGURE 20.8 Dependence of poly-2-methyl-5-vinylpyridine hydrogel swelling coefficient in presence of polymethacrylic acid hydrogel in 10 water lanthanum sulfate solution from time.

In parallel there is an occurrence of 3rd reaction as a result of which free proton is binded by pyridines ring and positively charged ions concentration in solution decreases sharply.

Lanthanum sulfate dissociation leads to La^{3+} ions binding by acid hydrogel and SO_4^{2-} ions have no counter ions, and, as a result, value of electric conductivity increases.

Hydrogels remote interaction effect provides significant changes in polyacid swelling coefficient (Figure 20.7). Intensive increase of polymethacrylic acid swelling is observed with poly-2-methyl-5-vinylpyridine share increase. As it can be seen from figure at gPMAA:gP2M5VP ratio 1:5 at all time of remote interaction intensive swelling of polyacid occurs. Dissociation of acid groups leads to repulsion of similarly charged –COO– groups, and, as a result, polymethacrylic acid polymer chain unfolds.

After hydrogels mutual activation poly-2-methyl-5-vinylpyridine hydrogel swelling coefficient increases significantly. With polyacid share increase there is a strong increase of gP2M5VP. The main reason of it is ionization of structure links of polybasis by cleaved proton. Maximum swelling of basic hydrogel is observed at gPMAA:gP2M5VP ratio 5:1. General type of curves shows that K_{sw} increase occurs with polyacid share increase. However, in time interval 1–6 hours there is maximum, which point to significant increase of polybasis swelling. Such effect may be due to destruction of intermolecular associates $\equiv NH...N\equiv$, which were formed by $\equiv NH$ groups.

20.4 CONCLUSIONS

1. Based on obtained results of specific electric conductivity, pH of solutions and K_{sw} of hydrogels it can be concluded that there is a result of remote interaction of hydrogels a significant change in their electrochemical, conformational and sorption properties occur.
2. Result of hydrogels long-range effect is additional activation of hydrogels due to what internode links obtain additional charge without counter ions.
3. Additional hydrogels swelling occur due conformational changes in internode chains during remote interaction of polymer networks.
4. Based on results of electric conductivity and solutions pH it can be assumed that maximum sorption capacity for intergel system gPMAA:gP2M5VP appears at ratio 5:1.

KEYWORDS

- intergel system
- La^{3+} ions
- poly-2-methyl-5-vinylpyridine hydrogels
- polymethacrylic acid hydrogel
- sorption

REFERENCES

1. Jumadilov, T., Shaltykova, D., Suleimenov, I., (2013). Anomalous ion exchange phenomenon. *Proc. of Austrian-Slovenian Polymer Meeting*, Abstr. *S5*, 51.
2. Alimbekova, B. T., Korganbayeva Zh. K., Himersen, H., Kondaurov, R. G., & Jumadilov, T. K. (2014). Features of Polymethacrylic Acid and Poly-2-Methyl-5-Vinyl-pyridine Hydrogels Remote Interaction in an Aqueous Medium. *J. Chem. Chem. Eng., 8*, 265–269.
3. Jumadilov, T. K., Kaldayeva, S. S., Kondaurov, R. G., Erzhan, B., Erzhet, B., (2014). Mutual activation and high selectivity of polymeric structures in intergel systems. Proceedings of symposium ICSP & AM Nova Science Publishers, 191–196.
4. Jumadilov, T. K., Himersen, H., Kaldayeva, S. S., Kondaurov, R. G. (2014). Features of Electrochemical and Conformational Behavior of Intergel System Based on Polyacrylic Acid and Poly-4-Vinylpyridine Hydrogels in an Aqueous Medium. *J. Mat. Sci. Eng. B., 4*, 209–213.
5. Bekturov, E. A., Dzhumadilov, T. K. (2009). New approaches to study hydrogels interaction remote effect. News of NAS RK, 1, 86–87.
6. Bekturov, E. A., Dzhumadilov, T. K., Korganbaeva, Zh. K. (2010). Long-range effect in polymer systems. KazNU herald, chem. series, *3*, 108–110.
7. Jumadilov, T. K. (2014). Electrochemical and conformational behavior of intergel systems based on the rare cross-linked polyacid and polyvynilpyridines. "Chemistry and Chemical Technology," *Proc. of Int. Conf. Lith. Chem. Soc.* 226–229.

CHAPTER 21

INVESTIGATION OF A COMPLEX FORMATION PROCESS OF COPPER WITH FULVIC ACIDS

T. MAKHARADZE, G. SUPATASHVILI, and G. MAKHARADZE

Ivane Javakhishvili Tbilisi State University, Chavchavadze Avenue 1, Tbilisi, 0128, Georgia, E-mail: gogigiorgi@mail.ru

CONTENTS

ABSTRACT

Fulvic acids were separated from the water of the river by the adsorption-chromotographic method. The charcoal was used as a sorbent. The complex formation process of fulvate complexes of copper were studied by using sephadex G-25 on different values of pH. The parameters of sephadex G-25: the mass of dry gel – 17 g, the height of swelled layer of gel – 42 sm, the inner diameter of column – 16 sm. It was calculated the average meanings of the conditional stability constants of fulvate complexes of copper. It was established that $\beta(pH9) \approx \beta(pH8) > \beta(pH6) > \beta(pH5)$.

21.1   INTRODUCTION

Fulvic acids are one of the first geopolymers, which were found in natural waters. They have functional groups and that's why they take an active place in complex formation and sorption processes and stipulates migration forms of toxic metals [1–16].

In spite of researches, experimental data on stability constants of complex compounds of fulvic acids are heterogeneous and they differ in several lines from each other. Therefore, it's difficult to investigate complex formation processes, taking place in natural waters, identify migration forms of heavy metals and evaluate and assess chemical-ecological condition of natural waters.

Our objective was to obtain the pure samples of fulvic acids, the investigation of complex formation processes between the pure samples of fulvic acids and Cu(II) on different pH by the gel-chromotographic method. According to the data, found in the literature about the fulvate complexes of copper are heterogeneous. Value of lgβ changes from 3.8 to 9.3 [8–15].

21.2   EXPERIMENTAL PART

To obtain pure samples of fulvic acids the authors concentrated the water of the river Mtkvari by the frozen method. Filtered water samples were acidified to pH 2 and were put for 2 hours on water bath at 60°C, for coagulation of humin acids. Then, the solution was centrifuged for 10 min at 8,000 rpm. To isolate FA from centrifugate, the adsorption- chromotographic method was used [16, 17]. The charcoal (BAU, Russia) was used as a sorbent. Desorption of amino acids and carbohydrates were performed by means of 0.1N HCl. For the desorption of polyphenols, the 90% acetone water solution was used. The elution of FA fraction was performed with 0.1N NaOH solution. The obtained alkali solution of FA for the purification was passed through a cation-exchanger (KU-2–8, Russia) and dried under the vacuum until the constant mass was obtained. Then, model solutions of FA was prepared.

The main solution of fulvic complexes were obtained by the solubility method. The same quantity of standard solution of fulvic acids and suspensions of $Cu(OH)_2$, were placed in fluoroplast cylinders. The constant ionic strength was made by adding 0.1M potassium nitrate. The final volume was 40 ml.

The concentration of hydrogen ions was regulated by 0.01M potassium hydroxide and 0.01M nitric acid, pH = 8 in model solutions (pH meter pH_2O06). The solution was put on a mechanical mixer for 60 hours (until the balance was achieved) and then suspension was filtered through the membrane filters (Sinpor N6). In filtrates, the concentration of copper was measured by Atomic Absorption Spectrophotometer (Perkin almer 200).

The main solution of obtained fulvate complexes was divided into 4 parts, which differs from each other only by the value of pH (9.02; 8.00; 6.03; 5.01). The formation of fulvate complexes were studied on different pH by the gel filtration method. For the investigation of complex formation process, taking account the associates of fulvic acids, various pH and the average molecular weight, for the optimal determination we used sephadex G-25(the limits of Fractionating 100–5000).

The parameters of sephadex G-25: the mass of dry gel - -17 g, the height of swelled layer of gel – 42 sm, the inner diameter of column – 1,6 sm. For the calibration of sephadex G-25 was used blue dextran, polyethylene glycols with molecular weights 300, 600, 1000 and glucose. The titer of standard substances – 1 mg/ml, transmission speed 3 ml/min, apply volume of solution – 2 ml.

The results of calibration: free volume 31–36 ml (I fraction), releasing volume of standard substances according to molecular weights (Mw) 1000 = 36–46 ml (II fraction), (Mw) 600 = 46–51 ml (III fraction), (600) Mw 300 = 51–56 (IV fraction), Mw 300 = 56–61 ml.(V fraction) Mw 180 = 61–67 ml (VI fraction).

We took the aliquots of different solutions with different pH (2–2 ml) which were placed in the top part of the column. The elution process was done by bidistilled water that has the same pH as the aliquots of solution.

We determined the quantity of metals connected with fulvic acids in I-V fractions. These are the fractions, which releasing volume fits substances with molecular weight $300 \leq Mw > 5000$. For this reason, I-V fraction was all gathered, then were concentrated up to 10 ml and the quantity of metals was measured by Atomic Absorption Spectrophotometer.

21.3   RESULTS AND DISCUSSIONS

Taking account the conditional diagrams of copper [10], the formation reactions of fulvate complexes of copper and their stability constants on different pH can be written in the following way:

$$Me^{2+} + mFA^{2-} \leftrightarrows [Me\,(FA)m]^{2m-2} \quad pH = 5\text{–}6 \tag{1}$$

$$\beta = [Me\,(FA)m]^{2m-2}/\{\,[\,Me^{2+}]\,[FA^{2-}]^{m}\} \tag{2}$$

$$[Me(OH)2]^{0} + mFA^{2-} \leftrightarrows [Me(OH)2\,(FA)m]^{2m-} \qquad pH = 8\text{–}9 \tag{3}$$

$$\beta = [Me(OH)2\,(FA)m]^{2m-}/\{\,[\,Me(OH)2^{0}]\,[FA^{2-}]^{m}\} \tag{4}$$

In solution, the concentration of fulvate complex (C''_{Me}) equals to the difference between final ($\sum_{Me}$) and initial (C'_{Me}) concentrations of metal received after formation of complex. The concentration of metal as ionic and hydroxocomplex concentration equals to the initial concentration of metal in solution (C'_{Me}). As we have already mentioned [8], in given systems, complexes with structure 1:1, or the stechiometral coefficient (m) equals to 1. Therefore, we can write (2) and (4) equations as the one equation:

$$\beta = [\sum_{Me} - C'_{Me}]/\{[\,C'_{Me}][FA]\} = [C''_{Me}]/\{[C'_{Me}][FA^{2-}]\} \tag{5}$$

During the gelchromotographic investigation of fulvate complexes, these characteristics will be calculated very easy: $C'_{Me} = \sum_{Me} C''_{Me}[FA] = C_{FA} - C''_{Me}$ where C'_{Me} is the quantity of metals which were not being in complex mol/l; $\sum_{Me}$ is the total quantity in main solution, mol/l, C''_{Me} is the quantity of metals connected with fulvic acids determined in the fractions $300 \leq$ Mw>5000, mol/l; [FA] – free ligand, mol/l; C_{FA}. The total quantity of fulvic acids in the system, mol/l.

The results of investigation fulvate complexes by the gel filtration method are given in Table 21.1.

21.4   CONCLUSIONS

The pure samples of fulvic acids were separated from the water of the river by the adsorption-chromotographic method. By the using of sephadex G-25 it was studied the complex formation process between copper and fulvic

TABLE 21.1 The Conditional Stability Constants Calculated on the Basis of the Results, Obtained by the Gel filtration Method $\beta = [MeFA]/\{[Me][FA]\} = [C\text{ "}_{Me}]/\{[C'_{Me}][FA^{2-}]\}$

pH	C'Cu	C"$_{Cu}$	[FA]	β
9.02	0.38×10^{-5}	1.65×10^{-5}	1.50×10^{-4}	2.89×10^4
8.00	0.33×10^{-5}	1.7×10^{-5}	1.50×10^{-4}	3.43×10^4
6.03	0.93×10^{-5}	1.1×10^{-5}	1.56×10^{-4}	7.18×10^3
5.01	1.48×10^{-5}	0.55×10^{-5}	1.62×10^{-4}	2.22×10^3

acids on various pH. It was studied the average meanings of stability constants of fulvate complexes of copper. It was established, that on pH = 9.02, $\beta = 2.89 \times 10^4$, on pH = 8.00 $\beta = 3.43 \times 10^4$, on pH = 6.03, $\beta = 7.18 \times 10^3$; on pH = 5.01, $\beta = 2.22 \times 10^3$.

ACKNOWLEDGMENTS

The financial support of Georgian Research and Development Foundation (#A60768) and Shota Rustaveli National Science Foundation (04/35) is gratefully acknowledged.

KEYWORDS

- average stability constant
- copper
- fulvic acids
- geopolymers
- molecular weight
- sephadex

REFERENCES

1. Raewvn, M., Town, R. M., Van Leeuwen, H. P., Buffle, J., (2012). Chemo dynamics of Soft Nanoparticulate Complexes: Cu(II) and Ni(II) Complexes with FA and Aquatic Humic Acids. *Environmental Science and Technology,* 46(19), 10487–10498.
2. Glaus, M. A., Hummel, W., Van Loon, L. R. (2000). Trace Metal-Humate Interaction. Experimental Determination of Conditional Stability Constants. *Applied Geochemistry,* 15(7), 953–973.

3. Rey-Castro, C., Mongin, S., Huidobro, C., David, C., Salvador, J., Garces, J., (2009). Effective Affinity Distribution for the Binding of Metal Ions to a Generic Fulvic Acid in Natural Waters. *Environmental Science and Technology,* 43(19), 7184–7191.

4. Claret, F., Schaver, T., Rabung, T., Wolf, M., Bauer, A., Buckau, G. (2005). Differences in properties and Cm(III) complexation behavior of isolated humic and fulvic acid derived from. Opalinus clay and Callovo-Oxfordian argillite. *Applied Geochemistry, 20,* 1158–1168.

5. Joris, W. J., Van, S., Dan, B. K., Jon, P. G. (2010). Acid-Base and Copper-Binding Properties of Three Organic Matter Fractions Isolated from a Forest Floor Soil Solution. *Geochimica at Cosmochimica Acta, 74,* 1391–1406.

6. Orsetti, S., Marco-Brown, J. L., Andrade, E. M., Molina, F. V. (2013). Pb(II) Binding to Humic Substances: An Equilibrium and Spectroscopic Study. *Environmental Science and Technology,* 47(15), 8325–8333.

7. Tinacher, R. M., Begg, J. D., Mason, H., Ranville, J., Powel, B. A., Wong J. C., Karsting, A. B., Zavarin, M., (2015). Effect of Fulvic Acid Surface Coatings on Plutonium Sorption and Desorption Kinetics on Goethite. *Environ. Sci. Technol., 49*(5), 2776–2785.

8. Mantoura, R. F. C., Dixon, A., Rilly, J. P. (1978). The Speciation of Trace Metals with Humic Compounds in Natural Waters. *Thalassia Jugoslavica, 14*(1–2), 127–145.

9. Qing, D., Zhong, X. S., Willis, F., Tang, H. X. (1999).Complexation in Illite-Fulvic Acid Cu^{2+} Systems. *Water Res., 33,* 693–706.

10. Makharadze, G., Makharadze, T., (2014). Method of Calculation of Stability Constants of Fulvic Complexes on the Example of Copper. *Journal of Chemistry and Chemical Engineering, USA, 8,* 108–111.

11. Gamble, D. S., Schnitzen, M., Hoffman, I. (1970). Cu^{2+}-Fulvic Acid Chelation Equilibrium in 0.1N KCl at 25°C. *Can. J. Chemistry, 48,* 3197–3204.

12. Schnitzer, M., Hansen, E. H. (1970). An Evaluation of Methods for the Determination of Stability Constants of Metal-Fulvic Acid Complexes. *Soil Science, 109,* 333–340.

13. Cheam, V., Gamble, D. S. (1974). Metal-Fulvic Acid Chelation Equilibrium in Aqueous NaNO$_3$ Solution Hg(II), Cd(II) and Cu(II) Fulvate Complexes. *Can. J. Soil Science, 54,* 413–417.

14. Shizuko, H., (1981). Stability Constants for the Complexes of Transition-Metal Ions with Fulvic and Humic Acids in Sediments Measured by Gel-Filtration. *Talanta, 28*(11), 809–815.

15. Schnitzer, M., Skinner, S. J. M. (1966). Organo-metallic interactions in soil. Stability constants of Cu^{2+}, Fe^{2+} and Zn^{2+} fulvic acid complexes. *Soil Science, 102,* 361–365.

16. Varshal, G. M., Kosheeva I. Y., Sirotkina, I. S., Velukhanova, T. K., Intskirveli, L. N., Zamokina, N. C. (1979). The Study of Organic Substances in Surface Waters and Their Interaction with Metal Ions. *Geochemistry* (Russia), *4,* 598–607.

17. Revia, R., Makharadze, G., (1999). Cloud-Point Preconcentration of Fulvic and Humicacids. *Talanta, 48*(2), 409–413.

CHAPTER 22

HgBa$_2$Ca$_2$Cu$_3$O$_Y$ SUPERCONDUCTOR PREPARED BY VAPOR DIFFUSION PROCESS

T. LOBZHANIDZE,[1] I. METSKHVARISHVILI,[2] G. DGEBUADZE,[2]
B. BENDELIANI,[2] M. METSKHVARISHVILI,[3] and V. GABUNIA[2]

[1]*Iv. Javakhishvili Tbilisi State University, Department of Chemistry, I. Chavchavadze Ave. 1, 0179, Tbilisi, Georgia*

[2]*Ilia Vekua Sukhumi Institute of Physics and Technology, Laboratory of Cryogenic Technique and Technologies, Mindeli Str. 7, 0186 Tbilisi, Georgia, E-mail: metskhv@yahoo.com*

[3]*Georgian Technical University, Department of Engineering Physics, Kostava Str. 77, 0175 Tbilisi, Georgia*

CONTENTS

ABSTRACT

Arsenic free and arsenic doped $HgBa_2Ca_2Cu_3As_xO_y$ superconductors were synthesized in sealed silica tubes by the solid-state reaction method. The Hg vapor pressure can be controlled by a small addition of As_2O_3 in the reactant bars. A series of experimental results obtained from the As-doped and As-free superconductors by ac susceptibility measurement techniques shows that a small addition of As_2O_3 increases the volume fraction of the (HgAs)-1223 superconducting phase and thermal stability of this superconducting phase.

22.1   INTRODUCTION

Among all of the known high-Tc superconductors the Hg-based cuprate superconductors $HgBa_2Ca_2Cu_3O_{8+\delta}$(Hg-1223) have the highest critical temperature Tc≈135 K, if these samples are prepared in normal conditions Tc≈135 K [1, 2] and Tc≈165 K under the external pressure (30 GPa) [3, 4]. This peculiar property makes the Hg-1223 phase as an invaluable material for its practical utilization.

Additives or substituent of the Hg-based superconductors by the appropriate metals or oxides can improve the nature of the grain boundaries, as well as promote the formation of the superconducting phase or introduce the effective pinning centers. Several scientific groups have studied the effect of the partial substitution of Hg by Sb ($Hg_{1-x}Sb_x$) [5, 6] in Hg-1223. They have founded that small substitution of Hg by Sb, improved the intergrain critical current density in Hg-1223. Besides, partial substitution of Hg by Pb($Hg_{1-x}Pb_x$) expedites the formation of nearly single-phase of Hg-1223 compound and enhances the irreversible magnetic fields [7–9]. Li et al. [10] have studied and optimized an influence of the synthesis and processing parameters on the phase purity and grain growth of Pb-doped $HgBa_2Ca_2Cu_3O_{8+\delta}$ cuprate superconductor. Goto [11] showed that in case of F-doped $HgBa_2Ca_2Cu_3Re_{0.2}O_y$ prepared by the diffusion method the maximal significance of critical current density J can reach $J_c≈10^4$ A/cm^2 at 77 K in zero fields.

Unlike the above-mentioned works, the aims of our studies are investigations of the effect of As_2O_3 addition on superconductive properties of Hg-1223 system. As far as we know, the effects of doping of arsenic oxide have not been studied yet for Hg-based polycrystalline superconductor.

22.2   EXPERIMENTAL PART

Arsenic free and arsenic doped HgBa$_2$Ca$_2$Cu$_3$As$_x$O$_y$ (x = 0.0–08 wt.%) were prepared by the two-step solid-state reaction method. In the first step highly pure powders of BaO (99% Oxford Chem Serve), CaCO$_3$ (99.9% Oxford Chem Serve) and CuO (98% Sigma-Aldrich) were mixed in the stoichiometric ratio Ba:Ca:Cu = 2:2:3 and then they were mixed and ground carefully in an agate mortar. The resulting powder mixture was calcined in an alumina crucible in air by using of the temperature programmed muffle furnace (KSL-1100X-S, MTI-corporation, temperature accuracy ±1°C) with four intermediate grindings at 900°C for 60 h.

The resulting powders were ground and pressed into the pellets by a hydraulic press (Holzmann-Maschinen, Type: WP10H) with about 500 MPa. The obtained pellet is annealed in home made a programmable tube type furnace (temperature accuracy ±1°C) at 900°C in flowing oxygen partial pressure of 0.3 bar for 15 h. In the second step the Ba–Ca–Cu–O precursor was mixed with HgO and As$_2$O$_3$ powders according to the composition HgBa$_2$Ca$_2$Cu$_3$As$_x$O$_{8+\delta}$ (the concentrations of As$_2$O$_3$ were in the range of x = 0.000, x = 0.004 and x = 0.008 identified as 0.0, 0.4, and 0.8 wt.%.) and After final grinding the powder was pressed into a disc-shaped pellet 6 mm in diameter, and 5 mm thick, by using a hydraulic press under a pressure of 400 MPa. The pellets were put in a quartz tube and from quartz tube was evacuated up to 10^{-3} Torr and sealed.

Thereafter, a quartz tube was inserted into a programmed muffle furnace. The temperature of the furnace was raised at a rate of 300°C/h up to 700°C and thereafter at a rate of 120°C/h up to 860°C and held at this temperature for 15 h. The furnace was cooled at the rate of 60°C/h to room temperature. Finally, the samples were oxygenated in tube type furnace at 300°C in flowing oxygen for 15 h.

The prepared patterns were characterized by X-ray diffraction (XRD, Dron-3M) with CuKα radiation. The phase method was used to study the real parts of the linear susceptibility [12]. The errors in the determination of χ' at higher frequencies than 1 kHz does not exceed 1%. For the measurements of intergranular critical current densities we used the method of high harmonics [13].

22.3   RESULTS AND DISCUSSION

An important step of the preparation of high pure phases HgBa$_2$Ca$_2$Cu$_3$O$_y$ superconductors is the sintering of the Ba$_2$Ca$_2$Cu$_3$O$_y$ multiphase precursor. To obtain the superconducting phase with optimal properties there are two, critical precursor parameters: the cation homogeneity and the oxygen content. The thermal annealing at the oxygen partial pressure provides homogeneity and stipulates an elimination of the carbonates, remained in the multiphase precursor sample [14]. The X-ray diffraction pattern of $Ba_2Ca_2Cu_3O_y$ precursor powder is plotted in Figure 22.1. As one can see, the sample consists only two phases of $BaCuO_2$ and Ca_2CuO_3. The existence of $BaCuO_2$ and Ca_2CuO_3 phases only is a good indicator for qualitative preparation precursor powder with Ba:Ca:Cu = 2:2:3 cation ration [15].

Figure 22.2 shows the temperature dependences of the real $(-4\pi\chi')$ part of ac susceptibility for un-doped and doped samples, measured on zero magnetic fields $(H = 0)$ at $h = 1$ Oe, $f = 20$ kHz. The diamagnetic onset temperature of the superconducting transition for 0.0 wt. %, 0.4 wt. % and 0.8 wt. % samples are $T\approx131$ K, $T\approx135$ K and $T\approx125$ K, respectively.

As we see un-doped sample clearly show a two-step decrease with T, which reflects the flux shielding from and among the grains. The high-temperature peak could be caused by the penetration of the field into the crystallites and be associated with intragranular critical current density, j_{cg}, while the low-temperature maximum is due to the penetration only into the Josephson medium formed by weak links between the crystallites. The latter

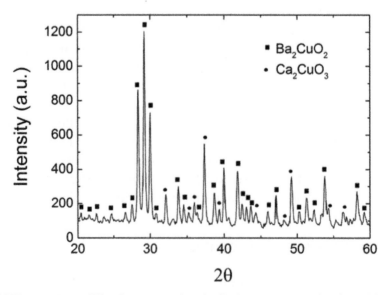

FIGURE 22.1   X-ray diffraction pattern of $Ba_2Ca_2Cu_3O_y$ precursor powder after calcinations at 900°C at flowing oxygen partial pressure of 0.3 bar for 15 h.

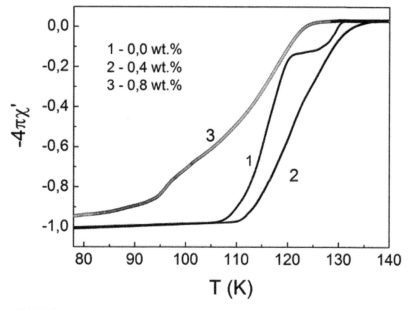

FIGURE 22.2   Temperature dependences of the real χ' parts of *ac* susceptibility.

is associated with intergranular critical current density j_{cJ}. Unlike the un-doped sample the Arsenic doped samples do not show two steps and full screening of applied *ac* magnetic fields for samples, whose As$_2$O$_3$ doping is in the range 0.4 wt.%, is observed at $T\sim110$ K.

By high harmonics methods we measured value of intergranular critical current densities [13] in liquid nitrogen temperature at $h = 1$ Oe and $H = 0$. We find for un-doped sample $j_{cJ}\approx100$ A/cm^2 and for As-doped samples: 0.4 wt.%–165 A/cm^2 and 0.8 wt.%–40 A/cm^2.

22.4 CONCLUSIONS

The effect of As$_2$O$_3$ doping on Hg-1223 system has been studied by the low field's *ac* susceptibility and the high harmonic response. Temperature dependences of χ' show a two-step process, for un-doped sample. It is important to note that the As-doped samples do not show two steps. The absence of two-step behavior could be explained in terms of smaller grain size in the As-doping samples, so that grains are fully penetrated at a lower field value. We found that in the low level doped sample, Arsenic enhances the value of the transport critical current densities from 110 A/cm^2 (0.0 wt.%) to 170 A/cm^2 (0.4 wt.%).

ACKNOWLEDGMENTS

This work has been fulfilled by financial support of the Shota Rustaveli National Science Foundation, Grant No. FR/423/6–260/12, contract number 31/36.

KEYWORDS

- ac susceptibility
- arsenic trioxide
- critical current
- Hg-based superconductors
- high harmonic
- solid-state reaction method
- vapor diffusion

REFERENCES

1. Schiling, A., Cantoni, M., Guo, J. D., Ott, H. R. (1993). Superconductivity above 130 K in the Hg–Ba–Ca–Cu–O system. *Nature, 363*, 56–58.
2. Antipov, E. V., Loureiro, S. M., Chaillout, C., Capponi, J. J., Bordet, P., Tholence, J. L., Putilin, S. N., Marezio, M., (1993). The synthesis and characterization of the HgBa$_2$Ca$_2$Cu$_3$O$_{8+\delta}$ and HgBa$_2$Ca$_3$Cu$_4$O$_{10+\delta}$ phases. *Physica C: Superconductivity, 215*, 1–10.
3. Chu, C. W., Gao, L., Chen, F., Huang, Z. J., Meng, R. L., & Xue, Y. Y. (1993). Superconductivity above 150 K in HgBa$_2$Ca$_2$Cu$_3$O$_{8+\delta}$ at high pressures. *Nature (London) 365*, 323–325.
4. Gao, L., Xue, Y. Y., Chen, F., Xiong, Q., Meng, R. L., Ramirez, D., Chu, C. W., Eggert, J. H., Mao, H. K. (1994). Superconductivity up to 164 K in HgBa$_{2Cam-1Cum}$O$_{2m+2+\delta}$ (m=1, 2, and 3) under quasihydrostatic pressures. *Phys. Rev. B 50,* 4260(R).
5. Li, J. Q., Lam, C. C., Hung, K. C., Shen, L. J. (1998). Enhancement of critical current density in HgBa$_2$Ca$_2$Cu$_3$O$_{8+\delta}$ superconductor doped by Sb. Physica C 304, 133–145.
6. Li, J. Q., Lam, C. C., Peacock, G. B., Hyatt, N. C., Gameson, I., Edwards, P. P., Shields, T. C., Abell, J. S. (2000). (HgSb)Ba$_2$Ca$_2$Cu$_3$O$_{8+\delta}$ thick films on YSZ substrates. *Supercond. Sci. Technol. 13*, 169–172.

7. Isawa, K., Yamamoto, A. T., Itoh, M., Adachi, S., Yamauchi, H., (1993). The effect of Pb doping in HgBa$_2$Ca$_2$Cu$_3$O$_{8+\delta}$ superconductor. *Physica C 217*, 11–15.
8. Isawa, K., Machi, T., Yamamoto, A. T., Adachi, S., Murakami, M., Yamauchi, H., (1994). Pb-doping effect on irreversibility fields of HgBa$_2$Ca$_2$Cu$_3$O$_{8+\delta}$ superconductors. *Appl. Phys. Lett. 65*, 2105–2107.
9. Shao, H. M., Lam, C. C., Fung, P. C. W., Wu, X. S., Du, J. H., Shen, G. J., Chow, J. C. L., Ho, S. L., Hung, K. C., Yao, X. X. (1995). The synthesis and characterization of HgBa$_2$Ca$_2$Cu$_3$O$_{8+\delta}$ superconductors with substitution of Hg by Pb. *Physica C 246*, 207–215.
10. Li, Y., Sastry, P. V. P. S. S., Knoll, D. C., Perterson, S. C., Schwartz, J., (1999). Synthesis of HgPb1223 superconductor. *IEEE Trans. Appl. Supercon. 9*, 1767–1770.
11. Goto, T., (1997). Partial melting of F-doped Hg$_x$Ba$_2$Ca$_2$Cu$_3$Re$_{0.2}$O$_y$ filament. *Physica C 282–287*, 891–892.
12. Metskhvarishvili, I. R., Dgebuadze, G. N., Bendeliani, B. G., Metskhvarishvili, M. R., Lobzhanidze, T. E., Mumladze, G. N. (2013). Low Field ac Susceptibility and High Harmonics Studies in PbMo6S8 Polycrystalline Superconductor. *J. Low Temp. Phys. 170*, 68–74.
13. Metskhvarishvili, I. R., Dgebuadze, G. N., Bendeliani, B. G., Metskhvarishvili, M. R., Lobzhanidze, T. E., Gugulashvili, L. T. (2015). Low-Field High-Harmonic Studies in Hg-1223 High-Temperature Polycrystalline Superconductor. *J. Supercond Nov Magn 28*, 1491–1494.
14. Sin, A., Fàbrega, L., Orlando, M. T. D., Cunha, A. G., Pinol, S., Bagio-Saitovich, E., Obradors, X., (1999). Improvement of superconducting (Hg, Re)-1223 ceramics synthesized by the sealed quartz tube technique. *Physica C 328*, 80–88.
15. Przybylski, K., Brylewski, T., Morawski, A., Lada, T., (2001). Ba$_2$Ca$_2$Cu$_3$O$_x$ precursor effects on synthesis of (Hg, Pb)Ba$_2$Ca$_2$Cu$_3$O$_{8+d}$ superconductor. *Journal of Thermal Analysis and Calorimetry 65*, 391–398.

CHAPTER 23

STUDY OF TEMPERATURE DEPENDENCE OF SPECIFIC RESISTANCE OF NICKEL-TUNGSTEN COATING ON COPPER

T. MARSAGISHVILI,[1] G. MAMNIASHVILI,[2] G. TATISHVILI,[1]
N. ANANIASHVILI,[1] M. GACHECHILADZE[1], and J. METREVELI[1]

[1]Ivane Javakhishvili Tbilisi State University, Rafiel Agladze Institute of Inorganic Chemistry and Electrochemistry, Mindeli Str., 11, 0186, Tbilisi, Georgia, E-mail: tamazmarsagishvili@gmail.com

[2]Ivane Javakhishvili Tbilisi State University, E. Andronikashvili Institute of physics, Tamarashvili Str., 6, 0162, Tbilisi, Georgia, E-mail: iph@iphac.ge

CONTENTS

ABSTRACT

Nickel-tungsten coatings on copper are obtained by electrodeposition method. Dependences of specific resistance on temperature for samples before and after thermal treatment are studied.

23.1   INTRODUCTION

Copper is widely used among metal structural materials in many industrial sectors. Copper is characterized by high thermo- and electro conductivity, but often during operation such properties are needed from copper, which this metal does not have. So, frequently copper must be protected by special galvanic coatings, which does not change appreciably mechanical characteristics of basic metal, but gives it additional functional signification.

In case, when during operation, besides high electro conductivity high thermo stability is needed for copper, coating of copper by nickel may be used (melting temperature of copper is 1085°C, specific resistance – 1.72×10^{-8} ohm·m; melting temperature of nickel is 1455°C, specific resistance – 6.84×10^{-8} ohm·m).

At the present time, one can see trend of supplanting of individual metals by their alloys, obtained by electrolytic method, which often have much higher characteristics in comparison with pure components and even metallurgical alloys.

Introduction of rare metals, particularly tungsten (melting temperature of which is 3422°C, and specific resistance is 5.5×10^{-8} ohm·m), in alloys of non-ferrous metals ensures obtaining of more heat-resistant materials having also high mechanical strength and other assets.

Tungsten is most refractory metal. It has also one of the highest density among metals. That endows tungsten with radiation-protective properties. High-melting and high density – these two basic properties determined its extraordinary importance in modern technologies and direction of its use.

Electro conductivity of tungsten can't be compared with electro conductivity of copper, but it is impossible to use copper contacts at high temperatures.

Electrolytic alloys are used in various fields of science and technology at present (tungsten alloys are widely used for fabrication of electrical contacts and components of production of defense industry), at that, range of their use permanently increases, what is explained by special physical and chemical properties of rare metals.

Tungsten does not precipitate from aqueous solutions. However, in presence of other metals, especially elements of subgroup of iron, tungsten precipitates together with them inconsiderable quantities.

23.2   EXPERIMENTAL PART

Study of electrical properties of nickel-tungsten alloy, obtained by electrochemical deposition on copper base before and after thermal treatment is proposed in the present work.

Pyrophosphate electrolyte [1], the main advantage of which is stability of composition and absence of complex former on the anode, was selected for deposition of nickel-tungsten alloy. Given electrolyte was prepared by definite methodology [2]. Copper plates of needed sizes, which were cut out on electro sparking machine tool with special accuracy were used as cathodes for obtaining of nickel coatings.

Standard four-point contact method was used for study of electrical properties of obtained samples. Specific resistance of substrate material and samples coated by nickel-tungsten was measured at different temperatures, before and after thermal treatment. The entity of burning was – impact thermal heating of the sample (during one minute) at the temperatures 600 and 950°C (in the atmosphere of helium). Afterwards the samples were cooled up to room temperature and measurement of temperature dependence of specific electrical resistance [$\rho = f(t)$] was carried out.

Content of tungsten and nickel in alloy was defined by roentgen-fluorescent spectral analysis method. Current yield was calculated by method adduced in Ref. [3].

23.3   RESULTS AND DISCUSSION

Obtained depositions contain 23–28% (weight) of tungsten, with current output 21–30%, thickness of coating is 12–15 mkm. Depositions are fine-crystalline, dense, without visible cracks.

Temperature dependences of specific resistances of copper [4], nickel [5] and tungsten [4] are adduced on Figure 23.1.

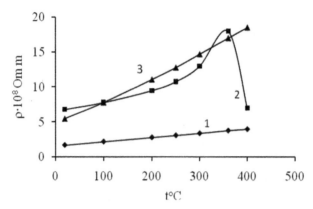

FIGURE 23.1 Changes of specific resistances of copper (1), nickel (2) and tungsten (3) at heating.

Specific resistance of nickel behaves rather unusually, at first it increases, but only up to temperature 358°C (Curie point), and then sharply decreases and at temperature 400°C becomes less than at room temperature [5].

One can see from Figure 23.2, that specific resistance of copper sample, coated by Ni-W at room temperature is slightly more than specific resistance of pure copper and at 130°C is equal to 6.5×10^{-8} ohm·m.

After heat treatment of the same sample this dependence extremely differs from previous one and is analogous to $\rho = f(t)$ of nickel. At heating from 20°C up to 130°C specific resistance permanently increases. At further

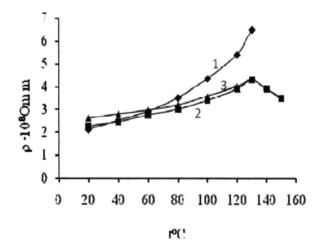

FIGURE 23.2 Temperature dependence for copper samples coated by Ni-W. (1) Without thermal treatment; (2) after burning at 600°C; (3) after burning at 950°C.

increase of temperature (up to 160°C) specific resistance decreases, probably because of relaxation of internal tensions, which promotes increase of electro conductivity.

23.4 CONCLUSIONS

Electro conductivity of copper sample coated by nickel-tungsten alloy is nearer to electro conductivity of copper. The obtained samples stand high temperature (previous to melt state ~1,000°C) and are characterized by good adhesion.

KEYWORDS

- **adhesion**
- **electrode position**
- **heat resistance**
- **nickel-tungsten coating**
- **specific resistance**
- **thermal treatment**

REFERENCES

1. Krasikov, A. V. (2012). Electrodeposition of nickel-tungsten alloy from pyrophosphate electrolyte. Abstract of a Thesis. St. Petersburg.
2. Kariakin, J. V., & Angelov, I. I. (1974). In: "Pure chemical substances" Moscow, Publishing house "Khimia" 408 p.
3. Flerov, V. N. (1987). Collected tasks of applied electrochemistry. Moscow, "High School," 293p.
4. Bogoroditski, N. P., Pasinkov, V. V., & Taraseev, B. M. (1985). Electro technical materials. "Energoatomizdat." Leningrad. 304p.
5. http://electrik.info/main/fakty/298-kak-izmeryaetsya-soprotivlenie-pri-nagreve-metallov.html.

CHAPTER 24

TOPOLOGY OF FORMATION OF LATEXES IN HETEROGENEOUS STATIC MONOMER-WATER SYSTEMS

A. A. HOVHANNISYAN,[1] M. KHADDAZH,[2] N. G. GRIGORYAN,[1] G. K. GRIGORYAN,[1] and O. A. ZHURAVLEVA[2]

[1]*The Scientific Technological Centre of Organic and Pharmaceutical Chemistry NAS RA, Institute of Organic Chemistry, 26, Azatutyan Str, 0014, Yerevan, Armenia, E-mail: hovarnos@gmail.com*

[2]*Peoples' Friendship University of Russia, Scientific-Educational Centre of Nanotechnology, Miklukho-Maklaya 10/2, 117198, Moscow, Russia*

CONTENTS

ABSTRACT

In the static model system styrene–aqueous solution of potassium persulfate was shown that the boundary layer of the monomer phase is one of the areas of the emulsion polymerization processes in which monomer phase is dispersed. It is proposed that the high rate of conversion of the monomer in the polymerization of monomers in emulsions is also associated with the polymerization reaction of the monomer in the boundary layer area.

24.1 INTRODUCTION

Emulsion polymerization (EP) has at least two particular properties which strongly distinguish it from other polymerization processes: (i) the high rate of the process, and (ii) a direct relationship between the rate and the degree of polymerization [1, 2]. A classical example is EP in micellar styrene-water emulsion where polymerization is initiated efficiently by a water-soluble initiator. In the initial stage of EP micelles begin to disappear [3]. At the same time monomer microdroplets begin to appear in the system. These microdroplets transfer to particles containing a polymer molecule and the growing radical. These droplets are then called polymer-monomer particles (PMP). After 5–10% of monomer conversion micellar emulsion is transformed to a dispersion consisting of two sets of particles of the monomer droplets and PMP, the diameter of which reaches 20–50 nm. After the disappearance of the micelles the quantity of PMP does not change, and their diameter will grow steadily to reach at 60% monomer conversion around 100 nm.

The fact of the disappearance of the micelles and the constancy of PMP after their disappearance became the basis for the creation of micellar model of EP [3], according to which the PMP originate in the aqueous phase when primary or oligomeric radicals falling into the micelles.

Other mechanisms of PMP initiation, according to which polymer molecules are formed in the aqueous phase as nuclei of the new phase, were also suggested by chemists [4–6]. Further, the polymer molecules associated and dissolving monomer converted into PMP. Such a homogeneous nucleation mechanism of particulate matter is considered to be more likely in the cases of EP of polar monomers (e.g., vinyl acetate). Disappearance of micelles in this case is explained by adsorption of molecules of emulsifier from the aqueous phase to the surface of dispersed particles. However, it is easy to show that the growth of the chain polymerization of styrene in the aqueous phase is up to a maximum of 3 or 4 monomer units. These surface-active oligomers with sulfate ion end groups are accumulated in the water phase, and that soon leads to the emergence of new micelles, rather than the disappearance of the existing ones.

The disappearance of the micelles is not justified thermodynamically in Harkins' micellar model of EP.

According to the theory of micelle formation micelles in water appear when the concentration of dissolved molecular emulsifier is what is called the critical micelle concentration. A necessary condition for the disappearance of the micelles is the creation of a new phase separation interface and adsorption of the emulsifier from the aqueous phase on that surface takes place.

Both micellar and homogeneous models do not take into account the heterogeneity of EP system and require a sophisticated phase boundary, which has bulk properties. Polymerization in this zone of the heterogeneous system may significantly influence both the kinetics of EP, and the process of dispersed particles forming.

According to the theory of heterogeneous systems, new surface interface can be created by elastic deformation of the interface, transfer a certain amount of substances from one phase to another and creating at the interface protuberances or depressions. New surface can also be obtained by dividing each of the phases into small particles. If both phases are liquid ones, the minimum unit of work to create a surface with all the above methods is the same – it is determined only by the temperature and the chemical potential of the contacting phases. It follows that if the released heat of polymerization is able to transfer a certain amount of the monomer from the interface to the volume of water, it can be assumed that the polymerization reaction can also deform the interface and fragment the system. Fragmentation of the system will inevitably lead to the disappearance of the micelles.

By means of numerical of calculations, it is shown that the heat of the polymerization reaction is sufficient for breaking heterogeneous system monomer – water [7]. In these calculations the value of the surface tension (γ) at the interface between the monomer-water is used. For the styrene – water system:

$$\gamma = 33 \ \text{mJ/m}^2 \tag{2}$$

The area occupied by the molecule at the interface is equal to 1/4 of the surface [8], therefore, the surface free energy per molecule of styrene is equal to

$$E = \gamma/4\pi r^2 \approx 4 \times 10^{-18} \, \text{mJ} = 4 \times 10^{-14} \, \text{erg}, \tag{3}$$

where r – radius of styrene molecules $\approx 2 \times 10^{-10}$ m.

For the styrene molecule to leave the interface surface and to diffuse into the aqueous phase an energy three times higher than 4×10^{-14} erg (its remaining portion is 3/4 the area associated with the styrene molecule adjacent molecules in the bulk monomer phase) is required. Radical polymerization evolves heat of reaction which is approximately equal to 20 kcal/mol, or about 1.4×10^{-12} erg per one molecule. Thus, in the emulsion at each occurrence of polymerization active center at the monomer – water interface, about 10 monomer molecules leave the surface and can get into saturated monomer aqueous phase. In water non-polar hydrocarbons molecules tend to the association and super saturation is necessary for the emergence of sustainable small droplets. Therefore, the polymerization reaction at the interface will certainly lead to the ingress of new monomer molecules in water and the formation of stable microdroplets.

Different mechanisms of fragmentation are possible. But in all cases, the work of fragmentation is spend on the polymerization reaction

In the following we demonstrate an experiment indicating that heterogeneous systems monomer – water polymerization reaction occurring at the monomer – water interface can disperse the monomer phase and form PMP.

24.2   EXPERIMENT AND DISCUSSION

On an aqueous solution of potassium persulfate in the tube we carefully layer styrene and then thermostat the system, after a certain time, we can watch the turbidity of the system, which subsequently converted into stable latex. Such a system is essentially a model of monomer droplets in the aqueous phase of the monomer – water emulsion.

Figure 24.1 shows a schematic latex formation in a static system styrene – water in the absence of an emulsifier [7–9].

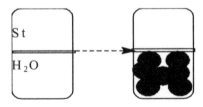

FIGURE 24.1   Schematic representation of the conversion of the monomer- water static system in the latex.

FIGURE 24.2 Topological picture of turbidity of the aqueous phase in the solution of styrene in ethanol – 0.4% aqueous solution of potassium persulfate.

To determine the nature of the system introduced in the disperse particles, we have paid attention to the fact that after the lamination the aqueous monomer phase, the diffusion process begins in the system, the density of the aqueous phase drops upwards. The diffusion rate of dispersed particles is much less than the rate of diffusion of the molecules. A density gradient along the aqueous phase and accumulation of particles can be localized in the zone of the aqueous phase below of which the medium density is higher than the density of dispersed particles. In this way, by measuring the density of the aqueous phase we can estimate the density of dispersed particles. The

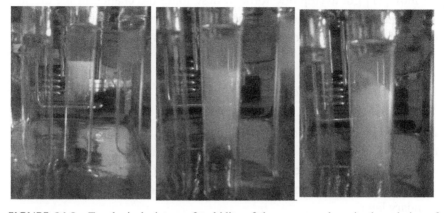

FIGURE 24.3 Topological picture of turbidity of the aqueous phase in the solution of styrene in ethanol – 0.4% aqueous solution of potassium persulfate in different time intervals of observation (1 – 30 min., 2 – 90 min., 3 – 120 min).

density gradient is created by introducing liquid alcohol (methanol, ethanol) to the monomer phase.

Polymerization was conducted in thermostatic tubes. Temperature of experiments was 60°C. Static monomer system – 0.4% aqueous solution of potassium persulfate was set by gently layering an aqueous monomer solution in ethanol. Volumes of monomer, ethanol, and aqueous phases were 2, 5, and 30 ml, respectively.

Topological picture of turbidity of the aqueous phase is shown in Figure 24.2.

Figure 24.3 clearly shows the dynamics of the process: the dispersed particles, created during the polymerization process are initially localized in a narrow zone of water phase near the monomer – water interface, then gradually sink deep into the aqueous phase.

At the time of turbidity, density of the upper, middle and lower zones of the aqueous phase were respectively 0.86, 0.92, 0.98 g/cm^3. Comparing densities in different zones of the aqueous phase with the values of the density of polystyrene (1.06 g/cm^3) we can notice that in the initial stages of formation the density of the dispersed particles is much less than the density of polymers, and therefore, they are the monomer microdroplets containing a certain amount of polymeric molecules.

In the boundary layer parameters of the monomer are substantially different from its bulk values [10, 11]. The thickness of the boundary layer can be determined by various physical parameters and it is obvious that it will depend on the selected parameter. Using the method of statistical mechanics, Rusanov [11] got asymptotic formula describing the change in the density of the liquid on the distance from the boundary surface and he used this formula to determine the thickness of the boundary layer (Δh)

$$\rho = \rho_0 + \frac{\pi \rho_0^2 x_0 \left(B' \rho' - B \rho_0 \right)}{6} * \frac{1}{\left(\Delta h \right)^3} \tag{1}$$

where ρ_0 and X_0 are, respectively, the density and isothermal compressibility of a liquid, B and B^1 are van der Waals interaction constants of liquid molecules with each other and with molecules of a neighboring phase, ρ' is the density of the neighboring phases, ρ is the density of the liquid in the boundary layer.

When $B\rho' > B\rho_0$ monomer at the interface – monomer density of water must be larger than its bulk density and polymerization of this layer should occur with a high speed. It is obvious that the kinetic effect is expected at a significantly small radius of particles, which is achieved in micro emulsions.

Almost all monomers are capable of being polymerized in an emulsion have a permanent dipole moment and in contact with water dipole–dipole interaction between the molecules of the monomer and water should lead to such an arrangement of molecules corresponding to a maximum of their interaction energy. Such an attraction between monomer and water molecules prevents thermal motion, but the tendency to orient the molecules of liquids at the interface is one of the qualitative differences between molecules in the above-mentioned case and those in bulk phases. It follows that in the polymerization of micro emulsions we can also find significant orientation effects.

24.3   CONCLUSIONS

The experimental results show that in a heterogeneous system monomer – water boundary layer monomer at the interfacial is a separate polymerization area and, under the influence of elementary acts of polymerization in this area monomer phase is dispersed.

Obviously, the presence of the surfactant in the water will flow dispersion irreversibly, leading to the formation of stable monomers of microdroplets.

It can be assumed that EP high speed is also associated with the polymerization reaction of the monomer in the boundary layers, because the density of the monomer in this area higher than its bulk value.

ACKNOWLEDGMENTS

The financial support of the National Academy of Sciences of Armenia is gratefully acknowledged.

KEYWORDS

- **dipole–dipole interaction**
- **microdroplets**
- **monomers**
- **thermostatic tubes**

REFERENCES

1. Bovey, F. A., Kolthoff, I., M, Medalia, A. I., Meehan, E. J. (1955). *Emulsion Polymerization*, New York, London, p. 165.
2. Peter, A., Lovell, El-Aasser, M. S., (1997). *Emulsion Polymerization and Emulsion Polymers,* ISBN: Hardcover, p. 218.
3. Harkins, W. D. (1950). General theory of mechanism of emulsion polymerization. *J. Polym. Sci., 5,* 217–251.
4. Hansen, F. K., Ugelstad, J., (1978). Particle nucleation in emulsion polymerization. I. Theory for homogeneous nucleation. *J. Polym. Sci. Polym. Chem. Ed. 16*(8), 1953–1979.
5. Fitch, R. M., Ross, B., Tsai, C. H. (1970). Homogeneous nucleation of polymer colloids: prediction of the absolute number of particles. *Amer. Chem. Soc. Polym. Prepr. II*(2), 807–810.
6. Fitch, R. M. (1980). Latex particle nucleation and growth. *Amer. Chem. Soc. Polym. Prepr. 21*(2), 286.
7. Oganesyan, A. A. (1986). Doctoral (Chem.) Dissertation, Moscow: *Inst. of Fine Chemical Technology* (In Russian).
8. Oganesyan, A. A., Gukasyan, A. V., Matsoyan, S. G., Boyajyan ,V. G., Gritskova, I. A., Pravednikov, A. N. (1985). *DAN USSR, 281*(5), 1145 (in Russian).
9. Oganesyan, A. A., Grigoryan, G. K., (2003). Effect of monomer nature on the mechanism of nucleation of emulsified latex particles in a two-phase monomer-water system., *Armenian Chemical, J., 56*(4), 121–124 (In Russian).
10. Good, R. J. (1957). Surface entropy and surface orientation of polar liquids. *J. Phys. Chem., 61,* 810–813
11. Rusanov, A. I. (1967). Phase Equilibrium and Surface Phenomena. Leningrad, Khimia, 387.

CHAPTER 25

SYNTHESIS AND RESEARCH OF SOME PHTALOCYANINIC DYES WITH CU, ZN FOR TESTING IN THE SYSTEMS OF SOLAR CELLS TYPE

S. ROBU,[1] G. DRAGALINA,[1] A. POPUSOI,[1] N. NASEDKINA,[2] A. COVALI,[2] and T. POTLOG[2]

[1]*Faculty of Chemistry and Chemical Technology, Moldova State University, 60 A. Mateevici Str., MD, 2009, Chisinau, Moldova, E-mail: s.v.robu@mail.ru*

[2]*Faculty of Physics and Engineering, Moldova State University, 60 A. Mateevici Str., MD, 2009, Chisinau, Moldova*

CONTENTS

ABSTRACT

In this chapter, copper (Cu) and zinc (Zn) phthalocianines was synthesized in nitrobenzene solution at a temperature of 200°C to 300°C. Firstly, Cu and

Zn salts-acetates were subject to dehydratation and then were purified by recrystallization from mixture of nitrobenzene and methanol. On the basis of the synthesized (Cu, Zn) Pc powders were obtained thin films by spin coating and thermal evaporation in high vacuum. The poliepoxipropil carbazole (PEPC) was used to improve the photoconductivity of the (Cu, Zn)Pc thin films. Annealing is a common method to improve the physical properties of thin films. We investigated the morphology and optical properties of (Cu, Zn) Pc thin films before and after thermal annealing using scanning electron microscope (SEM) and UV-Vis spectroscopy. Current–voltage characteristics demonstrated the formation of the ITO/CuPc hetero junctions.

25.1 INTRODUCTION

The actuality of these researches consists in the development of new methods of synthesis or optimization of described methods and their application in the synthesis of phthalocyaninic derivatives and carbazole containing polymers, to be used as the basis for producing luminescent and photoconductive materials. They are of concern for photonic, medicinal technique (x-ray films), optical fibers.

Phthalocyanines are of great interest as organic nanomaterials because of their exceptional properties [1, 2]. However, their application is limited due to their low solubility in organic solvents. The difficulties can be overtaking if in the aromatic ring are placed various derivatives, able to enhance the solubility of phthalocyanines and accordingly modify their properties [3–6].

25.2 EXPERIMENTAL PART

At each researches stage were applied various methods of organic synthesis, chromatography, modern methods of purifying of substances, optimal and economical ways of synthesis of copolimers, confirming the structure of the intermediates and final compounds, including elemental analysis, infrared spectroscopy, NMR spectroscopy, and X-ray analysis.

Research on synthesis of some derivatives of metal phthalocyanines showed that the introduction of electron or donor substitutes has a positive effect on the physical process of dissolution. For example, the introduction of hydroxyl or alkyl groups has allowed an increase in metallophtalocyanines solubility in organic solvents, but also ensuring their compatibility with certain polymers for the production of photosensitive materials.

For the purpose of developing new materials, photoconductive in the field of infrared and visible, we are synthesized metallophtalocyanines of transition metals Cu and Zn the compositions with different carbazolic polymers.

Synthesis of copper, cobalt and zinc phthalocianines (Figure 25.1) was achieved in nitrobenzene solution at a temperature of 200°C to 300°C. Firstly, Cu and Zn salts-acetates were subject to dehydratation. Phthalocianinic colors were purified by recrystallization from mixture of nitrobenzen and methanol.

From synthesized Cu, Zn phtalocyanines (Pc) powders were obtained CuPc and ZnPc thin films with thickness varied in the (2,0–10) μm interval. Photopolymer-layers were deposited by spin-coating method on glass substrates covered with a conductive layer of SnO_2 with resistance ~100 Ω. The layers were dried in the air, and then annealed in a vacuum at T ~ (60–80)°C. Simultaneous were obtained CuPc-thin films by thermal deposition in high vacuum with thicknesses of about 3–4 μm.

25.3 RESULTS AND DISCUSSIONS

The properties of metal phthalocyanines strongly depend on the structure of their molecules. As it is known, a change in the molecular structure as a result of the chemical modification of a peripheral part of the molecule leads to significant changes in the physic-chemical properties of compounds. It is known that poliepoxipropil carbazole (PEPC) is the most widely used compound among carbazole containing oligomeric photoconductors. (Cu, Zn)Pc organic compounds phthalocyanine presents a particular interest and they are perspective to be introducing in PEPC and to form a composite film

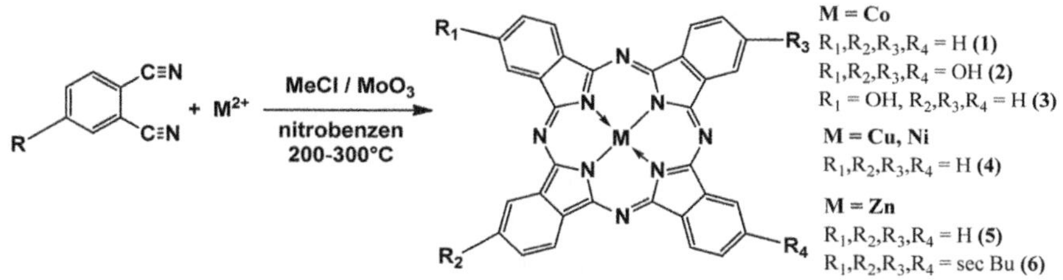

FIGURE 25.1   The structure of synthesized metallophtalocyanines.

with photosensitive properties due their structure with conjugated double bond [7] and to their strong absorption in ultraviolet and visible regions. Thin films of PEPC/(Cu, Zn) Pc nanocomposites were prepared from chemical solutions. CuPc and ZnPc thin films were obtained introducing PEPC and CuPc, ZnPc powders, respectively in proportion 1:1. The photoluminescence (PL) spectra of these nanocomposites were measured using the excitation with a nitrogen laser operating at 337 nm. All the PL measurements were performed at room temperature. As seen in Figure 25.2 the poliepoxipropil carbazol (PEPC) layer has photoluminescence peak at $\lambda = 442$ nm.

Spectra of photoluminescence of the nanocomposites in all cases show a well pronounced maximum which is considered to be attributed to the presence of the CH = CH double bond in the polymer and in CuPc and ZnPc compounds. Regretfully, the implementation of PEPC/CuPc and PEPC/ZnPc films obtained by spin coating method in fabrication of the photovoltaic devices was unsuccessfully. Therefore, we tried to obtain CuPc thin films by thermal evaporation in high vacuum. In our study an CuPc source and a substrate are separated by a 20 mm distance in a controlled atmosphere 5×10^{-6} Torr, and the source is maintained at a higher constant temperature at 670°C. The substrate temperature T_{sub} was fixed at ~50°C. The mean deposition time was approximately 6 minutes. The SEM micrographs of CuPc treated in vacuum at 90°C on glass/ITO substrate are shown in Figure 25.3. SEM image shows the branches of CuPc nanoribbons with different lengths. The UV-VIS optical transmission spectra results were analyzed in the strong absorption-edge region (Figure 25.4). Differences in the transmittance of the

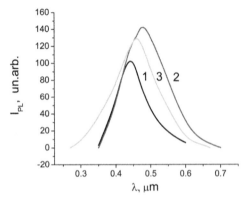

FIGURE 25.2 Photoluminescence spectra of (1) poliepoxipropil carbazol (PEPC) layer, (2) PEPC/ZnPc (1:1) and (3) PEPC/CuPc (1:1) layers.

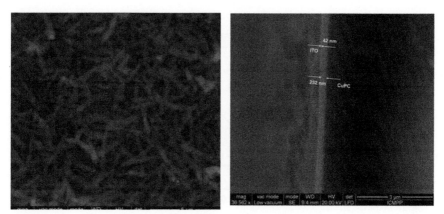

FIGURE 25.3 The morphology (left) and the cross section (right) of thermal treated CuPc film deposited on ITO/glass substrate by thermal evaporation in high vacuum.

films under examination can be attributed to differences in film thickness. CuPc thin films treated in vacuum show a typical absorption bands, namely the Q band in the visible region at 739.2 nm. The Q band can be attributed to the allowed highest occupied molecular orbital (HOMO)-lowest unoccupied molecular orbital (LUMO) (π–π^*) transitions.

A study on the photo response in the thin films was done by placing a monochromator between the lamp and the quartz window of the conductivity chamber. The photo response spectra of as-deposited and treated in vacuum at 150°C CuPc thin films are shown in Figure 25.5. Significant

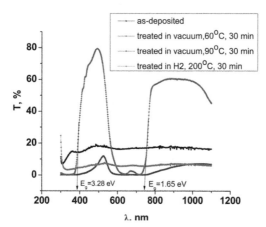

FIGURE 25.4 The transmittance of CuPc thin films deposited by thermal evaporation in high vacuum.

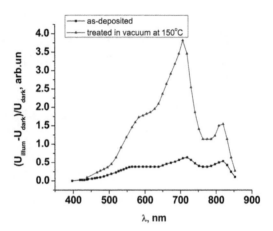

FIGURE 25.5 Photo response spectra of CuPc films deposited by thermal evaporation in high vacuum.

response is observed in the energy region, between 450 nm and 860 nm in both spectra, especially in the film treated in vacuum. The transition energies at the critical points which correspond to the maxima in the spectra are analyzed. The untreated CuPc thin film exhibits maxima at 1.51, 1.72, and 2.21 eV. When the film is subjected to thermal treatment in vacuum at 150°C the intensity of the photosensitivity increases and the position of the maxima located at 1.51, 1.72, and 2.21 eV shifted at 1.52 eV and 1.76 eV, respectively. The electron-hole pears are created. They drift under the influence of the applied field, giving rise to photo-current. The threshold of light absorption normally corresponds to exci-tation of electrons from valence band to conduction band which in turn corresponds to the maxima in the photo response spectrum. The photoac-tive transitions are assumed to take place within the π electron system of the phthalocyanine molecule. Since the CuPc thin film is p-type, their transport characteristics involve electron traps. So it is assumed that the net effect of photoconduction is to increase the population of holes in the valence band and each of the maxima corresponds to an energy level in the valence band [8].

The valence band is generally derived from the p-states of the constitu-ent elements, and the energy depends not only on the angular momentum, but also on the direction of the spin of the electrons. The interaction of the directed spin with the orbital angular momentum causes the valence bands to split [9].

The ITO/CuPc thin-film heterojunction photovoltaic devices were fabricated on glass substrates covered with ITO with transparency ~ 90% and a sheet resistance ~ 40 Ω. In this device the ITO acts as electron transport layer and CuPc as absorber. The Al back contact to the CuPc was deposited by the thermal evaporation. As ohmic contact to the ITO was used thermal evaporated indium. The dark and light current–voltage characteristics were measured by conventional methods. Photovoltaic characteristics were carried out on a solar simulator with the spectrum AM1.5. Figure 25.6 shows the dark and light current–voltage characteristics of the ITO/CuPc heterojunction solar cell at T = 300 K.

In heterojunctions, it is well known that, if the current transport is dominated by a thermal process, the current–voltage relationship takes the form:

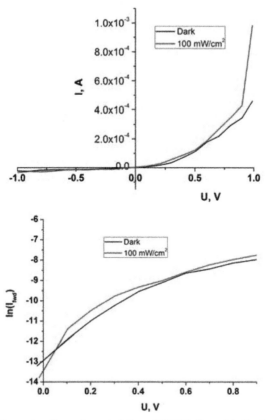

FIGURE 25.6 Current–voltage characteristics of ITO/CuPc and ln I = f(U) curves for this heterojunction solar cell.

TABLE 25.1   The Electrical Parameters of ITO/CuPc Heterojunction Solar Cell

Sample CuPc/ITO	K, rectification coefficient, (U = 1V)	I_o, µA	U_c, V	A	R_d, kΩ (U = 0.8V)
100 mW/cm²	27.59	1.52	0.379	1.95	1.26
Dark	19.83	2.41	0.165	3.91	1.55

$$I = I_0 (\exp \frac{qU}{AkT})$$

where I_o – the saturation current, A – the ideality factor. To estimate the saturation current I_o, the ideality factor A and the diffusion potential U_c for these characteristics was plotted $ln\ I$ versus U. I_o is obtained by extrapolating $ln\ I$ versus U curves to zero and the values of A and U_c are calculated from the slopes. The estimated parameters are illustrated in Table 25.1.

25.4   CONCLUSIONS

We have investigated the effect of the annealing on the optical properties and photoconductivity of (Zn, Cu) Pc thin films obtained from (Cu, Zn) Pc powders synthesized in the laboratory of the Moldova State University. An attempt to build the ITO/CuPc heterojunction solar cells from home (Cu, Zn) Pc powders synthesized was carried out. The obtained results can help us in future to design photoconductive or solar cells with a high efficiency of charge carrier generation by improving the photoelectrical properties of (Cu, Zn)Pc thin films. Therefore, studies are necessary for a clear decision about real photoconductivity of polymeric phthalocyanines.

KEYWORDS

- (Cu, Zn) phthalocyanines
- carbazolyl methyl acrylate
- copolymers
- photovoltaic devices
- thin films

REFERENCES

1. Gorduza, V. M., Tărăbăsanu, C., Athanasiuc, A., (2000). Unconventional applications of organic dyes, Ed. UNI-PRESS C-68 Bucharest, pp. 282–283.
2. Campbell, W. M., Jolley, K. W. (2007). Highly Efficient Porfirin sensitizers for due-sensitized solar cells, *Journal of Physical Chemistry Letters, 111*, 11760–11762.
3. Ali Erdoğmuş, Tebello Nyokong. (2002). New soluble methylendioxy-phenoxy-substituted zinc phthalocyanine derivatives: Synthesis, photo physical and photochemical studies, Department of Chemistry, Rhodes University, Grahams town 6140, South Africa, pp. 1–23.
4. Pulane, M. M., (2002). Synthesis of zinc ptalocyanine derivatives for possible use in photodynamic therapy. Thesis of Department of Chemistry Rhodes University, Grahams town, South Africa, pp. 1–131.
5. Rudolf Slot, Gabriela Dryad. (2012). Novel Lipophilic Lanthanide Bis-pthalocyanines Functionalized by Pentadecylphenoxy Groups: Synthesis, Characterization and UV-photo stability, Molecules, 17–10738–10753.
6. Imran Murtaza, (2011). Fabrication and Electrical Characterization of Organic semiconductor Phthalocyanine-Based, dissertation, Faculty of Engineering Sciences Ghulam Ishaq Khan Institute of Engineering Sciences and Technology, Pakistan, pp. 1–165.
7. Dementiev, s., Robu, S., Nasedchina, N. & Mitcov, D. (2010). The investigation of sensitization possibility of carbazole-containing polymeric photoconductors by transition metals phtalocyaninic dyes. 5th. International Conference on Materials Science and condensed Matter Physics, Moldova, p. 178.
8. Bodian, O., Verlan, V., Culeac, I., Iovu, M., Popusoi, A., & Dragalina, D., (2013). PPEPC/CoPc Nano composites, IC-NBME, Chisinău, Moldova.
9. Xavier, F. P., & Goldsmith, G. J. (1995). Role of metal phthalocyanine in redox complex conductivity. *Bull. Mater. Sci., 18*, 277.

CHAPTER 26

POLYMERIC RECORDING MEDIA BASED ON CARBAZOLE CONTAINING COPOLYMERS AND BENZOXYPHTHALOCYANINES OF TRANSITION METALS

S. ROBU,[1] M. ABADIE,[2] I. ANDRIES,[1] A. CHIRITA,[1] N. NASEDCHINA,[1] A. IVANCIC,[1] and O. CORSAC[1]

[1]Moldova State University, 60 A. Mateevici Str., MD, 2009, Chisinau, Moldova, E-mail: s.v.robu@mail.ru

[2]Université of Montpellier, Institute Charles Gerhardt of Montpellier – Aggregates, Interfaces & Materials for Energy (ICGM-AIME, UMR CNRS 5253), CC 1052, 34095 Montpellier Cedex 5, France

CONTENTS

ABSTRACT

In this work we study the problem of N-vinylcarbazole copolymers sensitization using classical sensitizers like 2,4,7-trinitrofluorenone and additionally metallo-phthalocyanine such as benzoxy phthalocyanine zinc, readily
soluble in organic solvents such as chloroform, toluene and others. Using
the composition of N-vinylcarbazole copolymer, 2,4,7-trinitrofluorenone
and benzoxyphthalocyanine copper were obtained photosensitive optical
transparent layers suitable for registration of holographic images in the visible and near-infrared spectrum. Studies of the developed photopolymers
photosensitivity have showed that the activation with 3–6 mass% of benzoxyphthalocyanine zinc does not reduce the photosensitivity in the range
$\lambda = 500$–700 nm, but shift the spectral photosensitivity curve to IR domain
850–900 nm. On these photopolymers coated with a thin layer (0.6–0.8 μm)
of thermoplastic were obtained high quality holographic gratings.

26.1   INTRODUCTION

Organic polymers are increasingly attractive alternatives to inorganic materials in optoelectronics [1, 2, 5–7]. They offer flexibility, low cost fabrication
and simplicity of application in various optical devices. Among the different
polymeric and co polymeric compounds the carbazole-containing polymers
enjoy a unique combination of properties well suited for optical applications such as electro photographic and holographic materials. The problem
of creation the new photosensitive media based on carbazole-containing
polymers, especially for registration of holographic information, are actual
and very necessary for solving of many technical (physics, photonics) and
economical problems, for example, protection of industrial and agrarian production from falsification.

   Wide distribution of carbazole-containing polymers in optical applications [1] is not only due to their technological, economical and ecological
properties. We consider more important being the appearance of large possibilities to manipulate their electrical and optical properties in a designed
manner. It can be achieved either by exploring large range of organic compounds or by exploring ways of doping them with small amounts of conductive metals. Finally, we have to note that photopolymers are the most
suitable materials for optical registration of information where materials
with high omic resistance are needed.

The recording functional materials predominantly are in form of thin-layer sets and functionally are of two kinds: multilayer compositions for photographic and holographic recording in presence of electric corona discharge and recording compositions of photo resistive materials. Separate thin layers can be of the following kind [1, 3, 4, 8]: optic transparent substrate layers (polyethylenetereftalat or glass); adhesive layers; metalized electrode layers (Cr, Ni, SnO_2); polymeric electro photoconductive, photo resistive and thermoplastic layers. The most important of them are electro photoconductive and thermoplastic layers for photo thermoplastic(PTP) recording process and photorefractive layer for photo structural transformations method. Materials for these layers can be synthesized from the following classes of organic carbazole-containing polymers: polymers with pendant carbazolyl groups, polymers containing electronically isolated carbazole moieties in the main chain, polymers with π-conjugated main chain, and σ-conjugated polymers.

In order to elaborate new photoresistive and electrophotographic materials with advanced properties in this chapter, several tercopolymers of N-vinylcarbazole (VC), 1-octene (OC-1) and n-octylmethacrylate (OMA) were synthesized and characterized. These compositions have good recording characteristics in visible and near-infrared region of spectrum. Studies have shown that activation them with a small addition of phthalocyanine dyes of metal (Zn, Cu, Co) does not reduce photosensitivity in visible range, but considerable shifts the spectral photosensitivity curve to IR domain. The detailed evaluation of this effect is given.

26.2 EXPERIMENTAL PART

In our search of electro photographic and holographic materials with advanced registering characteristics we have used binary copolymers of N-vinylcarbazole (N-VC) with 1-octene (OC-1) and ternary copolymers of N-VC:OC-1 and octyl methacrylate (OMA) with following structural formulas:

All copolymers were synthesized using the radical polymerization method in toluene. The polymerization was done in glass ampoule, sealed in inert argon atmosphere, at t ~ 80°C for 4–6 hours (Table 26.1) in the presence of 2 mol% initiator azo-bis-isobutyronitrile (AIBN). In the copolymers N-vinylcarbazole concentration ranged from 50 to 70 mol%.

TABLE 26.1   Composition and Some Characteristics of Synthesized Copolymers

Nr	Composition groups	Composition, mol %			Characteristic viscosity, Dl/g	Glass transition temperature, °C	Fluidity-temperature, °C
		N-VC	OC-1	OMA			
1	I	50	50	—	0.15	75–76	97–98
2	I	60	40	—	0.17	82–83	106–107
3	I	70	30	—	0.18	>90	—
4	II	50	30	20	0.14	72–73	92–93
5	II	60	20	20	0.15	78–79	96–97
6	II	70	20	10	0.19	86–87	>100

Obtained copolymers were purified by sedimentation in methanol and then dried in the vacuum oven (P~200 mmHg) at t ~ 40°C. For all copolymers was determined characteristic viscosity [η] and glass transition temperature (T_g). The composition and other characteristics of synthesized copolymers are presented in Table 26.1.

As we can see from the table the characteristic viscosity of copolymers ranges from 0.14 up to 0.19 Dl/g and the glass transition temperature from 72–73°C up to 90°C. For the development of bistratic carriers more accessible are copolymers of III group that have $T_g > 72°C$.

Sensitization of these polymers was done with 2,4,7-trinitrofluorenone (TNF). As an activating agent was used tetrabenzoxyphthalocyanine zinc

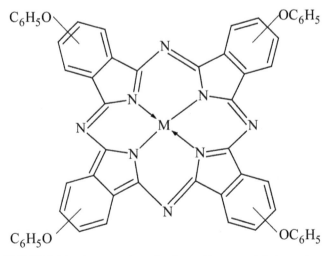

FIGURE 26.1   Structure formula of metal benzoxyphthalocyanines.

(TBO-Pc-Zn). For this purpose were synthesized a series of phthalocyanine dyes of zinc and cobalt benzoxyphthalocyanines with structure like that from the Figure 26.1

All synthesized dyes are well soluble in organic solvents: toluene, chloroform and others, which is why they can be used for obtaining nanocomposite and thin films (0,5–2,0 μm) from composite material with copolymers of N-vinylcarbazole and 1-octene and others.

From this copolymers prepared were solutions in chloroform of 10 mass% concentration with additions of 15% TNF and 0; 2.5; 5.0 and 10.0% of TBO-Pc-Zn. From the prepared solutions the photopolymer layers were deposited on flexible metalized support by the "meniscus" method. In turn, the optical glass support was coated with conductive layer through the "spin coating" method. The obtained layers thicknesses ranged ~ 2.0–2.5 μm. Semiconductor layers, also named as photopolymers (see Figure 26.2) were dried to the air and in the vacuum oven at t ~ 40°C, for 6–12 hours.

The recording layer (layer 4 in Figure 26.2) was made by the "spin coating" method from thermoplastic polymers. Recording layer thickness was 0.8–1.0 μm.

26.3   RESULTS AND DISCUSSIONS

The spectral photoconductivity of the photopolymer layers was investigated through an installation that allows variation of optical wavelength (λ) throughout the spectral range 400–800 nm. The results are shown in Figure 26.3.

As shown in Figure 26.3 all layers of the copolymer N-VC:OC-1 (II), sensitized with 12 mass% of TNF, possesses maximum photosensitivity in range $\lambda = 540$–600 nm (Figure 26.3, curve 1). In the some range maximum sensitivity show samples of copolymer II, additionally activated with 1.5–3.0 mass% phthalocyanine dye (curve 2). As shown in Figure 26.3, increasing

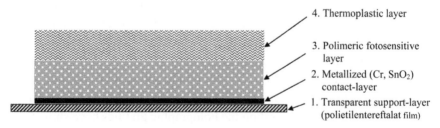

4. Thermoplastic layer

3. Polimeric fotosensitive layer

2. Metallized (Cr, SnO₂) contact-layer

1. Transparent support-layer (polietilentereftalat film)

FIGURE 26.2   A two-layer structure of the photo thermoplastic carrier.

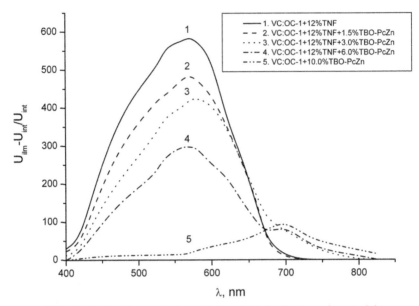

FIGURE 26.3   The spectral sensitivity of obtained polymeric materials.

the concentration of TB-Pc-Zn in semiconductor reduces the maximum value of spectral photosensitivity, but moves the curve $I = I(\lambda)$ to the infrared part of spectrum $\lambda = 800–850$ nm (curve 4). The most illustrative case we have in case of the copolymers sensitization only with phthalocyanine dye (Figure 26.3, curve 5). The maximum spectral photosensitivity evidenced in the range $\lambda = 680–720$ nm. Photosensitivity tail moves up to $\lambda = 850$ nm.

The appearance of photo sensitivity in the visible spectrum for these photopolymers when sensitizing them with 2,4,7-trinitrofluorenone related to the formation of charge-transfer complexes, was studied in detail by the authors [4, 9]. If layers are sensitizing only with phthalocyanine dyes, complex formation occurs too, which is confirmed by "Vis" spectra by appearance of a new absorption band at $\lambda \sim 700$ nm (Figure 26.4).

Recording of holographic data on the photo thermoplastic double-layer carrier from copolymers N-VC:OC-1 or N-VC:OC-1:OMA was done using equipment presented in the Figure 26.5.

Recording of holographic gratings on the PTP carrier with BMA-50 or PEPC recording layer was done via green light of laser $\lambda = 532$ nm at $t \sim 85–92°C$ and corona potential 9.0–9.5 kV. Charging time varied from 2 up to 5 seconds. Results of the recording the holographic gratings have shown that carriers photosensitivity was ranged $10^{-4}–10^{-5}$ J/cm^2; diffraction efficiency

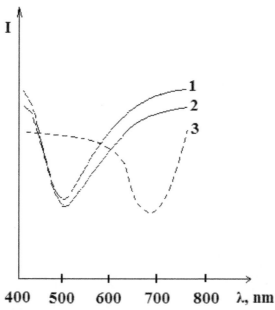

FIGURE 26.4 "Vis" transmission spectrum for solutions: 1 – copolymer II sensitized only with TNF; 2 – copolymer II sensitized with TNF+3 mas% TB-Pc-Zn; 3 – copolymer II sensitized only with TB-Pc-Zn.

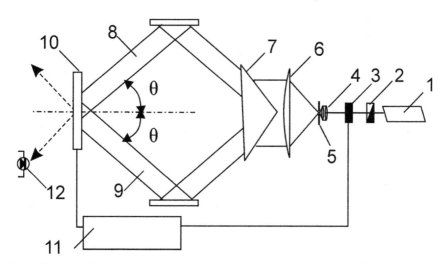

FIGURE 26.5 Optical scheme for electro photographic data recording: (1) laser ($\lambda = 532$ nm), (2) a set of neutral filters, (3) electro-optical shutter, (4) micro objective, (5) micro-aperture, (6) collimator, (7) prism, (8–9) plane-parallel laser beams, (10) PTP carrier, (11) PTP process control unit, and (12) photo-detector.

up to 8–10% and resolution $R \sim 2000$ mm^{-1}. Grooving deformation structure registered by the AFM method is shown in Figure 26.6. At the level of AFM resolution it can be seen that deformation structure has a pronounced trench character. This makes the optical material especially suitable for recording of the ruled objects, such as the holographic diffraction gratings.

Recording of holographic gratings was done also at $\lambda = 633$ nm. On the photo thermoplastic carriers of information with photosensitized layers of prepared compositions in the red light of He-Ne laser at $\lambda = 633$ nm, registered were holographic gratings with diffraction efficiency up to 10%, with resolution up to 1000 mm^{-1} and with low noise.

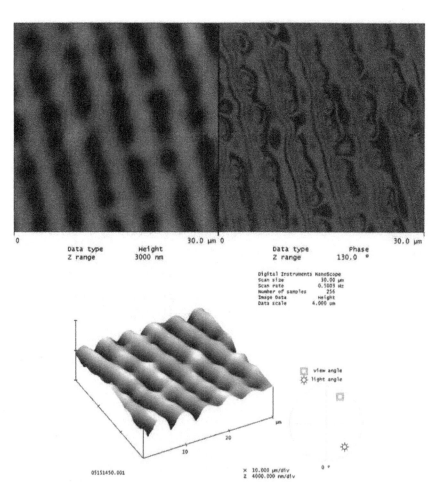

FIGURE 26.6   The thermoplastic surface deformation structure registered by the AFM.

26.4 CONCLUSIONS

1. Were synthesized and analyzed the binary and ternary compounds of N-vinylcarbazole copolymers with 1-octene and octylmethacrilate. The glass transition temperature of copolymers containing from 50 to 70 mol% of N-VC is above 70°C. Semiconductor layers obtained from these copolymers sensitized with different donor-acceptor additives are transparent in the spectral region 400–700 nm.

2. Photoconductive layers of the copolymers sensitized with 12 mass% of 2,4,7-trinitrofluorenone, possesses photosensitivity in large spectral range 400–700 nm with the maximum values in range $\lambda = 540$–600 nm.

3. It was shown that additional sensitization of photo polymeric layers with 10–50 mass% of copper tetrabenzoxyphthalocyanine leads to appearance of spectral photosensitivity in the range 400–800 nm.

4. Developed was photo thermoplastic belayed optical carrier from TNF and TB-Pc-Zn sensitized photopolymers, that allow holographic data recording through the laser red light. The microstructure of the surface layer deformation under the action of light in the FTP method of recording has a pronounced trench character, which is particularly important when recording holographic gratings.

ACKNOWLEDGMENTS

This research was partially supported by the Science and Technology Center in Ukraine (STCU Project #. 5808).

KEYWORDS

- bistratic photo thermoplastic carriers
- carbazole-containing polymers
- electro photographic and holographic materials
- holographic gratings
- photosensitivity
- phthalocyanines

REFERENCES

1. Kuvshinskij, N.G., Davidenko, N.A., KomKo, V.M. (1994). Physics of amorphous molecular semiconductors. Kiev, "Radio," p. 176.
2. Bodian, O., Verlan, V., Culeac, I., Iovu, M., Popusoi, A., Dragalina, G., (2013). PEPC/CoPc Nanocomposites: IC-NBME, Chisinau, Moldova.
3. Gainor, J., Afterguts, S., (1983). Photo thermoplastic method of image recording. *Photo. Sci. Eng. 37*(4), 200–213.
4. Negative Photographic Processes. Ed. A.l. Kortuzhanskogo. Leningrad, "Chemistry"1984, p. 376.
5. Erdogmus, A., Nyikong, T., (2002). New soluble methylendioxy-phenoxy-substituted zinc phthalocyanine derivatives: Synthesis, photo physical and photochemical studies, Department of Chemistry, Rhodes University, Grahams town 6140, South Africa. 1–23.
6. Murtaza, I., (2011). Fabrication and Electrical Characterization of Organic semiconductor Phthalocyanine-Based, dissertation, Faculty of Engineering Sciences Ghulam Ishaq Khan Institute of Engineering Sciences and Technology, Pakistan. 1–165.
7. Dementiev, I., Robu, S., Nasedchina, N., Mitcov, D., (2010). The investigation of sensitization possibility of carbazole-containing polymeric photoconductors by transition metals phtalocianine dyes. 5th International Conference on Materials Science and Condensed Matter Physics, Moldova. p. 178.
8. Avilov, G.V. (1985). Organic electrophotographic film – Moscow: Artp.125.
9. Vannikov, A.V., Grishin, A.D., (1984). Photochemistry polymer donor-acceptor complexes. Moscow: Sciences p. 261.

CHAPTER 27

BUCKMINSTERFULLERENE-PYRROLIDINES AS PROMISING ANTIOXIDANTS IN POLYMER MATERIALS

ELDAR B. ZEYNALOV and MATANAT Ya. MAGERRAMOVA

Nagiyev Institute of Catalysis and Inorganic Chemistry, Azerbaijan National Academy of Sciences, H. Javid Ave., 113, AZ1143 Baku, Azerbaijan, E-mail: zeynalov_2000@yahoo.com

CONTENTS

ABSTRACT

The intrinsic efficiency of fullerenes can be significantly activated by means of a connection with hydrogen donating groups of antioxidants such as phenol or secondary amine The developed system of conjugated s–p bonds arranged in the fullerene molecule in a closed shape can promote a strong resonance effect on the grafted units and increase the hydrogen atom abstraction efficiency. In this case the known ability of fullerene to trap alkyl radicals might be combined with an additional antioxidant functionality to afford a new class of antioxidants with bimodal action. Explorations in the field of fullerene derivatives on their antioxidant performance provide novel information on the potential stabilization properties of this type of molecular structure.

This chapter describes an anti-oxidative influence of fullerene-alkyl pyrrolidines on the model hydrocarbon and polymer oxidation environment. The antioxidant activity of the investigated derivatives was studied by measuring the inhibition rate constants for their reaction with alkyl and peroxy radicals in a model cumene initiated (2,2'-azobisisobutyronitrile, AIBN) oxidation experiments and compared to that recorded under identical experiments for buckminsterfullerene itself and commercial stabilizers.

The results indicate that linking the alkyl pyrrolidine groups directly to the fullerene core gives rise to an additional antioxidative power to the buckminsterfullerene: – they heighten the inherent rate constant of buckminsterfullerene for scavenging alkyl radicals due to the additional antioxidant contribution promoted by the radical-quenching ability of the formed nitroxyl intermediates.

This novel C60-amine conjugates may be considered as promising molecules for broad-spectrum radical scavenging antioxidants to use purposely in polymer materials, in particular in low-density polyethylene (LDPE) composite formulations.

27.1   INTRODUCTION

Pristine fullerenes C_{60} and C_{70} are now widely known as effective scavengers of carbon-centered radicals [1–10]. They are recommended in practical end-use applications as inhibitors of chain radical polymerization and thermal stabilizers for polymer materials [2, 5, 6, 11–27]. However, along with this advantage there is certain limit to the use fullerenes as antioxidants because of their insusceptibility towards proxy radicals [2, 28, 29]. In order to gain this important ability, essential for effective antioxidative functioning, fullerenes may be further modified, especially since the fullerene molecule, is essentially a fertile framework to obtain some derived molecular design with high antioxidative capacity.

So far there are only a few publications in the literature regarding the antioxidative efficiency of such fullerene derivatives. Notably this concerns the polyhydroxylated fullerene $C_{60}(OH)_{24}$ in biological systems [30–33]. The fullerenol has shown excellent efficiency to scavenging stable 2, 2'-diphenyl-1-picrylhydrazyl, reactive hydroxyl (OH)-O·, nitric oxide NO· radicals, superoxide anion-radicals O_2^-· and other reactive oxygen species [34–39].

The hydrogen atom donation was proved by ESR detection of the fullerenol radical $C_{60}(OH)_{23}$. In addition it was established that the interaction between a hydroxyl radical and fullerenol is also based on a radical-addition reaction of 2n (·OH) (n = 1–12) radicals to the remaining olefinic bonds of a fullerenol core.

Fullerene derivatives incorporating one or two 3,5-di-tert-butyl-4-hydroxyphenyl (BHT) units were synthesized and investigated for their antioxidative activity in controlled auto oxidation experiments. The results showed that grafting of the BHT structure does not qualitatively alter the thermo chemistry and kinetics of its reaction with peroxy radicals but the adducts exhibit an interesting dual mode of antioxidative action [28, 40].

Oil-soluble amine derivatives of buckminsterfullerene showed more antioxidative properties than zinc dialkyl dithiophosphate as described in Ref. [41].

The antioxidant assay of fullerene substituted phenyl aniline was determined in DMSO/PBS buffer to be significantly more potent than the alpha-tocopherol [42].

Thus, the field of antioxidative properties of fullerene derivatives may be envisioned as a new area with a promising future outlook.

In this work we report the results of the determination of inhibition rate constants of fullerene C_{60} amine derivatives by means of a model reaction of cumene-initiated oxidation. This model oxidation has repeatedly demonstrated high resolving power on the kinetic analysis of both chain-breaking antioxidants and radical scavengers. Nowadays the model reaction serves as one of the best tools for the preliminary quantification of antioxidant efficiency and allows basically the transfer of such knowledge to a polymer system. Next step of the investigation is devoted to a behavior of the (fullero)pyrrolidines in low-density polyethylene (LDPE) composite materials

27.2   THEORETICAL PREREQUISITES FOR THE MODEL OXIDATION

In our previous work we have clearly shown that the model oxidation relates to the onward assays which is effectively used for the determination of the kinetic parameters of both chain breaking antioxidants acting as acceptors of alkyl $R^{\cdot}$ and/or peroxy $RO_2^{\cdot}$ radicals [1, 43–48]. To make the determination of kinetic parameters the model reaction of cumene initiated (initiator is 2,2'-azo-bisisobutyronitrile (AIBN)) oxidation was designed to proceed under steady state conditions with long kinetic chains at moderate temperatures (40–80°C) where cumyl hydro peroxide does not contribute additionally to the initiation of oxidation and air oxygen pressure is sufficient not to limit the process [$Po_2 = 20$ kPa ($\approx 10^{-3}$ mol O_2 l^{-1})] [49, 50].

The scheme of oxidation represented for the chain breaking antioxidants [fullerene C_{60}, hindered amine stabilizer based on 2,2,'6,6'-tetramethylpiperidine (HALS)] has been shown to proceed according to the following mechanism:

Scheme 1. General scheme of the model cumene initiated oxidation in the presence of chain-breaking antioxidants (the generally accepted oxidation stage numbering is used:

Chain initiation: $AIBN \rightarrow r^{\cdot}$ $(rO_2) + RH \rightarrow R^{\cdot}$ (initiation rate is W_i)

$$R^{\cdot} + C_{60} \rightarrow {}^{\cdot}C_{60} R \text{ (rate constant } k_1) \tag{1}$$

$$R_{1\_}NH \rightarrow R_{1\_}NO^{\cdot} + R^{\cdot} \rightarrow R_{1\_}NOR + \text{inert products (rate constant } k_{1'}) \tag{1'}$$

Chain propagation:

$$R^{\cdot} + O_2 \rightarrow RO_2^{\cdot} + RH \rightarrow ROOH + R^{\cdot} \text{ (rate constant } k_2/k_3) \tag{2)–(3}$$

$$\textit{Chain termination:} \; R^{\cdot} + R^{\cdot} \rightarrow R - R \tag{4}$$

$$R^{\cdot} + RO_2^{\cdot} \rightarrow ROOR \tag{5}$$

$$2\,RO_2^{\cdot} \rightarrow \text{inactive products (rate constant } k_6) \tag{6}$$

where, RH: cumene, $R^{\cdot}$: cumyl alkyl radical, $RO_2^{\cdot}$: cumyl peroxy radical, ROOH: cumyl hydro peroxide), C_{60}: buckminsterfullerene, ${}^{\cdot}C_{60}$ R: radical adduct of buckminsterfullerene; R_1NH: hindered amine stabilizer (HALS), R_1-$NO^{\cdot}$: nitroxyl radical.

A system of derived kinetic expressions, fitting this scheme are as follows:

$$Wo_{2\,(C60)} = k_3 [RO_2^{\cdot}] [RH] = W_{i\,(1)}^{1/2} k_3 k_{6-}^{1/2}[RH] \tag{7}$$

$$Wo_{2 \, (HALS)} = k_3[RO_2\cdot] \, [RH] = W_{i \, (1')}{}^{1/2} \, k_3 \, k_6{}^{1/2} \, [RH] \qquad (8)$$

where $Wo_{2 \, (C60)}$, $Wo_{2(HALS)}$ – rates of oxidation in the presence buckminster-fullerene and sterically hindered amine antioxidants, respectively; $W_{i \, (1)}$, $W_{i \, (1')}$ – initiation rates promoted by AIBN in the presence of buckminsterfuller-ene and sterically hindered amine antioxidants, respectively; k_3, k_6 – rate constants of the chain propagation and termination, respectively.

$$W_{i \, (1)} = W_{i \, (AIBN)} - W_{(C60)} \qquad (9)$$

$$W_{(C60)} = k_{(1)}[R\cdot] \, [C_{60}] \qquad (10)$$

where /10/ is the rate of interaction between fullerene and alkyl radical.

$$W_{i \, (1')} = W_{i \, (AIBN)} - W_{(HALS)} \qquad (11)$$

$$W_{(HALS)} = k_{(1')} \, [R\cdot][HALS(-NO\cdot)] \qquad (12)$$

where /12/ is the rate of interaction between HALS and alkyl radical.

27.3   EXPERIMENTAL PART

27.3.1   AMINE DERIVATIVES OF BUCKMINSTERFULLERENE

The compounds used in this work were synthesized in the laboratory envi-ronment. The procedure of synthesis and their chemical structures are given in Table 27.1. The amine derivatives can be conditionally subdivided into two types: C_{60}-AmAO-1 and C_{60}-AmAO-2. And they are fullerene deriva-tives with grafted sterically hindered pyrrolidine fragments.

27.3.2   CHARACTERIZATION OF C_{60}-AMAO-1 AND C_{60}-AMAO-2 COMPOUNDS [51]

(a) C_{60}-AmAO-1: C_{60}: glycine: benzaldehyde
The infrared spectra shows a single peak around 3300 cm^{-1} which indicates the presence of a secondary amine group, with C-H stretching vibrations for the aromatic and aliphatic (–CH and –CH$_2$) parts of the pyrrolidine ring

TABLE 27.1   Features of Amine Derivatives of Buckminsterfullerene Used in the Work

Symbolic notation/ chemical name	Synthesis procedure	Chemical structures (molecular weight, MW)
1. C_{60}-AmAO-1 Fullerene-phenyl-pyrrolidine	C_{60}glycine: p-benzaldehyde = 0.07:0.14:0.35(mmol), under N_2, refluxed in toluene, purified in column chromatography	 MW = 879
2. C_{60}-AmAO-2 Fullerene-diphenyl pyrrolidine	C_{60}:phenylglycine:benzaldehyde = 0.07:0.14: 0.35(mmol), under N_2, refluxed in toluene, purified in column chromatographyw	 MW = 941

seen at 3020, 2921 and 2850 cm^{-1}, respectively. The skeleton absorptions of fullerene [60] have changed, however the peak at 527 cm^{-1} is clearly present showing that the fullerene cage is intact and that no additional groups have bonded to the cage and a mono adduct has been synthesized.

The UV spectrum of the derivative in cyclohexane shows a single descending curve with a small inflection near 260 nm, characteristic of covalent derivatives of C_{60}. There is also a shoulder present at around 420 nm due to the fullerene being present as an adduct rather than pristine [60] fullerene.

The ^{13}C NMR spectrum exhibits a set of well resolved lines in the 140–150 ppm region typical of a mono adduct, where the initial fullerene has a narrow peak at 143 ppm and poly adducts exhibiting a broad maximum at 140–150 ppm. The signals at 63–72 ppm, characteristic of sp^3 atoms of the modified fullerene, appear upon addition of amines at double bonds of the fullerene. Signals at 29 and 30 ppm are attributed to the carbon atoms of the CH$_3$ and CH$_2$ groups in the piperidine ring.

Analysis by MALDI mass spectrometry gave rise to a peak at m/z 839 which corresponds to the molecular ion with a peak at m/z 720 which is the M+ for [60] fullerene.

(b) C_{60}-AmAO-2: C_{60}: phenylglycine: benzaldehyde

Infrared via KBr disc, showed a single prominent peak at 3320 cm^{-1} indicating the presence of a secondary amine, C-H stretching vibrations attributed to the aromatic and aliphatic (-CH) portions of the pyrrolidine ring can be seen at 3028, 2901 and 2850 cm^{-1}, respectively. As expected for the presence of a mono adduct, the skeleton absorptions which are seen for pristine [60] fullerene at 1428, 1183, 577 and 527 cm^{-1}, have changed, with the only peak remaining unchanged being observed at 527 cm^{-1}, showing that the fullerene cage is intact, no additional groups have bonded to the cage, and a mono adduct has been synthesized.

Comparisons between the UV-Vis spectra of pristine [60] fullerene and 2,5-diphenylpyrrolidino [60] fullerene shows almost identical spectra's, with 2,5-diphenylpyrrolidino [60] fullerene exhibiting an additional characteristic peak around 430 nm which is attributed to the presence of a 6,6-closed mono-adduct.

The ^{13}C NMR spectrum exhibits a set of well resolved lines in the 140–150 ppm region typical of a mono adduct, where the initial fullerene has a narrow peak at 143 ppm and poly adducts exhibiting a broad maximum at 140–150 ppm. The signals at 63–75 ppm, characteristic of sp^3 atoms of the modified fullerene, appear upon addition of amines to the double bonds of fullerene. Signals at 21, and 27 ppm are attributed to the carbon atoms of the CH$_3$ and CH$_2$ groups in the piperidine ring.

Analysis by MALDI mass spectrometry gave rise to a peak at m/z 915 which corresponds to the molecular ion with a peak at m/z 720 which is the M+ for [60] fullerene.

27.3.3   MODEL REACTION OF CUMENE OXIDATION

To study the chain-breaking antioxidants the model cumene oxidation was undertaken at initiation rates: $W_i = 1.7 \times 10^{-8} – 6.8 \times 10^{-8} Ms^{-1}$, temperatures: 60 and 80 ($\pm$ 0.02)°C and oxygen pressure: $Po_2 = 20$ kPa (air). The employed cumene was 98% purity ("Aldrich").

2,2'-azobisisobutyronitrile (AIBN) was used as the initiator. The volume of the reaction mixture was 10 cm^3 (25°C). To achieve the assigned initiation rates 0.72–10 mg (at 60–80°C) of AIBN had to be added [49, 50, 52].

The rate constants of chain propagation and termination for the cumene oxidation at 60 and 80°C have the following values: $k_3 = 1.75$ and 4.05; $k_6 = 1.84 \times 10^{-5}$ and 3.08×10^{-5} $M^{-1}s^{-1}$, respectively; concentration of cumene [RH] $= 6.9$ (60°C) and 6.76 (80°C) mol l^{-1}. The rates of oxidation were evaluated by considering the amount of oxygen consumed, which was measured volumetrically with the simple equipment as described in [43, 50]. Oxidation rates were assessed both from slopes of the kinetic curves of oxygen consumption in the case of steady rate values and also by means of differentiating the curve in the case of an observed induction period.

Induction periods were graphically evaluated from kinetic curves [49, 50]. From the experimentally observed values of oxidation rates, using the known rate constants for cumene oxidation it is possible to determine inhibition rate constants k_1 (or $k_{1'}$) for the trapping cumylalkyl radicals by the fullerene or HALS type compounds:

$$k_1[R\cdot][C_{60}] \{k_{1'}[R\cdot][HALS]\} =$$
$$W_{i\,(AIBN)} - [Wo_{2\,(C60)}]^2 \{Wo_{2\,(HALS)}\} k_6 (k_3)^{-2} [RH]^{-2} \qquad (7)$$

Experiments were carried out at least in triplicate and the correctness in determining the kinetic parameters was within the range 1–10%.

Commercial LDPE (0.9185 g/cm³) was employed as polymer matrix for preparation of composites. The polyolefine is large-tonnage products and the problem of its stabilization is of a great practical importance.

The LDPE and the fullerene derivatives compounding was made on laboratory rollers at temperatures 150–160°C for 5–7 minutes.

The dried polymer samples with antioxidants have been studied by differential scanning calorimetry (DSC) and thermogravimetry (TGA).

The thermic analysis of composites was conducted in a dynamic mode on derivatograph Q-1500D of "Paulik-Paulik-Erdei" system, amount of a sample = 100 mg, rate of heating was 5°C/min., the reference sample was α-Al₂O₃.

27.4 RESULTS AND DISCUSSION

27.4.1 KINETIC STUDY OF C₆₀-AMAO-1 AND C₆₀-AMAO-2

The compound structures are composed of a directly grafted buckminster-fullerene with pyrrolidine fragments having sterically hindered bulky phenyl

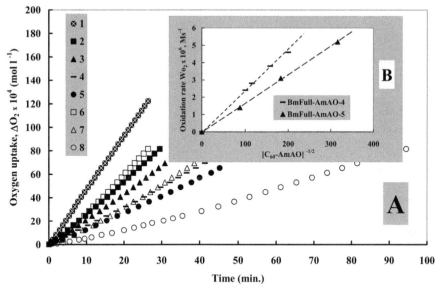

FIGURE 27.1   Kinetic dependences of oxygen oxygen-uptake (A) for initiated oxidation of cumene in the absence (1) and presence of different amounts of amine derivatives of fullerene C_{60} (2–8); B: is plot of induction period versus concentration of the compounds.

substituents. Apparently such propriety head to tail structure is a good realization for the mechanism of alkyl radical scavenging. Figure 27.1 represents the kinetic dependences of oxygen uptake in the model oxidation reaction in the presence of C_{60}-AmAO-1 and C_{60}-AmAO-2.

The initiator is 2,2′-azobisisobutyronitrile (AIBN), initiation rate: $W_i = 6.8 \times 10^{-8}$ Ms⁻¹, reaction mixture volume: 10 ml, oxygen pressure: $Po_2 = 20$ kPa (air), temperature: 60°C.

Concentration of the fullerene derivatives – $[C_{60}$-**AmAO-1**]: (1) = 0, (2) $= 2.5 \times 10^{-5}$, (3) $= 4 \times 10^{-5}$, (4) $= 7.5 \times 10^{-5}$, (5) $= 1.0 \times 10^{-4}$, mol/l; $[C_{60}$-**AmAO-2**]: (1) = 0, (6) $= 1.0 \times 10^{-5}$, (7) $= 3 \times 10^{-5}$, (8) $= 1.3 \times 10^{-4}$, mol/l.

It is seen from Figure 27.1 that over a sufficient wide range of concentrations of added amine derivatives the kinetic curves of oxygen consumption do not exhibit any induction period and the model oxidation proceeds with a retarded steady oxidation rate. Such a behavior pattern is consistent with scheme 1 and the related Eqs. **(1)**, **(2)** and **(8)**: $Wo_{2 \,(C60-AmAO-1/2)} = k_3 [RO_2\cdot] [RH] = W_{i \,(C60-AmAO-1/2)}^{1/2} k_3 k_{6-}^{1/2}[RH]$ **(8)**. The dependence plotted in the Figure 27.1(B) manifests quite a good linearity between the experimentally observed oxidation rates $Wo_{2 \,(C60-AmAO-1/2)}$ and the square root of

TABLE 27.2　Rate Constants for the Addition of Cumyl Alkyl Radicals to the Buckminsterfullerene, Amine Derivatives of Buckminsterfullerene and Some Alkyl Radical Accepting Stabilizers

Antioxidant/stabilizer	The inhibition rate constant, $k_{(333\ K)}$ M^{-1} s^{-1}	Reference
C_{60}	$(1.9 \pm 0.2) \times 10^8$	[1]
C_{60}-AmAO-1	$(3.4 \pm 0.5) \times 10^8$	[51]
C_{60}–AmAO-2	$(5.4 \pm 0.8) \times 10^8$	[51]
Cyasorb 3529	$(2.0 \pm 0.8) \times 10^8$	[46]
(1,6-Hexanediamine, N,N'-bis (2,2,6,6-tetramethyl-4-piperidinyl)-, Polymers with morpholine-2,4,6, -trichloro-1,3,5,-triazine)		
Chimassorb 119	$(1.2 \pm 0.2) \times 10^8$	[45]
(1,3,5-Triazine-2,4,6, -triamine, N,N'''-[1,2-ethane-diyl-bis [[[4,6-bis-[butyl (1,2,2,6,6-pentamethyl-4-piperidinyl) amino]-1,3,5-triazine-2yl-]imino]-3,1-propanediyl]]bis [N,' N''-dibutyl-N,'N''-bis (1,2,2,6,6-pentamethyl-4-piperidinyl)		
Chimassorb 119FL	$(1.4 \pm 0.2) \times 10^8$	[45]
Chimassorb 2020	$(1.5 \pm 0.2) \times 10^7$	[43]
(1,6-Hexanediamine, N,N'-bis (2,2,6,6-tetramethyl-4-piperidinyl)-polymer with 2,4,6-trichloro-1,3,5- triazine, reaction products with N-butyl-1-butanamine an N-butyl-2,2,6,6-tetramethyl-4-piperidinamine)		

the concentration of the compounds over the range $1 \times 10^{-5} - 1.3 \times 10^{-4}$ mol/l, e.g., $Wo_{2\ (C60-AmAO-1/2)} \sim [C_{60}\text{-AmAO-1/2}]^{1/2}$.

The inhibition rate constants for the C_{60}-AmAO-1 and C_{60}-AmAO-2 may be calculated from Eq. **(9)**:

$$k_{(C60-AmAO-1/2)} = 10^{-6}\{ W_{i\ (AIBN0)} - [Wo_{2\ (C60-AmAO-1/2)}]^2 k_6 (k_3)^{-2} [RH]^{-2}\}$$
$$[C_{60}\text{-AmAO-1/2}]^{-1} [Wo_{2\ (C60-AmAO-1/2)}]^{-1} \tag{9}$$

Values of the inhibition rate constants found for the buckminsterfullerene amine derivatives C_{60}-AmAO-1 and C_{60}-AmAO-2over the concentration range $1 \times 10^{-5} - 1.3 \times 10^{-4}$ mol/l are given in Table 27.2.

The evaluated rate constants appear to be considerably higher than those of fullerene and given commercials stabilizers, confirming the existence of a synergistic effect promoted by the fullerene component and the tethered amine units. It is possible also to infer that the more the steric hindrance in the grafted cyclic amines the higher the values of inhibition rate constants.

TABLE 27.3 Heat-Oxy Resistance of Different LDPE Composites

No	LDPE composites with 0.5 wt% additive LDPE	Weight loss at %:								
		220	225	250	275	300	325	350	375	400
0	Without additives	0	0	2.8	3.8	4.5	7.0	14	22	57
1	Fullerene C_{60}/C_{70}	0	0	0	2	3	6	15	35	53
2	Irganox 1010	0	0	0	1	1.5	3	6	10	37
3	**Agerite White**	0	0	1	1.5	3.5	6	10	15	42
4		0	0	0	0	1.0	1.2	4	12	42
5		0	0	0	0	0.5	1.0	3	10	40

27.4.2 THERMAL ANALYSIS OF LDPE COMPOSITES

Results of the model oxidation obtained are verified by thermo-gravimetric analysis data where decomposition temperatures onset, and points of 10, 20, and 50% weight loss as well as half-decay period $\tau/2$ of composites are adduced (Table 27.3).

The data given in the Table 27.3 specify (i) LDPE resistance to oxidation is low and the polymer decomposition starting point is near 200°C; and (ii) the stabilizing activity of the fullerene-pyrrolidines observed is purely comparable with that for the commercial stabilizers.

27.5 CONCLUSIONS

Results obtained in the work are evidence of the fact that the buckminster-fullerene derivatives containing grafted antioxidative units exhibit higher efficiency than conventional antioxidants. The found values of inhibition rate constants also corroborate that the grafted alkyl pyrrolidine groups are able to iteratively increase the inherent radical scavenging efficiency of the fullerene.

Thermo gravimetric and differential thermal analysis of the LDPE composites showed that fullerene-alkyl pyrrolidines exhibit high stabilizing activity comparable with action of strong commercial basic stabilizers-antioxidants.

Thus the fullerene derivatives may be recognized as a new promising class of antioxidants. However, many challenges are pending and therefore, the formulated inference requires further thorough explorations.

KEYWORDS

- amine derivatives
- antioxidative efficiency
- buckminsterfullerene
- fuller pyrrolidines
- inhibition rate constant
- model oxidation
- physical-mechanical properties
- thermal analysis

REFERENCES

1. Zeynalov, E. B., Allen, N. S., Salmanova, N. I. (2009). Radical scavenging efficiency of different fullerenes C60-C70 and fullerene soot. *Polymer Degradation and Stability, 94*(8), 1183–1189.
2. Zeynalov, E. B., Friedrich, J. F. (2007). Anti-radical activity of fullerenes and carbon nanotubes in reactions of radical polymerization and polymer thermal/thermo oxidative degradation: A review. Materials Testing (Materials and Components, Technology and Application) Section. *Polymer Materials, 49*(5), 265–270.
3. Guldi, D. M., Hungerbuhler, H., Janata, E., Asmus, K. D. (1993). Radical-induced redox and addition-reactions with C-60 studied by pulse-radiolysis. *Journal of the Chemical Society-Chemical Communications, 1*, 84–86.
4. Dimitrijevic, N. M., Kamat, P. V., Fessenden, R. W. (1993). Radical adducts of fullerenes C-60 and C-70 studied by laser flash-photolysis and pulse-radiolysis. *Journal of Physical Chemistry, 97*(3), 615–618.
5. Zeinalov, E. B., Koβmehl, G. (2001). Fullerene C_{60} as an antioxidant for polymers. *Polymer Degradation and Stability, 71*(2), 197–202.
6. Zeynalov, E. B., Magerramova, M. Y., Ischenko, N. Y. (2004). Fullerenes C-60/C-70 and C-70 as antioxidants for polystyrene. *Iranian Polymer Journal, 13*(2), 143–148.
7. Gasanov, R. G., Kalina, O. G., Bashilov, V. V., Tumanskii, B. L. (1999). Addition of carbon-centered radicals to C-60. Determination of the rate constants by the spin trap method. *Russian Chemical Bulletin, 48*(12), 2344–2346.
8. Gasanov, R. G., Tumanskii, B. L. (2002). Addition of (Me2CCN)-C-center dot, (Me2CPh)-C-center dot, and (CCl3CH2CHPh)-C-center dot radicals to fullerene C-60. *Russian Chemical Bulletin, 51*(2), 240–242.
9. Walbiner, M., Fischer, H. (1993). Rate constants for the addition of benzyl radical to C-60 in solution. *Journal of Physical Chemistry, 97*(19), 4880–4881.
10. Ghosh, H. N., Pal, H., Sapre, A. V., Mukherjee, T., Mittal, J. P. (1996). Formation of radical adducts of C-60 with alkyl and halo-alkyl radicals – Transient absorption and emission characteristics of the adducts. *Journal of the Chemical Society-Faraday Transactions, 92*(6), 941–944.
11. Camp, A. G., Lary, A., Ford, W. T. (1995). Free-radical polymerization of methyl-methacrylate and styrene with C(60). *Macromolecules, 28*(23), 7959–7961.
12. Stewart, D., Imrie, C. T. (1996). Role of C-60 in the free radical polymerization of styrene. *Chemical Communications, 11*, 1383–1384.
13. Arsalani, N., Geckeler, K. E. (1996). Radical bulk polymerization of styrene in the presence of fullerene [60]. *Fullerene Science and Technology, 4*(5), 897–912.
14. Cao, T., Webber, S. E. (1996). Free radical copolymerization of styrene and C_{60}. *Macromolecules, 29*(11), 3826–3830.
15. Chen, Y., Lin, K. C. (1999). Radical polymerization of styrene in the presence of C-60. *Journal of Polymer Science Part A- Polymer Chemistry 37*(15), 2969–2975.
16. Kirkwood, K., Stewart, D., Imrie, C. T. (1997). Role of C_{60} in the free radical polymerization of methyl methacrylate. *Journal of Polymer Science Part A: Polymer Chemistry, 35*(15), 3323–25,
17. Seno, M., Fukunaga, H., Sato, T. (1998). Kinetic and ESR studies on radical polymerization of methyl methacrylate in the presence of fullerene. Journal of Polymer Science, Part A, *Polymer Chemistry, 36*(16), 2905–2912.

18. Seno, M., Maeda, M., & Sato, T. (2000). Effect of fullerene on radical polymerization of vinyl acetate Journal of Polymer Science, Part A, *Polymer Chemistry, 38*(14), 2572–2578.
19. Pabin-Szafko, B., Wisniewska, E., Szafko, J. (2006). Carbon nanotubes and fullerene in the solution polymerization of acrylonitrile. *European Polymer Journal, 42*(7), 1516–1520.
20. Shibaev, L. A., Egorov, V. M., Zgonnik, V. N., Antonova, T. A., Vinogradova, L. V., Melenevskaya, E. Y., Bershtein, V. A. (2001). An enhanced thermal stability of poly(2, 6-dimethyl-1, 4-phenylene oxide) in the presence of small additives of C_{60} and C_{70}. *Polymer Science, A., 43*(2), 101–105.
21. Cataldo, F. (2001). On the reactivity of C-60 fullerene with diene rubber macro radicals. I. The case of natural and synthetic cis-1, 4-polyisoprene under anaerobic and thermo-oxidative degradation conditions. *Fullerene Science and Technology, 9*(4), 407–513.
22. Jipa, S., Zaharescu, T., Santos, C., Gigante, B., Setnescu, R., Setnescu, T., Dumitru, M., Kappel, W., Gorghiu, L. M., Mihalcea, I., Olteanu, R. L. (2002). The antioxidant effect of some carbon materials in polypropylene. *Material Plastic, 39*(1), 67–72.
23. Troitskii, B. B., Troitskaya, L. S., Dmitriev, A. A., Yakhnov, A. S. (2000). Inhibition of thermo-oxidative degradation of poly(methyl methacrylate) and polystyrene by C-60. *European Polymer Journal, 36*(5), 1073–1084.
24. Troitskii, B. B., Domrachev, G. A., Semchikov, Y. D., Khokhlova, L. V., Anikina, L. L., Denisova, V. N., Novikova, M. A., Marsenova, Y. A., Yashchuk, L. M. (2002). Fullerene-C-60, a new effective inhibitor of high-temperature thermo oxidative degradation of methyl methacrylate copolymers. *Russian Journal of General Chemistry, 72*(8), 1276–1281.
25. Ginzburg, B. M., Shibaev, L. A., Ugolkov, V. L., Bulatov, V. P. (2003). Influence of C-60 fullerene on the oxidative degradation of a free radical poly(methyl methacrylate. *Journal of Macromolecular Science – Physics, B., 42*(1), 139–166.
26. Zuev, V. V., Bertini, F., Audisio, G. (2005). Fullerene C_{60} as stabilizer for acrylic polymers. *Polymer Degradation and Stability, 90*(1), 28–33.
27. Kelar, K., (2006). Polyamide 6 modified with fullerenes, prepared via anionic polymerization of epsilon-caprolactam. *Polymer, 51*(6), 415–424.
28. Enes, R. F., Tome, A. C., Cavaleiro, J. A. S., Amorati, R., Fumo, M. G., Pedulli, G. F., Valgimigli, L. (2006). Synthesis and antioxidant activity of [60]fullerene-BHT conjugates. *Chemistry-A European Journal, 12*(17), 4646–4653.
29. Bulgakov, R. G., Ponomareva Yu. G., Maslennikov, S. I., Nevyadovsky, E., Yu, Antipina, S. V. (2005). Inertness of C60 fullerene toward RO_2˙ peroxy radicals. *Russian Chemical Bulletin, International Edition, 54*(8), 1862–1865.
30. Djordjevich, A., Bogdanovich, G., Dobric, S. (2006). Fullerenes in biomedicine. *J. Buon, 11*(4), 391–404.
31. Tsai, M. C., Chen, Y. H., Chiang, L. Y. (1997). Polyhydroxylated C-60, fullerenol, a novel free-radical trapper, prevented hydrogen peroxide – and cumene hydro peroxide-elicited changes in rat hippocampus in-vitro. *Journal of Pharmacy and Pharmacology, 49*(4), 438–445.
32. Shi, Z. Q., Li, Y. L., Wang, S., Fang, H. J., Zhu, D. B. (2001). Synthesis and antioxidative properties of polyphenol-fullerenes. *Chinese Science Bulletin, 46*(21), 1790–1792.
33. Sun, D. Y., Zhu, Y. S., Liu, Z. Y., Liu, G. Z., Guo, X. H., Zhan, R. Y., Liu, X. Y. (1997). Active oxygen radical scavenging ability of water-soluble fullerenols. *Chinese Science Bulletin, 42*(9), 748–752.

34. Djordjevic, A., Canadanovic-Brunet, J. M., Vojinovic-Miloradov, M., Bogdanovic, G. (2004). Antioxidant properties and hypothetic radical mechanism of fullerenol C-60(OH)(24) *Oxidation Communications, 27*(4), 806–812.
35. Mirkov, S. M., Djordjevic, A. N., Andrich, N. L., Andrich, S. A., Kostic, T. S., Bogdanovic, G. M., Vojinovic-Miloradov, M. B., Kovacevic, R. Z. (2004). Nitric oxide-scavenging activity of poly hydroxylated fullerenol C-60(OH)(24). *Nitric Oxide-Biology and Chemistry, 11*(2), 210–217.
36. Yu, C., Bhonsle, J. B., Wang, L. Y., Lin, J. G., Chen, B. J. Chiang, L. Y. (1997). Synthetic aspects and free-radical scavenging efficiency of polyhydroxylated C-60 *Fullerene Science and Technology, 5*(7), 1407–1421.
37. Zhu, Y. S., Sun, D. Y., Liu, G. Z., Liu, Z. Y., Zhan, R. Y., Liu, S. Y. (1996). An ESR study on (OH)-O-center dot-radical scavenging activity of water-soluble fullerenols. *Chemical Journal of Chinese Universities-Chinese, 17*(7), 1127–1129.
38. Chiang, L. Y., Lu, F. J., Lin, J. T. (1995) Free radical scavenging activity of water-soluble fullerenols. *Journal of the Chemical Society-Chemical Communications, 12*, 1283–1284.
39. Yin, J. L., Lao, F., Fu, P. P., Wamer, W. G., Zhao, Y. L., Wang, P. C., Oiu, Y., Sun, B. Y., Xing, G. M. Dong, J. Q., Liang, X. J., Chen, C. Y. (2009). The scavenging of reactive oxygen species and the potential for cell protection by functionalized fullerene materials. *Biomaterials, 30*(4), 611–621.
40. Enes, R. F., Farinha, A. S. F., Tome, A. C., Cavaleiro, J. A. S., Amorati, R., Petrucci, S., Pedulli, G. F. (2009). Synthesis and antioxidant activity of fullerene-flavonoid conjugates. *Tetrahedron, 65*(1), 253–262.
41. Ghanbari, B., Khailli, A. A., Taheri, Z., Mohajerani, B., Jamarani, M., Soleymani (2007). The effect of fullerene C-60 and its amine derivative on the ZDDP antioxidative functionality. *Fullerenes, Nanotubes and Carbon Nanostructures, 15*(6), 439–443.
42. Yang, J. Z., Alemany, L. B., Driver, J., Hartgerink, J. D., Barron, A. R. (2007). Fullerene-derivatized amino acids: synthesis, characterization, antioxidant properties, and solid-phase peptide synthesis. *Chemistry: A European Journal, 13*(9), 2530–2545.
43. Zeynalov, E. B., Allen, N. S. (2004). Simultaneous determination of the content and activity of sterically hindered phenolic and amine stabilizers by means of an oxidative model reaction. *Polymer Degradation and Stability, 85*(2), 847–853.
44. Zeynalov, E. B., Allen, N. S. (2004). An influence of micron and nano-particle titanium dioxides on the efficiency of antioxidant Irganox 1010 in a model oxidative reaction. *Polymer Degradation and Stability, 86*(1), 115–120.
45. Zeynalov, E. B., Allen, N. S. (2006). Effect of micron and nano-grade titanium dioxides on the efficiency of hindered piperidine stabilizers in a model oxidative reaction. *Polymer Degradation and Stability, 91*(4), 931–939.
46. Zeynalov, E. B., Allen, N. S. (2006). Modeling light stabilizers as thermal antioxidants. *Polymer Degradation and Stability, 91*(12), 3390–3396.
47. Zeynalov, E. B., Allen, N. S., Calvet, N. L., Stratton, J., (2007). Impact of stabilizers on the thermal catalytic activity of micro- and nano-particulate titanium dioxide in oxidizing condensed mediums. *Dyes and Pigments, 75*(2), 315–327.
48. Allen, N. S., Zeynalov, E. B., Sanchez, K. T., Edge, M., Kabetkina, Yu. P., Johnson, B. (2010). Comparative evaluation of the efficiency of a series of commercial antioxidants studied by kinetic modeling in a liquid phase and during the melt processing of different polyethylenes. *Journal of Vinyl & Additive Technology, 16*(1), 1–14.

49. Tsepalov, V. F., Kharitonova, A. A., Gladyshev, G. P., Emanuel, N. M. (1977). Determination of the rate constants and inhibition coefficients of phenol antioxidants with the aid of model chain reactions/Determination of rate constants and inhibition coefficients of inhibitors using a model chain reaction. *Kinetics and Catalysis, 18*(5), 1034–41/*18*(6), 1142–1151.

50. Zeynalov, E. B., Vasnetsova, O. A. (1993). Kinetic screening of inhibitors of radical reactions. Baku: Elm.

51. Allen, N. S., Zeynalov, E. B., Taylor, K., Birkett, P. (2009). Antioxidant capacity of novel amine derivatives of buckminsterfullerene: determination of inhibition rate constants in a model oxidation system. *Polymer Degradation and Stability, 94*(11), 1932–1940.

52. Van Hook, J. P., Tobolsky, A. V. (1958). The thermal decomposition of 2,2'-Azo-bisisobutyro nitrile. *Journal of the American Chemical Society 80*, 779–782.

CHAPTER 28

DESIGN AND PHOTOCATALYTIC PROPERTIES OF SEMICONDUCTOR/ DYE/POLYMER THIN FILM PHOTOCATALYSTS

O. LINNIK,[1] O. NADTOKA,[2] N. CHORNA,[1] N.SMIRNOVA,[1] and V. SYROMYATNIKOV[2]

[1]Chuiko Institute of Surface Chemistry, National Academy of Science of Ukraine, General Naumov Str. 17, 03164, Kyiv, Ukraine, E-mail: okslinnik@yahoo.co.uk

[2]Taras Shevchenko National University of Kyiv, Department of Chemistry, Volodymyrska Str. 60, 01033, Kyiv,Ukraine

CONTENTS

ABSTRACT

The semi conductive films contained 10 and 30% of TiO_2 in titania/silica nanocomposite were synthesized by sol-gel method using concentrated anatase titania colloid. Acridine Yellow (AY) was adsorbed onto the surface from its aqueous solution. The film with adsorbed AY was covered by polyepoxypropylcarbazole(PEPC) dissolved in benzene. The semiconductor/dye/polymer composites were tested for the stability in several solvents (water, ethanol/water mixtures and pure ethanol). The amount of adsorbed AY as well the concentration of PEPC influenced on the level of desorbed AY from the composites. Photo catalytic performance in the process of dichromate ions reduction and tetracycline degradation under UV and visible light was performed. Among tested materials, the most perspective photo catalytic system was the composite containing 10% $TiO_2/SiO_2/AY/PEPC$.

28.1 INTRODUCTION

Surface modification of titania or other semiconductors places a key role on the yield of photo induced electron transfer processes. A great attention is paid for a creation of photosensitized semi conductive materials as they find applications as a photoactive material in photovoltaic, optoelectronics and photo catalysis. Since 1980s the photovoltaic investigation of nanostructured TiO_2 observing the efficient charge injection from dyes into a conduction band of the semiconductor has been reported [1]. It was a basis for development of dye-sensitized solar cells (DSC)that directly convert solar energy to electrical energy without any permanent chemical transformation[2]. The principle is based on the photo excitation of*tris*(2,2-bipyridyl-4,4-carboxylate) ruthenium(II) resulting in the injection of an electron into the conduction band of the titania. In turn, the dye is restored by electrons from the electrolyte(the iodide/triiodide couple). The iodide is regenerated by the reduction of triiodide at the counter electrode the circuit being completed via electron migration through the external load. The voltage generated under illumination corresponds to the difference between the Fermi level of the electron in the solid and the redox potential of the electrolyte. Such sensitized electrochemical photovoltaic device exhibited a conversion efficiency of 7.1% under solar illumination. The evolution in the DSC research was intensively continued reaching the conversion efficiency of 15% employing an organic-inorganic hybrid perovskite($CH_3NH_3PbI_3$) as a sensitizing dye [3].

Two dyes were used to construct dye-sensitized solid-state solar cell exhibiting higher efficiency than the cells based on these individual dyes.

The monolayer of two dyes wereseparated by a thin layer of a semiconductor transparent to visible light. It is avoided the quenching and the carriers injected by the photo excited dyes tunnel through the thin barrier to n- and p-regions [4]. Later, a semiconductor—dye heterostructure containing n-TiO$_2$/Fast Green/p-CuSCN/Rhodamine6G/p-CuSCN/Acridine Yellow/p-CuSCN layers was found to generate photovoltaic response to light absorption by all the three dyes. The mechanism involved is suggested to be transfer of electrons to n-TiO$_2$ and holes to p-CuSCN via tunneling. The positioning of the excited level of Acridine Yellow allowing the electron tunneling to TiO$_2$ through a thicker barrier was more suitable comparing with Rhodamine6G. This technique could be a strategy to broaden the spectral response and enhance the efficiency of dye-sensitized solar cells [5].

Several polymethine dyes with indoline end groups were used to saturate Degussa P-25 titania with the following coating by polyepoxypropylcarbazole to protect the dye molecules to be removed from the surface of titania. The calculated LUMO energies of the tested dyes are higher than the conduction band edge of anatase TiO$_2$. The most active sensitizer was the dye without any substituent in the chain during photo catalytic oxidation of iodide to triiodide anions [6].

Titania is the most wide-spreadphoto catalyst absorbing only 4% of solar light and it is completely inert in the whole visible range of the solar spectrum. In addition, titania in anatase modification offers appropriate optical properties meaning the enhanced light trapping, it is also cheap and nontoxic. The challenge is therefore to extend the sensitivity of TiO$_2$ towards the visible range of the spectrum by modifying of TiO$_2$ with inorganic ions (metal or/and nonmetal)[7, 8] or organic compounds (dyes)[9].

Mixed oxide composite materials can often be more efficient photo catalysts than pure substances as a result of the generation of new active sites and improved mechanical strength, thermal stability and surface area of titania[10, 11]. The active surface area and adsorption capacity to the molecular sensitizer (dye) are greatly increased in such systems resulting in an effective light absorption [12, 13]. When Acridine Yellow molecule is adsorbed, its pyridine-type nitrogen atom forms bonds with both Brønsted and Lewis surface acidic sites of the substances mentioned above. The recently reported results of calculations [14] show that the highest adsorption energy of the molecular state of AY was obtained for the complex with titania–silica where Lewis acidic sites are present while the lowest one being related to the complex with titania corresponded to the experimental data [15].

To utilize the broad spectrum of visible region of solar radiation, we constructed a photo catalytic system based on TiO_2/SiO_2 mesoporous thin films with adsorbed Acridine Yellow (AY) dye and covered with polyepoxypropylcarbazole(PEPC). To investigate the photo catalytic performance of the systems, the processes of dichromate reduction and destruction of antibiotics tetracycline hydrochloride ($TiO_2/SiO_2/AY/PEPC$) were used. It shown [16–19] that the photo catalytic efficiency of pollutant transformations to the less toxic and dangerous compounds is strongly depended on the numerous points such as level of dopant in TiO_2, its nature, particle size and others.

28.2 EXPERIMENTAL PART

28.2.1 MATERIALS

Unless otherwise specified the reagents were obtained from Aldrich and used as received without further treatment. Carbazole(analytical grade) was used after recrystallization. Poly-(N-epoxypropylcarbazole) was synthesized by the reaction N-epoxypropylcarbazole with anhydrous N,N-dimethylformamide and further anion polymerization according to the literature [20].

28.2.1.1   Synthesis of N-EpoxypropylCarbazole(EPC)

Carbazole(0.167 g, 1 mmol), sodium methoxide(0.16 g, 3 mmol) and 23 mL of anhydrous N,N-dimethylformamide were added to a three necked flask. The solution was heated to 40°C and drop wise added epichlorohydrin(1.28 g, 13 mmol), then stirred for 1 h and was allowed to cool to room temperature. After removal of the solvent, the remaining crude product was isolated by flash chromatography (silica gel, hexane/dichloromethane: 5/l) to isolate 1 as a white solid in 53% yield. ^{1}HNMR(CDCl$_3$, 400 MHz) δ (ppm) 8.12 (d, 2H); 7.48 (d, 4H); 7.24 (t, 2H); 4.66 (d, 1H); 4.40 (d, 1H); 3.35 (d, 1H); 2.81 (d, J = 4.0 Hz, 1H); 2.58 (t, J = 4.8 Hz, 1H).

28.2.1.2   Synthesis of Poly-(N-EpoxypropylCarbazole)(PEPC)

Potassium hydroxide (0.0176 g, 0.3 mmol), 18-crown-6 (0.1170 g, 0.4 mmol) and 8 mL methanol were added to a flask. When the solid dissolved, 8 mL toluene was added. The mixture was heated in vacuum until the volume of the distillate became half that of the volume of the original solution. Then a new portion of toluene was added and distilled again. This procedure was repeated five times. The solution was then cooled down to room temperature and filtered. Polymerization of EPC was carried out in a flask equipped with nitrogen purge. 8 mL of the mentioned solution and EPC(0.8741 g) were added to the flask. The solution was heated at 90°C for 12 h. The resulting polymers were isolated by precipitation in hexane or in methanol and dried in vacuum at ambient temperature.

28.2.2 OBJECTS OF STUDY

The chemical structure of used dye and semiconductor polymer is presented in Figure 28.1.

28.2.3 CHARACTERIZATION

The XRD spectrum of TiO_2 sol was recorded with DRON-4-07 diffractometers(CuK$_\alpha$ radiation). Calculation of apparent crystallite size for titania has been performed by Debye-Scherrer formula $\beta(2\theta) = 0.94 \lambda/(Dcos\theta)$, using (1–0 1) reflections employing the FWHM procedure.

Optical spectra of the composites were recorded with a double beam spectrophotometer (Lambda 35, PerkinElmer) within the wavelength range of (190–1200) nm.

28.2.4 METHODS

Acid catalyzed silica sols were obtained using tetraethoxysilane $Si(OC_2H_5)_4$, aqueous solution of hydrochloric acid (used as a catalyst), de-ionized

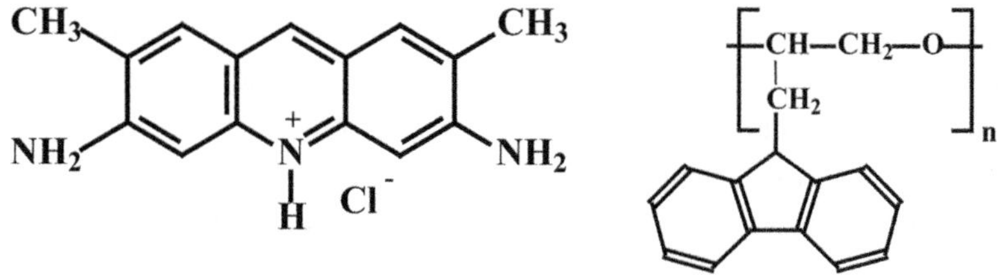

FIGURE 28.1   Structural formula of dye (AY)and semiconductor polymer (PEPC).

water for hydrolysis, and commercially available triblock copolymer polyethylene oxide – polypropylene oxide PluronixP123(BASF)(as template). To control hydrolysis-condensation reaction rates and to prevent oxides precipitation, acetylacetone was used as a complexing agent. Silica-titania films (10 and 30 wt% TiO_2) were prepared via addition of TiO_2 in anatase crystalline form (6–7 nm, Figure 28.2)[21] to SiO_2 sol. Dip-coating procedure was used for film formation on the clean and dried glass substrates (withdrawal rate was 1.5mm/s). Thermal treatment of films was performed at 350°C. Acridine Yellow (AY) dye (1×10^{-6}mol/L) was left for adsorption onto the films for 24 hours. After drying on the air, the polymer (5% PEPC in benzene) was twice covered onto the films with adsorbed AY. The composites were left for 48 hours at room temperature.

The investigated composites are sign as $10\%TiO_2/SiO_2/AY/PEPC$and $30\%TiO_2/SiO_2/AY/PEPC$ consisting of 10 or 30 wt% of titania in silica mesoporous thin films with adsorbed AY dye and covered with PEPC(the structural formula are given in Figure 28.1)

$TiO_2/SiO_2/AY/PEPC$ composite was immersed in 40 ml of an aqueous solution of potassium dichromate (in all experiments, the initial concentration of dichromate ions was 2×10^{-4} M) and the reducing agent (disodium salt of ethylenediaminetetraacetic acid (Na_2EDTA)) in the molar ratio 1:1 adjusted to pH$\geq$2 by perchloric acid. The change of Cr(VI) ions concentration was monitored with a Lambda 35 UV-vis spectrophotometer (Lambda 35, PerkinElmer) every 20 min at $\lambda = 350$ nm.

In the case of TC degradation, 40 ml of 2×10^{-5}mol/L of TC aqueous solution was used. The change of TC concentration was monitored with a Lambda 35UV-vis spectrophotometer (PerkinElmer) every 20 min at $\lambda = 357$ nm.

$TiO_2/SiO_2/AY/PEPC$ composite was immersed in the solution until complete adsorption in the dark occurred, and then irradiated by 1000 W middle-pressure mercury lamp for 90 min. For testing the visible light sensitivity, a filter transmitting light with $\lambda > 380$ nm was inserted in the photo catalytic setup. The distance lamp-reactor was set at 90 cm. The reaction temperature was kept constant (20°C). A blank experiment was carried out where a bare glass was used instead of the composite. The reaction conversion means the percentage of photo-reduced amount of dichromate ions or photo-degradated TC.

28.3   RESULTS AND DISCUSSION

In applied science, the photo catalysts in the film form are more convenient comparing to the powders due to the avoidance of the technical stage such as the sedimentation and/or ultra filtration of the solid form from liquid phase.

The titania sol used for the film synthesis is composed on the anatase particles due to the presence of 2θ peaks at 25.5, 37.8, 48.1, 54.3, and 63.2 as shown by XRD spectrum (Figure 28.2). The average crystallite size of the particle was 6–7 nm and was calculated from the most intense diffraction peak (101) using the Scherrer's equation through fitting the peak by a Lorentzian function.

Analyzing FTIR and XPS data of TiO_2, SiO_2 and TiO_2/SiO_2 we have concluded [22] that molecular scale mixing for sol–gel derived titanosilicates with formation of Ti-O-Si bonds is a key factor affecting surface properties such as level of surface hydroxylation and surface acidity as well as the catalytic function of silica–titania. Freshly prepared TiO_2/SiO_2 films showed high hydrophylic properties with the water contact angles being ca. 12 before and 5° after UV irradiation. In addition, the enhanced adsorption of AY on SiO_2/TiO_2 film (1.5–2 time higher)in comparison with that on the parent oxide: SiO_2or TiO_2 film surfaces, coincides with high acidity, surface area and hydrophilicity of synthesized films. Thus, strong adsorption of AY on

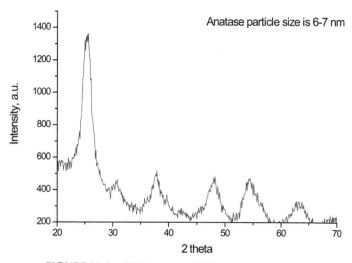

FIGURE 28.2   XRD pattern of TiO_2 precursor.

specific surface sites of SiO_2/TiO_2 films is responsible for its high efficiency of light absorption and its stability.

The semiconductor/dye/polymer composites were tested for the stability in several solvents (water, ethanol/water mixtures and pure ethanol). No penetration of AY molecules through PEPC layer in aqueous media was noted for the films with the amount of adsorbed AY corresponded to the intensity about 0.57–060 (Figure 28.3).

However, desorption of AY and consequently the decrease in the intensity of the absorption band at 443 nm was observed when the mixture of ethanol and water in the ratio 1:1 and pure ethanol were applied (not shown here). No stable systems were also obtained when the amount of adsorbed AY was doubled and when the $TiO_2/SiO_2/AY$ was covered by a single layer of PEPC.

Photo catalytic activity of the films was assessed via TC degradation and dichromate ions reduction. TC is one of the most frequently prescribed groups of antibiotics. Residues of TC and their metabolites were detected in eggs, meat and animals based on different exposure methods [23]. Their accumulation in human organism can produce arthropathy, nephropathy, central nervous system alterations, spermatogenesis anomalies, possible mutagenesis and photosensitivity in human beings. In turn, chromium-containing ions are recognized as a pollutant derived from the chrome-plating and electronic industries.

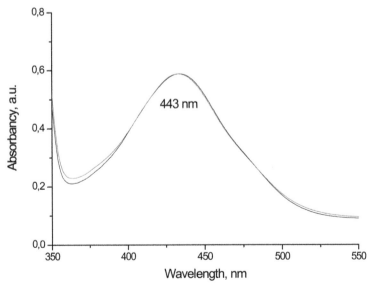

FIGURE 28.3   Absorption spectra of the 30%TiO_2/SiO_2/AY/PEPC composite before contact (black) and after 8 hours in water (red).

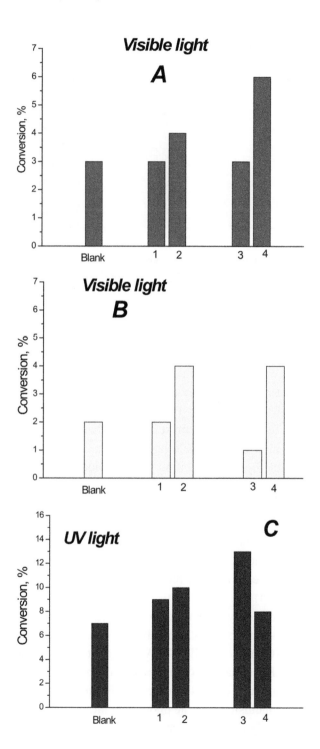

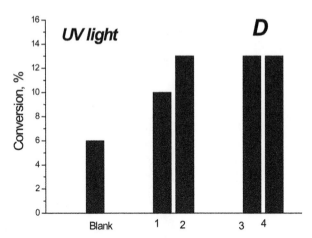

FIGURE 28.4 Photo catalytic activity of the composites tested in the reduction of dichromate ions (A and C) and degradation of TC (B and D): 1 – 10% TiO$_2$/SiO$_2$/PEPC, 2 – 10% TiO$_2$/SiO$_2$/AY/PEPC, 3 – 30% TiO$_2$/SiO$_2$/PEPC, 4 – 30% TiO$_2$/SiO$_2$/AY/PEPC.

When *visible light* was applied, the conversion percentage was increased slightly for 10% TiO$_2$/SiO$_2$/AY/PEPC and doubled for 30% TiO$_2$/SiO$_2$/AY/PEPC comparing with the AY free composites in dichromate reduction process (Figure 28.4a). The efficiency of the photo catalysts was enhanced in 2 and 4 times for 10% and 30% in TiO$_2$/SiO$_2$/AY/PEPC, respectively during TC destruction (Figure 28.4b). It must be noted that the AY free composites were completely inert under visible light (Figure 28.4 A and B columns 1 and 3). Upon visible light, the exciting of AY molecule takes place with the following charge carrier transfer from the excited (AY*) molecules to the CB of TiO$_2$ known as dye-sensitization process (Eqs. (1) and (2), and Figure 28.5, pathway 1.1). Further steps should implicate the reduction of the oxidized species (AY$^{\cdot+}$) by electron donor such as EDTA or TC(Eq. (3); Figure 28.5, pathways 1.3 and 1.5) and processes involving electrons from the CB as namely dichromate or oxygen reduction (Eqs. (5) and (7); Figure 28.5, pathways 1.2 and 1.4).

When light with energy higher than 3.2 eV interacts with titania particles, an electron transition from the valence band (VB) of semiconductor to the conduction band (CB) occurs (Eq. 4). As the photo catalyticactivity of *AY free* TiO$_2$/SiO$_2$/PEPC enhances comparing with the blank experiment and depends on TiO$_2$ amount for both processes (Figure 28.4 C, D columns 1, 3), it leads to the conclusion that the photo generated electrons became to be trapped and/or directly move through PEPC layer to the redox couple.

Dye-Sensitization Process

$$AY + h\nu(\leq 3.2eV) \rightarrow AY^* + e\text{-} \rightarrow AY^{\cdot+} \tag{1}$$

$$TiO_2 + e\text{-} \rightarrow Ti^{IV}O_2/Ti^{3+}(e\text{-}_{CB}) \tag{2}$$

$$AY^{\cdot+} + TC/EDTA \rightarrow AY + TC^{\cdot+}/EDTA^+ \tag{3}$$

Direct photo catalysis

$$TiO_2 + h\nu(\geq 3.2eV) \rightarrow TiO_2(e\text{-}_{CB} + h^+_{VB}) \tag{4}$$

Photo catalytic reduction of dichromate ions

$$Cr_2O_7^{2-} + 14 H^+ + 6e^-_{CB/trapped} \rightarrow 2Cr^{3+} + 7 H_2O \tag{5}$$

$$RCOOH(EDTA) + h^+_{VB/trapped} \rightarrow RCOO^{\cdot} + H^+(H_2O) \rightarrow CO_2 + R^{\cdot} \tag{6}$$

Photo catalytic destruction of TC

$$O_2 + e^-_{CB/trapped} \rightarrow O_2^{\cdot-} \rightarrow H_2O_2 + H^+ \tag{7}$$

$$TC + h^+_{VB/trapped} \rightarrow TC^{\cdot+} \rightarrow CO_2 + NH_3 + H_2O \tag{8}$$

Almost no influence on the photo catalytic performance under *UV light* was observed for $TiO_2/SiO_2/AY/PEPC$. Even the decrease in conversion percentage of dichromate reduction is noted (Figure 28.3 C, column 4). While the TC conversion is slightly raised or unchanged for $TiO_2/SiO_2/AY/PEPC$ (Figure 28.3 D, columns 2 and 4).

Figure 28.6 shows the change in the absorption spectra of 30% $TiO_2/SiO_2/AY/PEPC$ composites after light exposure. The intensity of adsorbed AY molecules is obviously lowered as a result of AY destruction by UV light. It is assumed that the photo formed electrons in CB of semiconductor are involved in the reaction with newly formed AY residues making the pathways 2.1 impossible. However, the destruction process with the holes participation follows freely (Figure 28.4D, columns 2 and 4; Figure 28.5, pathways 2.2). Based on this, we assumed the fate of the pathways 2.2 and 2.3 (Figure 28.5).

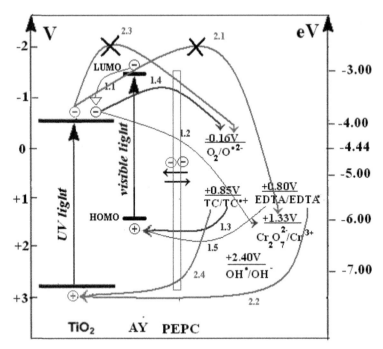

FIGURE 28.5 The proposed mechanism of dichromate reduction (pathways 1.2, 2.1, 1.5 and 2.2) and TC destruction (pathways 1.4, 2.3, 1.3 and 2.4) under visible (blue) and UV (red) light over $TiO_2/SiO_2/AY/PEPC$ composite.

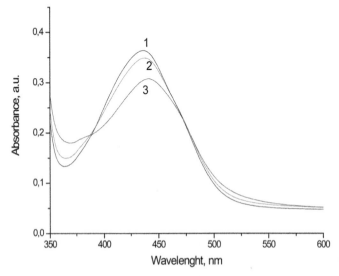

FIGURE 28.6 Absorption spectra of 30% $TiO_2/SiO_2/AY/PEPC$ composites: 1 – before irradiation; 2 – after 90 min of visible light exposure; 3 – after 90 min of full light irradiation.

28.4 CONCLUSIONS

Hence, the complex composites based on the mesoporous TiO_2/SiO_2 thin films with adsorbed AY dye and PEPC layer are designed. The double PEPC covering prevents the penetration of AY molecules to the liquid phase when aqueous solution is applied.

The presence of acidic surface sites in TiO_2/SiO_2 thin films, amount of adsorbed AY as well the number of PEPC layers influenced on the stability of the composites. The $TiO_2/SiO_2/AY/PEPC$ films can be used as the new type of photo catalytic systems in Ecological Photo catalysis as the significant increase in the photo activity under visible light for both reduction and destruction processes is reported.

The low photo catalytic efficiency of the tested composites is observed upon UV light. In turn, the new approaches have to be done to improve the photo catalytic activity and stability to the more aggressive media.

KEYWORDS

- composite
- dye
- photo catalysis
- polyepoxypropylcarbazole
- semiconductivity
- titania

REFERENCES

1. Vlachopoulos, N., Liska, P., Augustynski, J., &Gratzel, M., (1988). Very efficient visible light energy harvesting and conversion by spectral sensitization of high surface area polycrystalline titanium dioxide films. *J. Am. Chem. Soc., 110*(4), 1216–1220.
2. O'Regan, B. C., &Gratzel, M., (1991). A Low-Cost, High-Efficiency Solar Cell Based on Dye- sensitized Colloidal TiO_2 Films. *Nature, 353*, 737–740.
3. Burschka, J., Pellet, N., Moon, S., -J., Humphry-Baker, R., Gao, P., Nazeeruddin, M. K., Grätzel, M., (2013). Sequential deposition as a route to high-performance perovskite-sensitized solar cells. *Nature, 499*, 316–319.
4. Tennakone, K., Kumara, G. R. R. A., Kottegoda, I. R. M., Perera, V. P. S. (1999). An efficient dye-sensitized photo electrochemical solar cell made from oxides of tin and zinc.*Chem. Commun., 1*, 15–16.

5. Perera, V. P. S., Pitigala, P. K. D. D. P., Senevirathne, M. K. I., Tennakone, K., (2005). A solar cell sensitized with three different dyes. *Solar Energy Materials & Solar Cells 85,* 91–98.

6. Gusiak, N. B., Kobasa, I. M., Kurek, S. S. (2013). Nature inspired dyes for the sensitization of titanium dioxide photo catalyst, *Chemik, 67*(12), 1191–1198.

7. Linnik, O., Petrik, I., Smirnova, N., Kandyba, V., Korduban, O., Eremenko, A., Socol, G., Stefan, N., Ristoscu, C., Mihailescu, I. N., Sutan, C., Malinovski, V., Djokic, V., Janakovic, D., (2012). TiO_2/ZrO_2 thin films synthesized by PLD in low pressure N-, C- and/or O-containing gases: structural, optical and photo catalytic properties. Digest *Journal of Nanomaterials and Biostructures, 7*(3), 1343–1352.

8. Linnik, O., Shestopal, N., Smirnova, N., Eremenko, A., Korduban, O., Kandyba, V., Kryshchuk, T., Socol, G., Stefan, N., Popescu-Pelin, G., Ristoscu, C., Mihailescu, I. N. (2015). Correlation between electronic structure and photo catalytic properties of non-metal doped TiO_2/ZrO_2 thin films obtained by pulsed laser deposition method. *Vacuum, 114,* 166–171.

9. Zakharenko, V. S., Parmon, V. N. (1987). Efficiency of sensitization of titanium dioxide by adsorbed acridine yellow in photo catalytic hydrogen evolution.*React. Kinet. Catal. Lett., 4,* 389–394.

10. Bach, U., Lupo, D., Comte, P., Moser, J. E., Weissortel, F., Salbeck, J., Spreitzer, H., Gratzel, M., (1998). Solid-state dye-sensitized mesoporous TiO_2 solar cells with high photon-to-electron conversion efficiencies.*Nature, 395,* 583–585.

11. Kumar, D. A., Shyla, J. M., Xavier, F. P. (2012). Synthesis and characterization of TiO_2/ SiO_2 nanocomposites for solar cell applications.*Appl. Nanosci., 2,* 1–8.

12. Worrall, D. R., Williams, S., Eremenko, A., Smirnova, N., Yakimenko, O., Staruch, G. (2004). *Coll. and Surfaces A: Physico Chem. Eng. Aspects. 230,* 45–55.

13. Imamura, S., Ishida, S., Tarumoto, H., Saito, Y., Ito, T. (1993). Effect of the composition of titania-silica on its physical and photo catalytic properties.*J. Chem. Soc. Faraday Trans. 89,* 757–762.

14. Smirnova, N. P., Surovtseva, N. I., Fesenko, T. V., Demianenko, E. M., Grebenyuk, A. G., Eremenko, A. M. (2015). Photo degradation of dye acridine yellow on the surface of mesoporous TiO_2, SiO_2/TiO_2 and SiO_2 films: spectroscopic and theoretical studies. Journal of Nanostructure in Chemistry, DOI10.1007/s40097-015-0165-y.

15. Fesenko, T. V., Pokrovskiy, V. A., Surovtseva, N. I., Eremenko, A. M., Smirnova, N. P., Boryak, O. A., Selkovsky, V. S., Kosevich, M. V. (2009). Laser desorption/ionization mass-spectrometry of acridine dyes adsorbed on the surface of TiO_2 and SiO_2 films. *Surface, 1*(16), 125–135.

16. Smirnova, N., Vorobets, V., Linnik, O., Manuilov, E., Kolbasov, G. (2010). Photo electro chemical and photo catalytic properties of mesoporous TiO_2 films modified with silver and gold nanoparticles. *Sur. Interface Anal., 42,* 1205–1208.

17. Eremenko, A., Smirnova, N., Gnatiuk, I., Linnik, O., Vityuk, N., Mukha, Y., Korduban, A., (2011). Silver and Gold Nanoparticles on Sol-Gel TiO_2, ZrO_2, SiO_2 Surfaces: Optical Spectra, Photo catalytic Activity, Bactericide Properties. Chapter in Book 3: Composite Materials, 2–32.

18. Linnik, O., Smirnova, N., Korduban, O., Eremenko, A., (2013). Gold nanoparticles in $Ti_{1-x}Zn_xO_2$ Films: synthesis, structure and application. *Materials Chemistry and Physics. 142,* 318–324.

19. Linnik, O., Smirnova, N., Zhukovskiy, M., Orekhovskaya, T., Asharif, A., Borisenko, V., Gaponenko, N., (2013). Influence of Support Nature on Photo catalytic Activity of TiO_2 Film.*Advanced Science, Engineering and Medicine, 5,* 281–286.

20. Leiqiang Qin, Shimin Zhang, JingkunXu, Baoyang Lu, XueminDuan, Danhua Zhu, Yao Huang. (2013). Novel poly(ethylene oxide) grafted polycarbazole conjugated free-standing network films via anionic and electrochemical polymerization, *Int. J. Electrochem. Sci., 8,* 5299–5313.

21. Pavlova-Verevkina, O. B., Politova, E. D., Nazarov, V. V. (1999). Preparation and structure of stable dispersions of uniform TiO_2 nanoparticles. *Colloid Journal. 61,* 359–362.

22. Andrulevic̆ius, M., Tamulevic̆ius, S., Gnatyuk Yu, Vityuk, N., Smirnova, N., Eremenko, A., (2008). XPS investigation of TiO_2/ZrO_2/ SiO_2 films modified with Ag/Au nanoparticles. *Mater Sci. (Medziagotyra). 14,* 8–14.

23. Zurhelle, G., Muller-Seitz, E., Petz, M. J. (2000). Automated residue analysis of tetracyclines and their metabolites in whole egg, egg white, egg yolk and hen's plasma utilizing a modified ASTED system. *Chromatogr. B: Biomed. Sci. Appl., 739*(1), 191–203.

CHAPTER 29

GRADED ZONE STRETCHING OF THE LINEAR POLYMERS

L. NADAREISHVILI, N. TOPURIDZE, L. SHARASHIDZE, and I. PAVLENISHVILI

Georgian Technical University, V. Chavchanidze Institute of Cybernetics, 5 S. Euli St., 0186 Tbilisi, Georgia, E-mail: levannadar@yahoo.com

CONTENTS

ABSTRACT

A new method of uniaxial zone drawing of thermoplastic polymers-controlled graded zone stretching method is developed. The description and principles of operation of the appropriate device are given. The experimental data illustrating the possibilities of the proposed method are presented.

29.1   INTRODUCTION

Uniaxial oriented stretching is a widespread method of the structural modification of the linear polymers [1]. Various approaches of uniaxial orientation stretching are known: cold/hot drawing, solid state extrusion, rapid drawing from melts or solutions. Structural modification is possible also by zone stretching of polymers [2].

As a result of stretching at the above glass-transition temperature isotropic polymer passes into specific, so-called oriented state, which is characterized by preferential location of the structural elements (fragments of macromolecules and permolecular structures) in the stretching direction.

The oriented state can be quantitatively characterized by the degree of orientation which gives information about conformations and orientation of macromolecules. For these purposes research methods such as polarized IR-spectroscopy, birefringence, etc., are used.

The most common value used to quantitative characterization of the degree of orientation is the standard deviation of the macromolecules section's orientation – $\overline{\cos^2 \theta}$ (θ is the angle between stretching direction and section of the macromolecule). This value is changed in the range of $\frac{1}{3} \leq \cos^2 \theta$ $\theta \leq 1$. $\cos^2 \theta$ depends linearly on the value of the relative elongation, which can be written as

$$\overline{\cos^2 \theta} = f(\lambda)$$

where $\lambda = \Delta\iota/\iota$ ($\Delta\iota$ is the real elongation and l is the initial length of the sample. The degree of orientation depends uniquely on the extension value under the immutability of other factors of stretching mode (temperature, stretch velocity) [3–5].

Degree of orientation may be characterized also by the average angle of disorientation $\Theta_{1/2}$ and Herman's factor of orientation $F = (3 \overline{\cos^2 \theta} - 1)/2$. These values vary respectively in the range $0 \leq \Theta_{1/2} \leq \infty$ and $0 \leq F \leq 1$.

29.2   GRADUALLY ORIENTED (STRETCHED) STATE

Structural anisotropy arising as a result of orientation stretch leads to changes in many physical and mechanical properties. The samples with different properties may be obtained by means of the variation of the orientation degree depending on:

- stretching temperature;
- stretching rate;
- cooling rate of the heated polymer;
- relative elongation – λ.

Throughout the volume of the sample oriented in conventional (non-gradient) mode (such as films obtained in industry) anisotropy, and therefore the physical and mechanical properties are practically identical.

If in the stretching process a gradient of at least one of the four above mentioned parameters is provided, this should lead to the *gradient of anisotropy* (orientation degree) and hence to the gradient of physical and mechanical properties.

Earlier we have established the conception about *new structural state* of the thermoplastic polymers – about *gradually oriented (stretched) state* (GOS) [6–11].

In GOS $\overline{\cos^2 \theta}$ (as well as $\Theta_{1/2}$ and F) is not the same in throughout the entire volume of the oriented polymer sample. In GOS there are continuous sequence of changing values of $\cos^2 \theta$ that vary from one area of the sample to another in the $\frac{1}{3} \leq \overline{\cos^2 \theta} \geq 1$ range.

GOS may be characterized in terms of relative elongation. In GOS there are continuous sequence of changing values of λ that vary from one area of the sample to another in

$0 \leq \lambda \leq n$ range, wherein the maximum value of n for the currently existing thermoplastic polymers may be $8 \div 10$. For a complete description of the GOS is also necessary to specify the length and profile of these changes.

Therefore, quantitative parameters of graded stretched polymers (polymers in GOS) are:
- range of change in relative elongation;
- length of this change;
- profile of this change (linear, hyperbolic, parabolic, logarithmic, etc.).

According to this conception as a result of transformation of isotropic polymers and their composites to GOS (in other words – by graded oriented stretching) in materials the gradient of all the properties are generated that depend on the value of relative elongation/orientation's degree. These properties are: optical, electrical, magnetic, acoustic, thermal, mechanical, sorption, etc. The validity of this argument proved by the relationship between the birefringence and relative elongation:

$$Dn = n_1 - n_2 = gl$$

where n_1 and n_2 are respectively index of refraction of ordinary and extraordinary rays;

γ– optical coefficient of deformation; λ– relative elongation ($\Delta l/l$). From this equation it follows that gradient of relative elongation should causes gradient of birefringence.

Transition into GOS is realized by the graded oriented stretching – by action of inhomogeneous mechanical field on the isotropic linear polymers or by uniaxial managed graded zone stretching that is implemented on graded zone stretching device (GZSD) which is a standard tensile-testing machine equipped with additional technical units.

Both technical approaches (especially graded zone stretching method) ensures the predetermined changes in relative elongations at predetermined areas of the polymer samples, that is, allows to manage the such quantitative parameters of gradually oriented polymers as a range of change in relative elongation/orientation degree, length of this change and profile (linear, hyperbolic, parabolic, logarithmic, etc.).

29.3   GRADED ZONE STRETCHING DEVICE (GZSD)

In Figure 29.1, the GZSD scheme is presented (a – side view, b – front view). The clip 1 is immovable practically, the active clip 9 is connected with electric motor 10. The arrow indicates the displacement direction for the clip 9 at stretching.

The investigated sample, for example the film 2, is fixed between clips 1 and 9. The thermal insulators 4 and 13 are located equidistantly on both

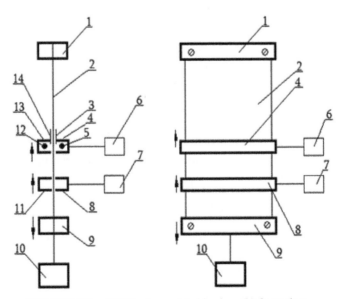

FIGURE 29.1   GZSD scheme:(a) side view, (b) front view.

sides of the film face to face perpendicularly to stretching direction. The tubular heaters 5 and 12 are fixed in insulators 4, 13. The thermal insulators are provided with heat flow directing channels, shape and size of which are regulated by the gap width regulators 3 and 14. The thermal isolators 4 and 13 and also the heaters 5 and 12 are fastened to the electric motor 6. The arrow indicates the displacement direction of thermal isolators and heaters at stretching. Between the thermal insulators 4 and 13 and active clip 9 on the both side of the film 2 the cooler tubes like air blowers 8 and 11 are located.. The blowers are connected with electric motor 7. The arrow indicates the displacement direction of the blowers 8, 11 at stretching of the film 2.

The GZSD works as follows: The selected distance between the clips is set. The dimensions and profile of the thermal channels are selected. The thermal insulators 4, 5 with inserted in them heaters 5, 12 and blowers 8, 11 are installed near clips 9 on the selected distances. They switch the heaters 5, 12. When the temperature reaches the desired value, the motor 6, blowers 8, 11, electric motors 7, 10 turn on. At this moment the temperature of the selected zone of film 2 is higher than one for other parts of the film and therefore the forces for stretching of this region of the film is minimal (in comparison with other regions of the film). Consequently the film 2 will be elongated practically only in the selected region. Because of shift of the

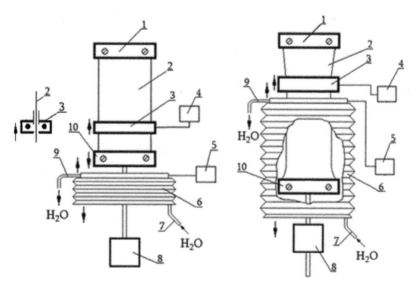

FIGURE 29.2 GZSD. Cooler – the sylphon bellows with liquid:(a) initial position;(b) stretching process.

heaters 6, 7 the selected zone in dependence on its height and rate of shift of heaters and clip 2 comes out from heated zone gradually and enters to the cold zone, where the film stretching is practically ceases (due to the high tensile yield point). At simultaneous shift of air blowers 8, 11 and clip 10 the described process spreads to heaters 5, 12 shift direction.

The process may be continued till reaching of heaters 5, 12 to the clip 1, or finished at any stage. The stretching regime (temperature and moving speed of heaters 5, 12, air blowers 8, 11 and clip 9) may be changed at any stage.

The Figure 29.2 shows the GZSD scheme, in which the sylphon bellows are used as a cooler. Sylphon bellows 6 is filled with liquid, for example with water. The clip 1 is practically immovable. The clip 10 is active and connected with electric engine 8. Investigated sample, for example film 2 is fixed in the clips 1 and 10. The arrow indicates the shift of clip 10. The clip 10 is fixed hermetically at the bottom of the sylphon bellows 6, which is connected with an electric motor 8. The arrow indicates the direction of movement of the bottom of sylphon bellows 6 and the clip 10, respectively, at stretching of film 2. On both sides of the film 2 perpendicularly to the stretching direction the heat insulators are placed.

Figure 29.2 shows only one heat insulator 3. In the heat insulators tubular heaters are fixed (as shown in Figure 29.1, the positions 4,5, 12,13). The thermal insulators are provided with heat flow directing channels, shape and size of which are regulated by the gap width regulators (as shown in Figure 29.1, the positions 3, 14). The thermal insulator 3 is connected with electric engine 4. The arrow indicates the replace direction of heat insulator 3 (heat insulators and heaters) at stretching. The sylphon bellows 6 is provided with pipes 7 and 9 for inputting and outputting the liquid. The upper part of the sylphon bellows 6 is connected with electric engine 5. Arrow indicates the displacing of the top part of the sylphon bellows 6 at stretching of sample 2. The sylphon bellows 6 is provided with pipes 7 and 9 for the liquid inlet and outlet.

The GZSD works as follows: The preliminary operations are the same as in the previous case. When the temperature reaches the desired value, the electric motors 4, 5 and 8 turn on. Temperature of the selected zone of the film 2 is higher than one for other parts and therefore the forces for stretching of this region of film is minimal (in comparison with other regions of the film). Consequently the film 2 will be elongated practically only in the selected region. Because of shift of the heaters the elongated zone comes out from heated zone gradually and enters the sylphon bellows 6, where it is cooled by running water. Due to the increasing tensile yield point strength elongation of the film is stopped. At simultaneous shift of open end of sylphon bellows 6, heaters and clip 10 the described process spread to heaters shift direction.

GZSD makes it possible to implement a very wide range of stretching modes, in particular:

- the possibility of creating different front of stretching;
- stretching of sample in the continuous and/or jumping regime with constant rate and/or acceleration;
- conducting of stretching in homogeneous or gradient regime;
- realization of uniaxial stretching of different selected region of the same sample in the different regime (temperature, rate and value of deformation) at temperature higher than room temperature without taking out of the sample from clips;
- in accordance with above item at temperature higher than room temperature it is possible to define the influence of deformation rate and temperature on such mechanical parameters of separate regions of the sample, investigation of which do not requires the destruction of whole sample. These characteristics are: mechanical stress and deformation according to the Hooke's region; modulus of elasticity; flow modulus; mechanical stress at the fixed deformation; relaxation of mechanical stresses.

The stretch degree of the sample is determined by the speed ratio of the displacement of active clamp and a heater. The gradient stretching mode is achieved by varying this ratio in stretching process. Preselected experiment mode is carried out via the control unit. Some illustrative materials are presented below.

Figure 29.3 shows the gradually oriented rectangular PVAL-films stretched at 358 K using GZSD. The overlay parallel lines are applied to register the elongation distribution of the film.

Figure 29.4 shows the distribution of relative elongations along the length of the same gradually oriented films. In Figure 29.4(a) the relative elongation over the length of the film varies linearly within 0 ÷ 225% in case of a total lengthening of 213%. In Figure 29.4(b) the distribution of relative elongation first increases and then decreases along the length of the film. In case of a total film lengthening of 195% the relative elongation varies over the length within 0 ÷ 220% ÷ 0.

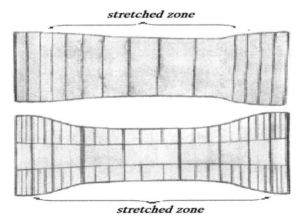

FIGURE 29.3 Gradually stretched rectangular PVAL-films. The arrow indicates the direction of propagation of the stretch front.

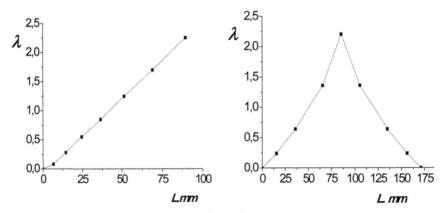

FIGURE 29.4 Relative elongations distribution in the gradually oriented rectangular film: (a) Figure 29.3(a); (b) Figure 29.3(b).

The uniaxial graded oriented stretching method can be considered as technological innovation for the creation of a new type of FGMs on the base of linear polymers/copolymers and its composites [12].

KEYWORDS

- graded zone stretching
- graded zone stretching device

- **gradient of physical and mechanical properties**
- **gradually oriented state**
- **oriented state**
- **thermoplastic polymers**

REFERENCES

1. Ward, I. M. (Ed.). (2013). Structure and Properties of Oriented Polymers. Springer Science & Business Media. pp. 500.
2. Kunugi, T. (2008). Preparation of highly oriented fibers or films with excellent mechanical properties by the zone – drawing/zone-annealing method, in: Oriented Polymer Materials, S. Fakirov, Ed. John Wiley & Sons. pp. 394–421.
3. Laius, L., Kouvshinsky, Ye. (1967). *Mechanics of Polymers, 3*, 455. (in Russian).
4. Laius, L., Kouvshinsky, Ye. (1967). *Mechanics of Polymers, 4*, 579. (in Russian).
5. Shishkin, N., Milagin, M., Gabarayeva, A. (1967).*Mechanics of Polymers, 6*, 1048. (in Russian).
6. Nadareishvili, L. Fabrication method and investigation of polymer films with a specified gradient of birefringence. *Georgian Engineering News, 2*, 73–77, 2001 (in Georgian).
7. Lekishvili, N., Nadareishvili, L., Zaikov, G. E., Khananashvili, L. (2002). New concepts in polymer science. Vygodsky, J. S., & Samsonia, Sh. A., (Eds.), *Polymers and Polymeric Materials for Fiber and Gradient Optics.* VSP Utrecht, Boston, Koln, Tokyo. 222 pp.
8. Nadareishvili, L., Gvatua, Sh., Blagidze, Y., Zaikov, G. E. (2004). (Gradient birefringence) – optics – a new direction of gradient optics, *J. Appl. Polym. Science, 9*, 489–493.
9. Nadareishvili, L., Lekishvili, N., Zaikov, C. E. (2005). Polymer medias with gradient of the optical properties, in: Zaikov, G. E. (Ed.), *Modern Advanced in Organic and Inorganic Chemistry.* Nova Science Publishers: New York, USA, pp. 31–134.
10. Nadareishvili, L., Wardosanide, Z. et al., (2012). Gradient Oriented State of Linear Polymers: Formation and Investigation, *Molecular Crystals and Liquid Crystals, 556*, 52–56.
11. Nadareishvili, L., Bakuradze, R., Kilosanidze, B., Topuridze, N., Sharashidze, L., Pavlenishvili, I. (2015). Graded Orientation of the Linear Polymers. XIII International Conference on Polymers, February 23–24, 2015. Paris, France. International Science Index, 1358–1362.
12. Nathan, J. Reinolds (Ed.). (2011). Functionally Graded Materials, Nova Science Publishers, New York, 324 pp.

CHAPTER 30

POSSIBILITIES OF CURRENT CARING SUPERCONDUCTING POLYMER-CERAMIC NANOCOMPOSITES OBTAINMENT

S. P. DAVTYAN and A. O. TONOYAN

Armenian National Polytechnic University, 105 Teryana Str., 375009, Yerevan, Armenia, E-mail: atonoyan@mail.ru

CONTENTS

ABSTRACT

With the use of yttrium and bismuth ceramics' powders ($Y_1Ba_2Cu_3O_{6.97}$ and $Bi_2aSr_2Ca_2Cu_2O_8$), on the basis of super-high molecular polyethylene binder, current-carrying polymer-ceramic nanocomposites are obtained. It is shown that the implementation of the proximity effect between the ceramics' grains leads to obtaining superconducting polymer-ceramic nanocomposites with a transport current. Additives of silver and aluminum nano-scale powders cause significant increase of current density, and the application of acoustic fields during the formation of samples allows to further increase the density to ~ 4–103A·cm^{-2}. Physical and mechanical properties of current-carrying polymer-ceramic nanocomposites have been investigated and it is shown that nano-sized aluminum additives increase the breaking strength and the elasticity modulus of the samples. The morphological features of current-carrying polymer-ceramic nano-composites' interphase layer on the basis of various binders (polymethylmethacrylate, polystyrene, super-high molecular polyethylene and isotactic polypropylene) have been studied.

30.1 INTRODUCTION

In the Refs.[1–10] in order to obtain high-temperature superconductive product with a transport current, the various metals as binders, particularly silver [1–3,10] and metals of transitive row Mo, Zn, Sr [8], etc. have been used.

For polymer-ceramic composites' obtainment in the literature there are descriptions of using as binders both thermoplastics [11,12] and reactoplastics [12]. In Ref.[11], polymer additives are used to protect high-temperature superconductors from moisture, whereas in [13–21] for obtaining polymer-ceramic samples of different geometries (rods, tubes, rings, etc.) both methods of hot pressing [13, 15, 16, 18, 19] and polymerization filling [14, 15, 17, 20, 21] were used. In the Refs.[14, 15, 17, 18, 20, 21] superconducting, physical-mechanical, thermo chemical, thermo physical properties of polymer-ceramic nanocomposites were studied. It was shown [14–21] that for high temperature SC nanocomposites, the critical temperature of transition to superconducting state is 2–30 higher than that of the original ceramics.

It becomes obvious that the burning of organic part of polymer-ceramic nanocomposites, their sintering and recovering at any particular thermal regime in the analogy with Ref. [11], as a rule, give rise to a transport current. However, polymer-ceramic nanocomposites with a superconducting (SC) transport current can be obtained without the stages of burning and sintering. This is achievable via so-called nearing effect thus effectively shielding dielectrization of near-to-surface layers of the ceramic's grains.

The purpose of this study—based on yttrium and bismuth ceramics, fine dispersed powder of super-high molecular polyethylene as a binder and nano-sized additives of aluminum and silver—is to obtain polymer-ceramic nanocomposites with a transport current.

30.2 EXPERIMENTAL PART

In this chapter, a high temperature as ceramic powder were used to synthesized yttrium and bismuth oxide ceramics ($Y_1Ba_2Cu_3O_{6.97}$ and $Bi_2Sr_2Ca_2Cu_2O_8$), with dispersion less than 50 μm and the critical transition temperature to superconducting state of 93 K and 109 K, respectively. As polymer binders in the form of fine dispersed powders were used super-high molecular (HMPE)(brand U506–000) and branched (BPE) polyethylene with melting temperature Tm = 128–135°C, 105–108°C, respectively, isotactic polypropylene (IPP) with Tm = 167–171°C, polymethylmethacrylate (PMMA) with the temperature of nitrification point Tv = 110–115°C and polystyrene(PS) with Tv = 98–102°C [22].

For finding-out the influence of nano-sized powders of aluminum and silver on the transport current of received composites were made experiments where in one case the powders of polymers (HMPE, BPE, IPP, PMMA, polyanilin) and superconducting ceramics ($Y_1Ba_2Cu_3O_{6.97}$ and $Bi_2Sr_2Ca_2Cu_2O_8$) were pre-mixed in an agate mill while in others to this mix nano-sized powders of aluminum (40 nm) and silver (3O nm) were added. From the obtained homogeneous mixtures by the method of hot pressing HMPE (~150°C under the pressure of 100 MPa) tablets with a diameter of 10 mm and thickness of 2 mm were molded. The golden contacts by thermo compression method were deposited on the tablets. Resistance of the samples was determined by the conventional 4-lead method. In some cases forming of the samples was carried out under the influence of acoustic fields, which causes deagglomeration of the agglomerated nano-particles (sonofication at 500W was kept on for the entire process of samples' formation). The start of the superconducting transition (Ts) and its width are determined by measuring the magnetic susceptibility at the frequency 1 kHz with amplitude of the magnetic field 10 me. Physical-mechanical properties and heat of sample fusion of SC polymer-ceramic nanocomposites were measured on the instruments: Differential Scanning Analyzer (DSA) and differential scanning calorimeter (DSC) made by the firm Perkin-Elmer.

30.3   RESULTS AND DISCUSSION

30.3.1   REGULATION OF THE CURRENT DENSITY IN SUPERCONDUCTING POLYMER-CERAMICS

Characteristic change of resistance of the samples from temperature (curve 1) for polymer-ceramic nanocomposites of $Y_1Ba_2Cu_3O_{6.97}$:HMPE:Al = 80:10:10 (% mass) is shown in Figure 30.1.

It is seen that up to the critical temperature of superconducting transition, the course of the resistance curve (curve 1) has a semi-conducting character. Further, at the SC transition temperature 96 K, a decrease in resistance is noticed, and at 78 K, the transition ends and the sample resistance turn to zero.

Unexpected results obtained during repeated thermal cycling of the very samples. The behavior of the samples resistance up to the critical transition temperature to the superconducting state again remains semi-conductive (Figure 30.1, curves 2,3).

However, as it can be seen the level of residual resistance increases with every next cycle, and after the transition to the superconducting state, zero resistance is not achieved. Here, there is a plateau, up to 8–5°K. Moreover, after four thermo-cycling SC transition disappears altogether (Figure 30.1, curve 4).

These results are explicable either by the accumulation of residual strains on the binder-ceramic phase boundary, as a result of the micro-cracks formation while thermo-cycling, or by the poor contacting which tend to deteriorate

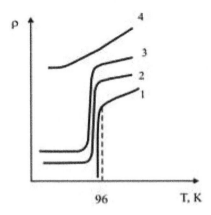

FIGURE 30.1   Change of the $Y_1Ba_2Cu_3O_{6.97}$ + HMPE nanocomposite's resistance with the additives of nano-sized aluminum powder.

as the cycling progresses. Developing current creates its own magnetic field, which eventually destroys the SC temperature transitions.

Similar results we obtained on the composites consisting of bismuth ceramic – polyaniline – HMPE with the addition of nano-sized silver powder. Consequently, to be stated that here, in the temperature range 109–115 K, "SC-step" is observed as well (Figure 30.2), without reaching resistive zero.

In the nano-composites with bismuth ceramics,' in contrast to the above discussed case, repeated thermal cycling (5–6 cycles) does not affect the curve of resistance from temperature. To be noted that in this case until the transition temperature to the superconducting state, the resistance curve's course depending on temperature is semi-conductive.

It is obviously that the increase in current density can be achieved by the implementation of proximity effect, for example, decrease in the share of the binder in SC polymer-ceramic nanocomposites. For this purpose, by hot pressing method were formed a number of polymer-ceramic nanocomposites, in which nanopowders weren't added, and the proportion HMPE was from 1 to 5 wt. %. The dependence of current density on the amount of binder is shown in Figure 30.3 (curve 1).

Indeed, as it is seen from the figure, the implementation of the proximity effect leads to a substantial increase in current density. In this case, with the increase of the share of binder the current density decreases monotonically and at 5% of the content of HMPE becomes zero. Additives in the same nanocomposite of nano aluminum powder in the proportions HMPE:

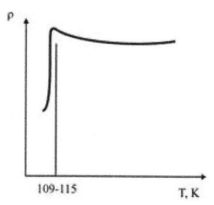

FIGURE 30.2 The change in resistance of nanocomposites with bismuth ceramics with aniline, with HMPE binder and with the addition of nano-sized silver powder.

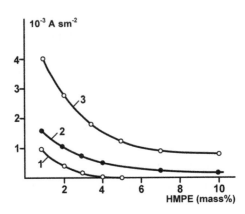

FIGURE 30.3 Dependence of current density on the amount of polymer binder.

nano-additives = 1:1 (wt%) increases the current density else more (Figure 30.3, curve 2).

As it is seen from curve 2 of Figure 30.3, increasing of the quantity of HMPE also reduces the current density from $-1.5 \cdot 103$ A·cm^{-2}(1% mass content of binder) and already at 10% content of HMPE it becomes equal to ~ 150 A· cm^{-2}. The critical transition temperature into superconducting state is 96 K and the width is 6°. Probably relatively low current density in the nanocomposites is determined by the agglomerated state of aluminum nanoparticles.

Therefore, the further shaping of the joint venture superconducting nanocomposites was carried out under the influence of acoustic fields. Indeed as it is seen from curve 3 in Figure 30.3 application of acoustic fields in the pressing process leads to rather sharp increase in current density, which is associated with deagglomeration of aggregated particles and their uniform distribution over the volume of samples. As it is seen from curve 3 in Figure 30.3, at 1% binder content of HMPE current density reaches $4 \cdot 10^3$A·cm^{-2}.

30.3.2 PHYSICAL-MECHANICAL PROPERTIES OF POLYMER-CERAMIC NANOCOMPOSITES WITH TRANSPORT CURRENT

Samples with low content of HMPE are quite fragile, in connection with which difficulties arose when applying contacts by thermo-compressing. For example, superconducting polymer-ceramic nanocomposites with 5% HMPE have the following physical and mechanical characteristics: ultimate

tensile strength of −0.7 MPa, modulus of elasticity of −1.4 MPa and elongation −3%.

It is interesting that the addition of nano-sized aluminum powder in
superconducting polymer-ceramic nano-composite with nearly constant
critical transition temperature into superconducting state (Meissner effect)
and of the current density leads to significant improvement in physical –
mechanical properties (Table 30.1).

Increasing of the critical transition temperature into superconducting
state is evidently associated with the processes of intercalation of binders'
macromolecule fragments into the layered structures of ceramic grains [19–
21]. As to noticeable improvement in physical – mechanical properties of
SC polymer-ceramic nanocomposites, it is possible that such a fact is associated with the introduction of nano aluminum particles into the polymer
matrix HMPE, as well as their uniform distribution in the volume of binder
and the formation of a rigid amorphous fraction of the polymer on the surface of nano-particles [23–26].

Thus, on the basis of these results we can conclude that the simultaneous
implementation of the proximity effect, additives of nano-sized powders of
aluminum and their uniform distribution over the samples' volume under the
influence of acoustic fields provide a current-carrying SC polymer-ceramic
nanocomposites with current density to $\sim 4 \cdot 10^3$ A·cm^{-2} with simultaneous
improvement of physical and mechanical properties.

TABLE 30.1 Impact of the Initial Composition of Polymer-Ceramic Nanocomposites on
Physical–Mechanical and Superconducting Properties

The weight ratio HMPE: nanoadditive ceramics, mass. %	σ, MPa	E, MPa	ε, %	Tc, K	Tk, K
1:1:98	0.27	0.4	2.0	95	88
2:2:96	0.33	0.56	3.5	95	88
3:3:94	0.68	1.2	5.0	96	88
4:4:92	1.00	2.0	7.0	96	88
5:5:95	1.50	5.0	9.0	95	88
10:10:80	3.80	9.0	11.0	96	88

30.3.3 MORPHOLOGIC PECULIARITIES OF CURRENT-CARRYING SC POLYMER-CERAMIC NANOCOMPOSITES

To study the morphological features of current-carrying polymer-ceramic nanocomposites the melting temperature (Tm) and the enthalpy (ΔHm) of nanocomposites containing 5% nano-sized aluminum (mass relative to HMPE) by differential scanning calorimetry are defined. A wide range of the nanocomposites composition (the ratio of HMPE with an oxide ceramic $Y_1Ba_2Cu_3O_{6.97}$) was studied. The results are shown in Table 30.2.

It is seen that with increasing of the filling degree in nano-composites the enthalpy of melting is increasing.

The observed increasing in enthalpy value is associated either with the degree of crystallization (Tables 30.2 and 30.3) or with a change in morphology of the binder in the surface layers of the phase.

However, on the basis of the obtained results it is impossible to establish unambiguously the decisive role of any of these factors. To do this, by DSC methods in regime of scanning temperature the influence of oxide ceramic quantities $Y_1Ba_2Cu_3O_{6.97}$ on the heat of melting and electron-microscopic investigations of samples of polymer-ceramic nanocomposites with transport current were carried out.

The dependence of the samples' melting heats from temperature for the SC polymer-ceramic nanocomposites, obtained by varying the initial temperature of samples' molding, and its content with a binder of BPE is shown in Figure 30.4.

From the curves on Figure 30.4 it is seen that the maximum values of the melting heats (curves 1–4, Figure 30.4) SC polymer-ceramic products are practically independent of the initial forming temperature (curves 1,2, 3, Figure 30.4), whereas quite strongly depend on the amount of ceramics in the

TABLE 30.2 Impact of Filling Degree on Temperature and Enthalpy of Melting of Binder in Nanocomposites HMPE + $Y_1Ba_2Cu_3O_{6.97}$

The weight ratio HMPE: $Y_1Ba_2Cu_3O_{6.97}$	Tm start	ΔHm, per gram HMPE	The degree of crystalline, %
100:0	140	115.0	39.1
80:15	151	121.5	41.4
45:50	140	128.5	43.7
10:85	138	130.0	44.2

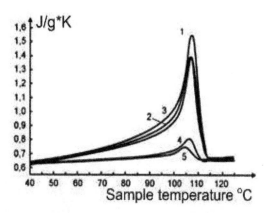

FIGURE 30.4 Impact of temperature on the variation of the melting heats for SC polymer-ceramic nanocomposites obtained at different initial temperatures (T0) and initial ratios of BPE with ceramics: T0, 0C 130 (1), 140 (2), 160 (3) $Y_1Ba_2Cu_3O_{6.97}$:BPE = 90:10 (1–3), 97:3 (4), 99:1 (5). The quantity of nano-sized aluminum 5% of the weight of $Y_1Ba_2Cu_3O_{6.97}$.

composite (curves 4–5, Figure 30.4). Temperature of melting and enthalpy determined from the data in Figure 30.4 are presented in Table 30.3.

Thus, in the case of a binder of BPE with the increase of the filler amount the enthalpy and the degree of crystalline are increasing. Therefore, based on obtained data we can suppose that a fairly strong increase of the value of ΔHm(calculated per gram HMPE and BPE) could be due to two reasons:

- increase of the amount of SC ceramics leads to an increase in the degree of crystalline;
- intercalation of fragments or individual elements of macromolecular binding into the interlayer spaces of the ceramic grains alters the morphology of super-high-molecular or branched polyethylene on the boundary of phase section of ceramic-binder, which is more likely [18–21].

Indeed, the study of the structural features of the current-carrying SC polymer-ceramic nanocomposites by scanning electronic microscopy in case of using of both amorphous and crystalline polymers, with complete and uniform enveloping of ceramic grain by polymer binders (Figure 30.5a, b), indicates that there is quite strong interaction on the boundary of ceramic-polymer's phase.

Interaction of macromolecular binder of PMMA and PS with the surface of the ceramic grains $Y_1Ba_2Cu_3O_{6.97}$ may contribute to the formation of rigid amorphous fraction (RAF) of polymer binder.

TABLE 30.3 Impact of Filling on the Temperature of Melting and Enthalpy in the Nanocomposites BPE + $Y_1Ba_2Cu_3O_{6.97}$

The weight ratio of BPE: $Y_1Ba_2Cu_3O_{6.97}$	Tm start	ΔHm, per gram BPE	The degree of crystalline, %
90:10	107	84	29
97:3	107	97	33
99:1	105	133	45

 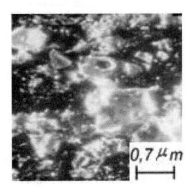

FIGURE 30.5 EM Microphotography of nanocomposite SC samples with binders of the PMMA (a) and PS (b). Ratio of binder:Ceramics = 15:85.

In current-carrying SC nanocomposites with binders HMPE, PP, regardless of the amount of SC-ceramics, by analogy with [23–26], there are fibers structures formations. For example, in Figure 30.6 are given electron microphotographs' photos of samples with the HMPE binder. As it can be seen from the figure the formation of fibers structures, not typical for polyethylene, takes place.

As it has already been noted in Refs. [18–21], fibers formations are the result of intercalation of macromolecule fragments of HMPE in the layered structure of ceramic grains. Such a binding of macromolecules affects the mobility of macro-chains of super-high molecular polyethylene, reduces their flexibility. Consequently, the crystallization of macromolecules, associated in this way, takes place via cooperative interaction between them.

It should be specially noted that in case of current-carrying polymer-ceramic nanocomposites with a binder of IPP, the above mentioned phenomena are more emphasized. Indeed, as it is seen from the change of the dependences of heat capacity from temperature (Figure 30.7), a splitting of peaks into two components is observed.

Possibly, the splitting is a result of the presence of two various structures in polymer-ceramic nanocomposites with IPP binder.

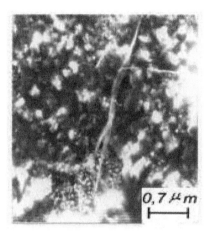

FIGURE 30.6 EM Microphotographs of polymer-ceramic nanocomposites with HMPE binder. Ratio of binder: ceramics: nanoaluminum – 5:90:5.

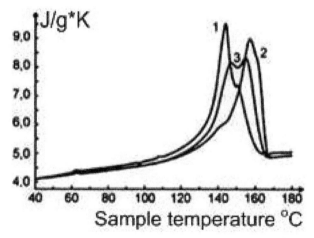

FIGURE 30.7 Temperature impact on the variation character of melting heats for the SC polymer-ceramic nanocomposites obtained at different initial ratios of PP with ceramics. $Y_1Ba_2Cu_3O_{6.97}$:PP = 85:15 (1) 70:30 (2), 50:50 (3).

As a proof of such an assertion can be the electron-microscopic investigations. Indeed, as it is seen from the microphotographs of samples with IPP binder (Figure 30.8), here the number of fibers formations is much larger than in HMPE binder.

A large number of formed fibers structures are probably to be the cause of splitting (Figure 30.8) peaks of melting temperatures in SC polymer-ceramic nano-composites with a binder of isotactic propylene.

Thus, the obtained results permit to conclude that the introduction of nano-sized aluminum powders allows to obtain SC polymer-ceramic

FIGURE 30.8 EM microphotography of polymer-ceramic nanocomposite with IPP binderof binder:Ceramics:nanoaluminum = 15:80:5.

nanocomposites with a transport current with current density up to ~ $4 \cdot 10^3 A \cdot cm^{-2}$. A natural question arises regarding the possibility of increasing the current density. The answer is positive, and it is possibly with use of other binders and nano-fillers of a different nature with varying average size of nano-particles.

KEYWORDS

- current density
- current-carrying
- high-temperature superconductors
- interphase space
- laminar structure
- physical-mechanical properties
- polymer-ceramic nanocomposites

REFERENCES

1. Jha, S. R., Reddy, Y. S., Sharma, R. G., (1987). *Chemistry and Materials Science, 31,* 167–169.

2. Hill, D. M., Gao, Y., Meyer III, H. M., Wagener, T. J., Weaver, J. H., Capone II, D. W., (1988). Phys. Rev., B 37, 511–514.

3. Meyer, H. M., Hill, D. M., Wagener, T. J., Weaver, J. H., Capone II, D. W., (1988). *Phys. Rev. B 38*, 6500–6507.

4. Gao, Y., Vitomirow, J. M., (1988). *Ibid. 37*, 3741–3746.

5. Romano, L. T., Wilshaw, P. R., Long, N. J., Grovenor, C. R. M., (1989). *Supercond. Sci. Tech., 1*, 285–289.

6. Lindberg, P. A. P., Shen, Z., -X., Wells, B. O., Dessau, D. S., Mitzi, D. R., Lindau, I., Spicer, W. E., Kapitulnik, A., (1989). *Phys. Rev., B 39*, 2890–2896.

7. Hikata, T., Sato, K., (1989). *Jap. J. Appl. Phys. 28*(1), 182–188.

8. Gafarov, S. F., Dzhafarov, T. D., (1990). *Technical Physics, 16*(9), 59–67.

9. Asoka Kumar, P. S., Mahamuni, S., Kulkarni, P., Mulla, I. S., Chandrachood, M., Sinha, A. P. B., Nigavekar, A. S., Kulkarni, S. K., (1990). *J. Appl. Phys. 67*, 3184.

10. Docenko, V. I., Braude, I. S., Ivanchenko, L. G., Kislyak, I. F., Puzanova, A. A., (1996). *Low Temperature Physics, 22*, 935–937.

11. Goto, T., Kada, M., (1987). *Jap. J. Appl. Phys., 26*, 1527–1531.

12. Golyamina, E. M., Lykov, A. N., (1989). *Superconductivity, Physics, Chemistry and Technology. 2*, 51–60.

13. Davtyan, S. P., Tonoyan, A. O., (2008). High-Temperature Superconductors. Superconducting Polymer-ceramic Nanocomposites, Publ. Limush, Yerevan.

14. Tonoyan, A. O., Davtyan, S. P., Martirosyan, S. A., Mamalis, A. G., (2001). *J. Materials Processing Technology, 108*, 201–204.

15. Hayrapetyan, S. M., Tonoyan, A. O., Arakelova, E. R., Davtyan, S. P., (2001). *Polymer Sci., A43*, 1814–1825.

16. Davtyan, S. P., Tonoyan, A. O., Hayrapetyan, S. M., Manukyan, L. S., (2005). *J. Materials Processing Technology, 160*, 306–312.

17. Davtyan, S. P., Tonoyan, A. O., Tataryan A. A., Schick Ch, (2006). *Composite Interfaces, 13*, 535–541.

18. Anahit Tonoyan, Christoph Schick, Sevan Davtyan, Materials 2009, vol. 2, doi:10.3390/ma2042154, PP. 2154-2187 and Online Materials 2010, ISSN 1996-1944 www.Mdpi.Com/journal/materials.

19. Davtyan, S. P., Tonoyan, A. O., Tataryan A. A., Schick Ch, Sargsyan, A. G., (2007). *J. Materials Processing Technology, 163*, 734–742.

20. Tonoyan, A. O., Davtyan, S. P., (2013). Ceramic Nanocomposites Chapter 9 "High-Temperature Superconducting Polymer Ceramic Nanocomposites," Woodhead Publishing Limited, 284–322.

21. Davtyan, S. P., Tonoyan, A. O., Schick, Ch., (2009). *Materials, 2*, 2154–2187.

22. Encyclopedia of Polymers. (1977). Moscow: Soviet *Encyclopedia,2*, 1005.

23. Sargsyan, A. G., Tonoyan, A. O., Davtyan, S. P., Schick, Ch., (2007). *European Polymer Journal, 8*, 3113.

24. Davtyan, S. P., Berlin, A. A., Tonoyan, A. O., Schick, Ch., Ragovina, S. Z., (2009). *Russian Nanotechnologies, 4*, 489.

25. Davtyan Sevan, TonoyanAnait, (2014). Theory and Practice of Adiabatic and Frontal Polymerization, Monography. p. 668, Palmarium Academic Publishing, Saarbrucken, Deutschland.

26. Davtyan, S. P., Tonoyan, A. O., Berlin, A. A. (2011). Advances and Problems of Frontal Polymerization Processes. *Review Journal of Chemistry, 1*, 56–92.

PART IV

GREEN CHEMISTRY AND RECYCLING

THE INFLUENCE OF RUBBER RECYCLATE MORPHOLOGY ON THE PROPERTIES OF RUBBER-ASPHALT COMPOSITE

M. SIENKIEWICZ, H. JANIK, K. BORZĘDOWSKA-LABUDA, and Sz. KONKEL

Polymer Technology Department, Chemical Faculty, Gdansk University of Technology, Gabriela Narutowicza Street 11/12, 80-233 Gdansk, Poland, E-mail: helena.janik@pg.gda.pl, macsienk@pg.gda.pl

CONTENTS

ABSTRACT

Rubber products, particularly those used in the automobile industry, are responsible for a vast amount of wastes, mostly in the form of used tires.

One of the methods of waste tires application use them as a modifier of asphalt. However, the properties of rubber-asphalt composites depend on the morphology of waste tiregranulate and kind of tire recycled (car/truck). In thischapter, the influence of grinding methods of waste tire on the properties of rubber-asphalt composites were studied. SEM analysis indicates that waste tire granulate obtained using granulator with flat-die granulation process has a higher specific surface area. In this case the higher quality of the rubber-asphalt composite may be achieved.

31.1 INTRODUCTION

Worldwide, the amounts of used polymer products are increasing by the year: most of them are created by automobile tires. Numerous postconsumer rubber products are a challenge not only for environmentalists, but also for engineers in the world. Rubber products are composed of long polymer chains, mostly cross-linked by sulfur bridges. Their lifetime is estimated for about 100 years. Reprocessing products due to their mechanical strength, thermal and solvent resistance, would be too laborious and therefore expensive [1–3].

The progress made in recent years in waste management of polymers means that postconsumer tires came to be seen as a potential source of valuable raw materials. Development of research on improving methods of recovery and recycling and regulations implementing the principle of extended producer responsibility in the field of life tires led to solutions that enable the transformation of rubber waste into energy or new polymer products [1, 4].

One of the methods of waste tires management use them as a modifier of asphalt. The addition of recycled rubber to asphalt improves tenacity, mechanical properties, fatigue life, flexibility, resistance to rutting, abrasion resistance, low temperature cracking, reduce noise during contact tire with the ground [5–12].

Rubber-asphalt composites obtained with the use of rubber wastes were described by many scientists [13–17]. Airley showed that the density of recycled rubber-composite is changing when dust rubber is added to asphalt. It was also proved that grains of rubber wastes swell when they are combined with asphalt and then the dust mass of the rubber wastes increases up to 140% by weight. This is due to absorption of aromatic hydrocarbon by the rubber granulate [14].

Absorption of light hydrocarbon fractions by rubber dust depends on asphalt content and the value of its penetration: the higher those parameters, the greater absorption. On the other hand, as it was noted before, rubber granulate is easily swelled by asphalt what causes also increasing the viscosity of the rubber-asphalt composite, compared to the viscosity of the virgin asphalt [15].

The interaction between the components of the rubber granules and asphalt depends on molecular weight of both components. The interaction is stronger when molecular weight of asphalt is lower. This is due to the possibility to spread the light fraction between the grains of the rubber dust [14].

Morphology of rubber-asphalt composites was also investigated. SEM analysis showed that pure asphalt is characterized by a smooth surface. Addition of recycled rubber changes the surface homogeneity: the larger particle of recycled rubber the higher heterogeneity[14, 15].

In this chapter, we present the results of research characterizing the influence of the grain surface morphology of recycled rubber for physical properties of rubber-asphalt composite. The studies were conducted on the example of asphalt modified by the rubber granules, which were obtained on the warm grinding processes of waste tires by using of a flat die pelleting press (modifier I) or knife granulator (modifier II).

31.2 EXPERIMENTAL PART

31.2.1 MATERIALS

Asphalt 70/100 used in our studies was delivered by LotosAsfalt Sp. z o.o. in Gdansk (Poland). Two types of rubber granulate were analyzed. Waste tire granulate grinded via innovative flat die pelleting press (modifier I) came from Gumeko Sp. z o.o. fromChwaszczyno(Poland). The granulate size was 0.2–08 mm. Waste tire granulate grinded via standard granulate process on the knife granulator (modifier II) was delivered by Orzeł S. A. from Lublin (Poland). The granulate size was 0–08 mm.

31.2.2 PREPARATION OF RUBBER-ASPHALT COMPOSITES

Rubber-asphalt composites were prepared with stirrer powered by "IKA RW 16 basic" in a metal container with a volume of approx. 1.2 dm^3. Firstly, pure asphalt was heated to a temperature of 180°C. Then appropriate quantities of the rubber granulate were added maintaining constant temperature to eliminate the destruction of the compounds present in the asphalt. Homogenization of the mixture was carried out for 1 hour at 180°C. The next stage of the study consisted a three-hour "maturation" of rubberized asphalt composition at 180°C in the oven. Table 31.1 shows the amounts of the individual ingredients used in all the performed modification processes using asphalt 70/100 and rubber granules.

31.2.3 SOFTENING POINT MEASUREMENTS

Samples were tested after "maturation" at 180°C for 3 h. The softening point was tested by"the ring and ball" method. The first step is filling two rings with rubber-asphalt composites then put balls on them. Then the rings are arranged on a special stand and put in a water which is heated at a rate of 5°C. The temperature at which the rubber-asphalt composite touches the base of the stand is considered as the softening temperature.

31.2.4 PENETRATION MEASUREMENTS

To determine the consistency of rubber-asphalt composites, penetration was determined; the harder composite, the higher value of penetration is noticed. Penetration measurements consist of insertion needle penetrometer into a sample with constant load for needle applied. Measurement time is 5 seconds; the measurement temperature is 25°C. Before testing, the samples should be placed at 25°C for at least 1h. In this study we used a semi-automatic penetrometer PNR12 from Petrotest.

TABLE 31.1   Ingredients used to Obtain Rubber-Asphalt Composite and Codes of the Samples

Type of composite	Type of granulate (modifier)	Amount of granulated rubber wastes [% by mass]	Amount of asphalt 70/100 [% by mass]
A	I	5	95
		10	90
B	II	5	95
		10	90

31.3   RESULTS AND DISCUSSION

31.3.1   MORPHOLOGY OF RUBBER GRANULATE

Figure 31.1 presents SEM micrographs of waste tire granulates grinded via innovative flat-die granulation process (modifier I) and standard granulation process (modifier II). The SEM micrographs of both granulates made at the corresponding magnification of the image show differences in the morphology of waste tire granulate.

In the case of waste tire granulate shredded via standard granulation process, more regular shapes of particles are observed. The walls of the grains are mostly completely smooth, what results in minimal specific surface area of the granules. There exist a few jags which are created probably by friction of rubber pieces during shredding of rubber wastes. However, they do not have much influence on development of the specific surface area of the grains.

In the case of waste tire granulate shredded via innovative flat die granulation process, spongy shaped particles are observed. The product has mostly jagged walls of particles comparing to the granules shredded via standard granulation process. This effect was achieved by using a granulator with a flat die and thickening rollers. As a result, the rubber material is squeezed through openings of a perforated plate granulator, allowing to obtain particles with high specific surface area.

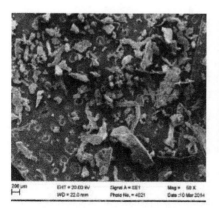 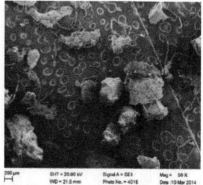

FIGURE 31.1   Morphology of surfaces of two types of waste tire granulates (shredded via flat die granulation process – right, shredded via standard granulation process – left).

TABLE 31.2   Softening Point of Rubber-Asphalt Composite

Type of composite	Granulation process	Amount of modifier [%]	Softening point [°C]
A	Flat die granulation process	0	45.7
		5	49.6
		10	54.8
B	Standard granulation process	0	45.7
		5	50.2
		10	55.2

31.3.2 SOFTENING POINT

Table 31.2 and Figure 31.2 present results of softening point of rubber-asphalt composite modificated by two kinds of waste tire granulate. It can be concluded that with an increasing amount of modifier content the softening point becomes higher. It should be noted that the increase in the softening temperature is relatively proportional to the increase of granulate content. As can be seen in Figure 31.2, there are rather slight differences in softening point value depending on type of granulates(I or II) used as asphalt modifiers.

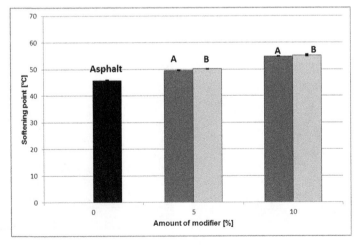

FIGURE 31.2   Softening point of rubber-asphalt composites (A – containing rubber granulate shredded via flat die granulation process, B – containing rubber granulate shredded via standard granulation process).

TABLE 31.3   Penetration of Rubber-Asphalt Composites

Type of composite	Granulation process	Amount of modifier [%]	Penetration [0,1 mm]
A	Flat die granulation process	0	83
		5	71
		10	68
B	Standard granulation process	0	83
		5	72
		10	64

31.3.3   PENETRATION

As can be seen from Table 31.3 and Figure 31.3, penetration in the rubber-asphalt-composite changes when rubber granulate is added. It is observed for both types of granulates. The more rubber granulate is added the lower penetration value is achieved. In addition, smaller decline in the penetration value was observed in compositions modified by granulate (I) compared to compositions modified by granulate (II). Thus waste tire granulatation

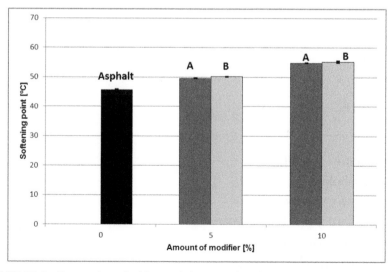

FIGURE 31.3   Penetration of rubber-asphalt composites (A – containing rubber granulate shredded via flat die granulation process, B – containing rubber granulate shredded via standard granulation process).

process via innovative flat die pelleting press appeared more effective to improve penetration value of rubber-asphalt composites.

31.4 CONCLUSIONS

It has been shown that the morphology of ground tire rubber used to prepare rubber-asphalt composite has the influence on the properties of the final product.

1. Both softening temperature and penetration were improved by adding to asphalt some modifiers obtained by grinding of waste tires.
2. Softening point of rubber-asphalt composites increased when rubber granulate was added to asphalt. Softening temperature measurements showed comparable improvement in properties of rubber-asphalt composites modified by granulate shredded via flat die granulation process (modifier I) and granulate shredded via standard granulation process (modifier II).
3. Penetration value studies revealed that more effective modifier to improve penetration of rubber-asphalt composites was granulate obtained via flat die granulation process (modifier I) especially in case of adding 10% of rubber granulate.

ACKNOWLEDGMENTS

Lotos Company from Gdansk (PL) is gratefully acknowledged for delivering asphalt for our studies.

KEYWORDS

- **modified asphalt**
- **morphology**
- **rubber granulates**
- **standard granulation process**
- **flat-die granulation process**
- **rubber-asphalt composite**
- **waste tires**

REFERENCES

1. Sienkiewicz, M., Kucinska-Lipka, J., Janik, H., Balas, A., (2012). *Waste Management. 32*, 1742–1751.
2. Karger-Kocsis, J., Meszaros, L., Barany, T., (2013). *J. Mater Sci. 48*, 1–38.
3. Wojciechowski, A., Michalski, R., Kaminska, E., (2012). *Polymer. 11*, 9.
4. Ivan, G., (2002). *Elastomer. 6*, 24–38.
5. Gronowicz, J., Kubiak, T., (2007). *Exploitation Problems. 2*, 5–7.
6. Xiao, F., Amirkhanian, S. N., (2010). *Materials and Structures. 43*, 223–233.
7. Weidong Cao, (2007). *Construction and Building Materials. 21*, 1011–1015.
8. Navarro, F. J., Partal, P., Martínez-Boza, F., Gallegos, C., (2004). *Fuel. 83*, 2041–2049.
9. Yue Huang, Bird, R. N., Heidrich, O., (2007). *Resources, Conservation and Recycling. 52*, 58–73.
10. Gawel, I., Kalabinska, M., Pilat, J., (2001). Asfalty drogowe (Road Asphalts), WKŁ, Warszawa.
11. Radziszewski, P., Kalabińska, M., Piłat, J., (2001). *Elastomery (Elastomers). 4*, 8–16.
12. Navarro, F. J., Partal, P., Martínez-Boza, F., Gallegos, C., (2010). *Polymer Testing. 29*, 588–595.
13. Huang, Y., Bird, N. R., Heidrich, O., (2007). *Conservation and Recycling. 52*, 56–73.
14. Airey, G., Rahman, M., Collop, A., (2003). The influence of crude source and penetration grade on the interaction of crumb rubber and bitumen ,International Asphalt and Rubber Conference, Brasilia, Brazil.
15. Ying, G., Rongji, C., (2010). *Journal of Wuhan University of Technology, Mater. 25*, 853–855.
16. Gaweł, I., Stepkowski, R., Czechowski, F., (2006). *Industrial and Engineering Chemistry Research. 45*, 3044–3049.
17. Radziszewski, P., Piłat, J., Sarnowski, M., Kowalski, K., Król, J., Ruttmar, I., Witczak, S., Heinrich, P., Skowroński, D., Krupa, Z., Nawracaj, M., Gil, K., (2011). *Nawierzchnieasfaltowe (Road Surfaces). 3*, 10–15.

CHAPTER 32

DEGRADATION OF MODIFIED TPS IN THE NATURAL AND INDUSTRIAL COMPOST

H. JANIK, M. SIENKIEWICZ, M. WAWROWSKA, K. WIECIERZYCKA, and A. PRZYBYTEK

Polymer Technology Department, Chemical Faculty, Gdansk University of Technology, Gabriela Narutowicza Street 11/12, 80–233 Gdansk, Poland, E-mail: helena.janik@pg.gda.pl, macsienk@pg.gda.pl

CONTENTS

ABSTRACT

Nowadays, there is a big interest in a new class of polymeric materials that can be degradable after usage in natural environment or industrial composting. Biodegradable/composting polymers are converted by the action of micro-organisms(bacteria, fungi, algae) to carbon dioxide, water and biomass. The aim of the study was to determine in the pilot studies the degree of decomposition of modified by us thermoplastic starch (TPS) by epoxidized soybean oil and gum arabic in the environment of compost consisting of sewage sludge. Commercial foils for packaging advertised as compostable (*Castorama, BioBag*) were tested for comparison as well The samples were placed in an open roofing at normal weather conditions. The degree of sample degradation was analyzed by examining the change in weight loss and visual view of the samples. The tests were carried out for 2 weeks in July, in the north of Poland (Władysławowo). For the comparison standardized test of industrial composting overworked for biodegradable polymers were done as well for new prepared foils. Studies have shown more visible changes in surface view in the samples of modified TPS by epoxidized soybean oil and arabic gum in comparison to commercial samples. Both new prepared foils and commercial ones did not change their mass after incubation in the compost after 7 days.

32.1 INTRODUCTION

Conventional plastic packages are post-consumer wastes which are difficult for disposal and constitute a burden to the environment, due to their large volume and inability to biodegrade or to compost [1–3]. Due to the current environmental requirements, in the mid-90s of the twentieth century, innovative technologies for obtaining polymer materials from renewable raw materials, and also technologies/biotechnologies leading to the manufacture of biodegradable or compostable plastics started to be implemented [4–12].

Starch obtained from potato, corn, wheat and rice is of industrial importance [2–4]. Starch may be used to modify many polymers in the form of thermoplastic starch (TPS). TPS is biodegradable, can be manufactured by means of conventional technologies (extrusion, thermoforming, high pressure injection) which are used for the manufacturing of the first generation, non-biodegradable, synthetic polymers.

In our studies, we used commercial compostable samples and new prepared in our laboratory foils [4] from potato TPS, epoxidized soybean oil (EOS) and Arabic gum. The introduction of epoxidized soybean oil – EOS, as an additive modifying TPS will allow in the future improving the compatibility of TPS with other polymers like PLA[4–6].

On the other hand, the introduction of the modifying additive, which is arabic gum (GA), being a natural thickening and emulsifying compound prevents flocculation and coalescence of epoxidized vegetable oils used in the modification of TPS. Arabic gum also often leads to formation of a durable coating on the surface of polymer composition, which provides a high barrier capacity for oxidation processes of its components and improvement of adhesion of the composition to overprints, without disturbing its glossiness.

The aim of the experiments was to determine the effect of natural additives on the degree of degradation of thermoplastic starch by registration of surface and mass changes of the samples incubated in degradation conditions and compare those results with some commercial samples denoted as compostable.

32.2 EXPERIMENTAL PART

32.2.1 MATERIALS

32.2.1.1 Commercial Foils

- Samples cut out from shopping bag (Figure 32.1a), sold in supermarket chains *Castorama*(thickness 0.020 mm)
- Samples cut out from the film produced by BioBag(thickness 0.031 mm, Figure 32.1b)
- The above products are claimed by the manufacturers as biodegradable in compost conditions
- (*Biobag.* after 45 days and *Castorama* after 42–63 days).

Laboratory foils *(samples obtained according to a new idea* [4] *of preparation polymer composition for biodegradable polymer material).* Each type of laboratory sample was characterized by a thickness of the order of 0.7 mm and was cut out from the following materials prepared in our laboratory:

FIGURE 32.1   The view of commercial foils (a)BioBag, (b)Castorama.

- thermoplastic starch (TPS);
- thermoplastic starch modified with epoxidized soybean oil – TPS/ EOS (0.5% and 1% w/w of modifier added to TPS);
- thermoplastic starch modified with epoxidized soybean oil and arabic gum (TPS/05% EOS were modified with 1% w/w of Arabic gum) – TPS/0.5%EOS/1% GA.

32.2.2 PREPARATION OF THE LABORATORY NEW FOILS

In a first step, native potato starch is mixed at room temperature, using a mechanical mixer, with a plasticizer in the form of glycerol and an additional modifying component [epoxidized vegetable oil (EOR) or/and arabic gum (GA)]. The resulting mixture is subjected to conditioning in closed vessels for a period of 6–24 hours. In a second step, the mixture from the first step is subjected to extrusion using a single-screw, while maintaining the head temperature in the range of 120–2109°C and the temperature of the heating zones of the plasticizing unit in the range of 100°C to 210°C, with simultaneous degassing of volatile matter. The thermoplastic extradite is cooled down with air to room temperature and subjected to granulation, thus obtaining a modified thermoplastic starch (mTPS) granulate with a diameter of approx. 1 mm. Then the proper foil is obtained by melt press procedure at the temperature of 180°C.

32.2.3 DEGRADATION TEST

32.2.3.1 Composition of the Compost

We used for the studies the compost from the sewage treatment plant "Water and Sewage Company Swarzewo"(Pomerania district in Poland). According to the definition proposed by the European Committee for Standardization (CEN) sludge is called as a mixture of water and solids separated from various types of water as a result of natural or artificial processes. The compost used for testing was characterized by the presence of micro-organisms like aerobic bacteria, nematodes, thermophilic bacteria, annelids and molds.pH was 6.7, moisture content 40–70%.

32.2.3.2 Procedure of Degradation Test

The specimens were placed in the plastic frames with a size of 10x7 cm free surface of polymer for studies and put to the compost at room temperature in sheltered outdoor conditions. Samples were covered with several centimeters thick layer of compost (Figure 32.2). Each type of sample was divided into 3 series. The samples from every series were drawn out from degradation environment after 2, 6 and 7 days. Then they were cleaned out from remnants of compost and analyzed.

Standardized test of composting was also carried out according to PN-EN 14806:2010.

32.2.4 VISUAL EVALUATION OF SAMPLE

To estimate the degree of degradation of polymer samples in the compost environment visual evaluation of the samples was performed twice: before inserting the samples into compost and after the proper time of incubation. It was estimated the general view, the color of the samples and the texture on the surface.

FIGURE 32.2 The view of the set prepared for incubation of the samples in the compost.

32.2.5   WEIGHT CHANGES AFTER THE INCUBATION OF THE SAMPLES IN THE COMPOST

Calculated weight changes of samples were based on the measured weights before and after incubation in the environment. Measurements were made using an analytical balance WPS 510/C/2. The mass loss in percentage was determined from the following formula:

$$\Delta m = \frac{(m_1 - m_2)}{m_1} \times 100\%$$

where:Δm – percentage of mass loss of the polymer sample [%]; m_1 – mass of the polymer sample before the start of the experiment [g]; m_2 – mass of the sample after a specified period of incubation time in the compost [g].

32.3   RESULTS AND DISCUSSION

32.3.1   VISUAL OBSERVATION OF COMMERCIAL SAMPLES

No clear changes were observed in the surface view after incubation in compost.

32.3.2   VISUAL EVALUATION OF THERMOPLASTIC STARCH (TPS) FOILS (FIGURE 32.3)

There were noted changes in texture and color of the material. There are visible numerous small cracks after 2 days of composting, which became bigger after farther incubation time. No mold on the sample surface was observed.

32.3.3   VISUAL EVALUATION OF TPS/0.5% EOS (FIGURE 32.4)

There were observed changes in the color and texture of the sample TPS/0.5% EOS starting from the second day of incubation. Big cracks appeared on the surface of the sample after 6 days of composting. Moreover the presence of mold was observed after 7 days of composting.

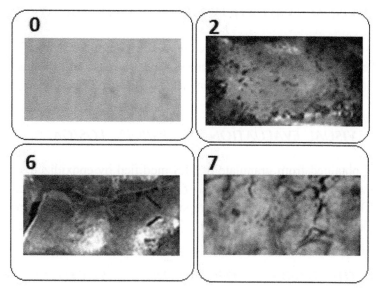

FIGURE 32.3   The TPS sample's view after incubation in the compost for a period of 2, 6 and 7 days.

FIGURE 32.4   The TPS/0.5%EOS samples' view after incubation in the compost for a period of 2, 6 and 7 days

32.3.4 VISUAL EVALUATION OF TPS/1% EOS (FIGURE 32.5)

There were observed changes in the color and texture of the sample TPS/1% EOS after 2 days of incubation. Huge cracks appeared on the surface after 6 days of incubation. The presence of mold was also visible after 7 days of composting.

32.3.5 VISUAL EVALUATION OF TPS/0.5% EOS/GA1%

The foil obtained from granulate of TPS modified by epoxidized soybean oil in a weight ratio of 0.5%, and 1% of gum arabic clearly lost the contemporary flexibility after 2 days of composting, many small cracks appeared after 6 days which became bigger after 7 days of composting (Figure 32.6).

32.3.6 THE DEGREE OF DEGRADATION – WEIGHT CHANGES

32.3.6.1 Commercial Samples

No weight changes were registered after the incubation in compost environment for both commercial samples studied.

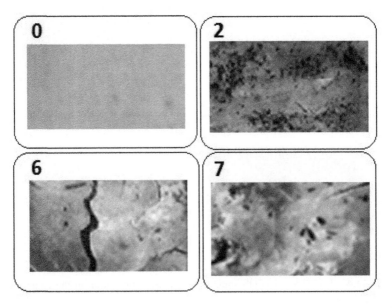

FIGURE 32.5 The TPS/1.0%EOS samples' view after incubation in the compost for a period of 2, 6 and 7 days.

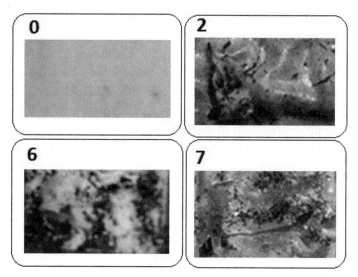

FIGURE 32.6   The TPS/0.5%, 0%EOS/1%GA samples' view after incubation in the compost for a period of 2, 6 and 7 days.

32.3.6.2   Laboratory Samples

No weight changes were registered after the incubation in compost environment for laboratory new foils prepared according to the procedure described in this chapter.

32.3.6.3   The Degree of Degradation – Standardized Test

The laboratory samples were composted in 100% according to the demand of PN-EN 14806:2010 (no residue on the sieves after 60 days).

32.4   CONCLUSIONS

1. All new laboratory prepared foils passed the test of composting according to PN-EN 14806:2010 (commercial composting).
2. No weight changes were noted for all samples (commercial and new laboratory foils) incubated in outdoor conditions for 7 days in the compost having sewage sludge in its composition.
3. No visible changes were observed in the surface view of commercial samples after the incubation for 7 days in compost environment (sewage sludge, outdoor conditions).

4. Clear changes in surface view were registered for all new laboratory foils even after 2 days of incubation in compost having sewage sludge in its composition (outdoor conditions). The modified TPS by epoxidized soybean oil (0.5%) and Arabic gum (1%) appeared the most vulnerable to degradation in compost having sewage sludge (outdoor conditions).

5. Epoxidized soybean oil and Arabic gum are promising modifiers to speed up degradation of biodegradable/compostable foils, which at present are mostly compostable at elevated temperature.

KEYWORDS

- **biodegradable/composting polymers**
- **biodegradation**
- **degradation**
- **environment of compost**
- **sewage sludge**
- **thermoplastic starch**

REFERENCES

1. Gregory, M. R., (2009). Environmental implications of plastic debris in marine settings—entanglement, ingestion, smothering, hangers-on, hitch-hiking and alien invasions, *Phil. Trans. R. Soc. B, 364,* 2013–2025.
2. Nafchi, A. M., Moradpour, M., (2013). Thermoplastic starches: Properties, challenges and prospects, *Starch, 65,* 61–72.
3. Belhassen, R., Vilaseca, F., Mutje, P., Boufi, S., (2014). Thermo plasticized starch modified by reactive blending with epoxidized soybean oil, *Industrial Crops and Products,58,* 261–267.
4. Janik, H., Sienkiewicz, M., Kucińska-Lipka, J., Bożek, K., Lachowicz, A., Stelmasik, A., EPO Application 15460007.6/EP 1540007, 4.03.2015.
5. TaydeSaurabh, Patnaik, M., Bhagt, S. L., Renge, V. C. (2011). Epoxidation of vegetable oils: a review. *International Journal of Advanced Engineering Technology, 2,* 491–501.
6. Xiong, Z., Yang, Y., Feng, J., Zhang, X., Zhang, C., Tang, Z., Zhu, J., (2013). Preparation and characterization of poly(lactic acid)/starch composites toughened with epoxidized soybean oil, *Carbohydrate Polymers, 92,* 810–816.
7. Carvalho, A. J. F., Curvelo, A. A. S., (2005). Surface chemical modification of thermoplastic starch: reactions with isocyanides, epoxy functions and stearoyl chloride, *Industrial Crops and Products, 21,* 331–336.

CHAPTER 33

RENEWABLE RESOURCES FOR POLYURETHANES AND POLYURETHANE COMPOSITES: A REVIEW

J. T. HAPONIUK, A. HEJNA, Ł. PISZCZYK, J. DATTA, and K. FORMELA

Polymer Technology Department, Chemical Faculty, Gdansk University of Technology, Gabriela Narutowicza Street 11/12, 80–233 Gdansk, Poland, E-mail: jozef.haponiuk@pg.gda.pl

CONTENTS

ABSTRACT

Each year, more than two million tons of polyurethane is produced in the EU by reacting isocyanides with polyols made from fossil fuel. In addition, there are appreciable quantities of petroleum based functional additives applied in the industry nowadays for both polyols and polyurethane materials. It is therefore of key importance to develop sustainable economically viable polyols with enhanced functionalities, and thereby reducing the need for polyols and additives made from non-renewable fossil fuel. By basing polyurethanes on biomaterials, the carbon footprint is reduced, and using biomass waste/byproducts will further improve the environmental profile. Relatively new materials from renewable resources that can be incorporated into polyurethane technology are glycerol and its polymeric form. Incorporation of these compounds is related to the production of biofuels, based on fatty acid methyl esters. Glycerol is a by-product in biodiesel production; 100 kg of glycerol is obtained for each tone of fuel. According to European Biodiesel Board, consumption of biodiesel in EU will exceeds yearly 15 Mt, what gives 1.5 Mt of waste glycerol to utilize which can be polymerized and successively used in PU. This work, based on the literature and personal research, presents the status of knowledge about application in the synthesis of polyurethanes of renewable raw materials and waste materials, mainly derived from natural oils, waste glycerol from biodiesel production and biomass.

33.1   INTRODUCTION

Nowadays, global trends on sustainable development and environmental protection are strongly related to the reduction of greenhouse effect through the limitation of harmful gases' emission. There has been implemented number of law regulations, which should help in achieving this goal, such as Kyoto Protocol, European Union Directives or EU Climate Package 3x20. All these regulations have one thing in common – they are created to fight the excessive emission of carbon dioxide and to promote the use of energy and raw materials from renewable resources. For example, EU Directive 2009/28/WE is obligating the member countries to reach the 10% share of biofuels in transport sector [1]. Climate Package 3x20 states that the emission of CO_2 in 2020 should be 20% lower than in 1990; the share of renewable energy should reach 20%; and consumption of energy should be decreased by 20% [2]. With such obligations, all the branches of chemical industry and many researchers have focused their attention and resources on the development of environmentally friendly technologies based on ecological materials. Such trend applies especially to oleo chemicals industry, which is rapidly growing over the last years. More and more chemicals derived from plant and animal fats are appearing on the market and they are not only ecologically, but also economically competitive to petrochemical analogs. Similar situation is observed in polymer industry, market offers increasing number of different products based on renewable materials; however the majority of polymer products are still manufactured using petrochemical materials.

Nowadays, polyurethane (PU) industry is strongly depending on petroleum, because the majority of polyols – main components used in PU manufacturing is petroleum-based. However, petroleum is not a renewable material, so its availability will become lower simultaneously increasing its price, which together with the previously mentioned global trends related to sustainable development and various law regulations will make companies pay more attention to polyols from renewable raw materials. Materials, which are less expensive and available almost all year round, because of their year-to-year regeneration. In terms of their chemical structure, biopolyols are generally esters of glycerol and unsaturated fatty acids [3]. Depending on the location, the most common oils used in biopolyols manufacturing are: in Europe – rapeseed and sunflower oils [4], in Asia – palm and coconut oil [5], and in USA – soybean oil [6–8]. Moreover, researchers were investigating application of other oils such as linseed oil or chestnut oil [9].

Biopolyols can be manufactured by number of various processes, including oxidation of unsaturated fatty acids, epoxidation of double bonds and further ring opening reactions, hydroformylation, ozonolysis and esterification of natural oils. Each of the processes shows some advantages and disadvantages.

Oxidation of fatty acids is generally simple process, although it is hard to control, except alcohols there are also obtained other products of oxidation, such as peroxides, aldehydes, ketones or carboxyl compounds, whose presence can have unfavorable influence on the properties of final products, such as discoloration, odor or uneven mechanical properties [10].

Epoxidation of natural oil guarantees better control over creation of hydroxyl groups in the molecules. Application of proper hydrogen donors can lead to high yield of process [12,13]. Polyols prepared through this process show hydroxyl values from tens to even 400 mg KOH/g and are quite resistant to oxidation and environmental influence due to the low content of double bonds, however it is often related to their high viscosity, which can be a problem from the technological point of view [14–16].

Petrovic et al. [17] prepared polyols by epoxidation of soybean oil and further ring opening assisted by methanol, hydrogen chloride and hydrogen bromide. Authors found out that chemical compound used for ring opening have significant influence on the density, mechanical properties, thermal stability and flammability of resulting polyurethane material. Incorporation of chlorine and bromine atoms to the structure of polyol led to preparation of self-extinguishing materials, without the incorporation of additional flame retardants.

Hydroformylation process includes reaction of oils with mixture of hydrogen and carbon oxide, and further conversion of aldehyde groups to hydroxyl groups by hydrogenation [18]. The main disadvantage of this process is the cost of used catalysts [19, 20]. The most effective catalysts are rhodium compounds allowing almost 100% conversion, however in order to use them in economically feasible way they need to be restored after reaction, which requires very expensive methods. Other, less expensive, potential catalysts are cobalt compounds, but they are not guarantying as high selectivity and yield of the process, it is estimated that they allow only 65% conversion [21]. Polyols obtained through hydroformylation are generally used for the manufacturing of rigid polyurethane foams or polyurethane coatings, however, their functionality can be decreased by partial esterification [22].

Ozonolysis is based on the conversion of unsaturated bonds into hydroxyl groups. This process allows almost quantitative conversion, although results also in generation of by-products – low molecular weight glycols and monohydroxyl alcohols. Alternatively, process can be also performed in the presence of glycols, causing simultaneous esterification of formed by-products with polyol[23, 24]. Generally, compounds prepared through the ozonolysis show noticeably lower molecular weight and viscosity comparing to polyols resulting from epoxidation or hydroformylation[25].

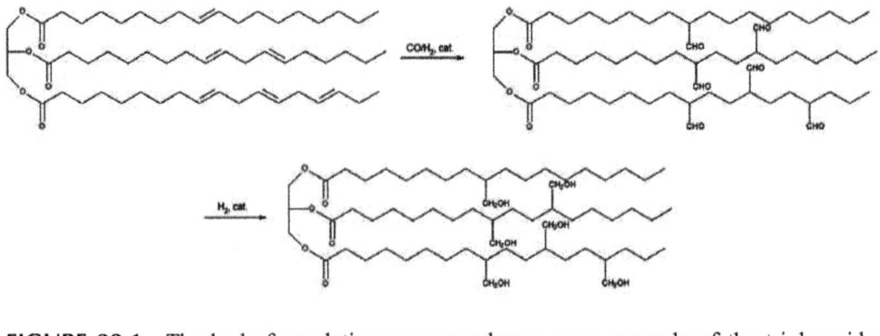

FIGURE 33.1   The hydroformylation process scheme as an example of the triglyceride-containing residues of oleic, linoleic and linolenic acids.

Although all methods mentioned above can result in polyols showing very satisfactory properties, the most popular process in manufacturing of polyols is esterification of natural oils with alcohols containing at least two hydroxyl groups, most of the with glycerol [26–28]. This method of production requires the use of catalysts in order to obtain high yields, however various low-cost and commonly available inorganic bases and acids might be applied [29]. Polyols obtained through esterification of glycerol with natural oils are often characterized by high hydroxyl values, which is very beneficial for manufacturing or rigid polyurethane foams, coatings and other materials with high crosslink density [30]. Moreover, except glycerol, different bifunctional glycols can be used, leading to bi-functional polyols suitable for production of flexible foams or thermoplastic polyurethane elastomers [31].

Although natural oils are important compounds in production of biopolyols, their production and technology is well known and analyzed. Nowadays more emphasis is put on the incorporation of biodiesel-based glycerol into production of biopolyols for polyurethane materials. Crude glycerol is a byproduct of biodiesel production, which becomes more and more popular. Diesel is considered one of the two most important or even the most important (e.g., in Europe) transportation fuel. Nowadays cars with diesel engine account for more than 55% of total numbers of produced cars [32]. Moreover, the European Union law regulations require the reduction of carbon dioxide emission for new cars to the level of 95 g/km in 2021 (there are proposals for further reduction to 68–78 g/km till 2025)[33]. Under those circumstances, popularity of the biodiesel as a replacement for petroleum-based fuel will be constantly growing. Over the last decade, European production of biodiesel growth from around 3 million tons to more than 9 million tons and according

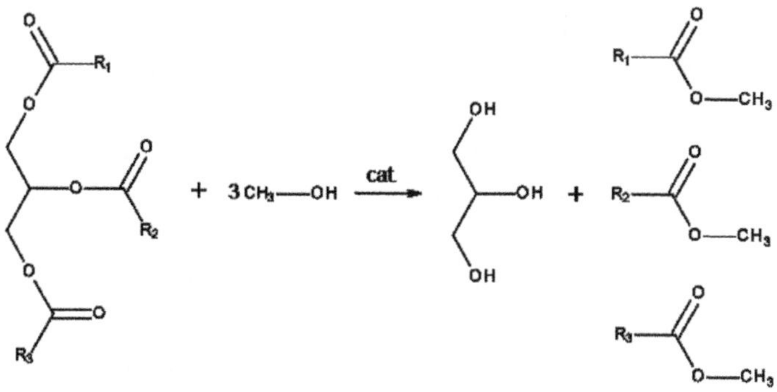

FIGURE 33.2   Transesterification of triglycerides resulting in crude glycerol and fatty acid methyl esters.

to forecasts it will exceed 15 million tons in 2020. Great enlivenment of bio-diesel market has been also observed outside of Europe [34, 35].

Around 100 kg of crude glycerol is obtained for each tone of produced biodiesel [36]; it means that almost 1 Mt of biodiesel-based glycerol is generated annually in Europe. Crude glycerol contains various impurities, such as methanol, water, soaps, residual catalyst, free fatty acids (FFA) or fatty acids methyl esters (FAME)[37, 38]. Detailed composition obvi-ously depends on the type of used oils, type and amount of catalyst and potential treatment applied to final product. Purification of crude glycerol is often very expensive and economically unavailable for smaller biodiesel producers, and at the same time is often essential in order to utilize it in various branches of industry, such as food or pharmaceuticals [39]. That is why, at this moment, it is very important to develop economically-effec-tive methods of purification or possibilities of crude glycerol utilization without complicated purification steps. Such actions could cause enliven-ment of the crude glycerol market and improve the quality of final product, simultaneously decreasing the cost of biodiesel production [40]. The most obvious way of crude glycerol utilization is the combustion resulting in energy production, however such method does not use the whole potential of this by-product. During the last few years, very popular topic among the researchers is the incorporation of biodiesel-based glycerol into polymer technology. Polymer industry, in contrast to food or pharmaceutical indus-try does not require complicated and often expensive purification processes in order to utilize crude glycerol [41, 42]. Among many types of plastics which can be manufactured with the use of crude glycerol the most popu-lar are polyurethanes. Glycerol obtained from biodiesel production might be incorporated into polyols and PU production by many different ways, indirectly, by the use of materials derived from glycerol, or directly, by incorporation glycerol or its polymeric form into processes. Indirect uses may include incorporation of low-molecular weight compounds such as propanediol or butanediol obtained through microbial conversion of crude glycerol [43, 44] or hydrogenolysis of glycerol [45, 46] and glycerol car-bonate derived from carboxylation of glycerol [47, 48], esterification of glycerol with organic carbonates [49, 50] or glycerolysis of urea [51, 52]. Direct ways of biodiesel-based glycerol incorporation into PU manufactur-ing include biomass liquefaction and oligomerization and polymerization of glycerol and those two methods will be described in more details in this chapter.

33.2   BIOMASS LIQUEFACTION

Liquefaction—next to combustion—pyrolysis and gasification are one of the most common method of biomass utilization. It is based on solvolysis reactions and results in generation of low molecular weight compounds, which can be soluble in water or other solvents and re-polymerize with themselves and with solvent particles leading to oil compounds with vast range of molecular weight [53]. Such compounds can be further incorporated into manufacturing of different polymeric materials, such as polyesters, epoxy resins, phenolic resins or polyurethanes, because of their high content of hydroxyl groups [54–56]. Various types of biomass have been subjected to liquefaction, such as wood [57, 58], corn bran, corn stover or corncobs [59–62], wheat straw [63], soybean straw [64], cork [65], bamboo residues [66–68], palm kernel cake [69] or biodiesel production solid residues [70]. Many researchers have been also analyzing the liquefaction of neat biomass components, such as cellulose or lignin [71–73]. The structure of biomaterial shows great influence on the liquefaction process, obviously the highest rate of liquefaction is observed for amorphous materials, such as lignin or amorphous cellulose and hemicellulose, which are due to the high entropy of their structure. The slowest part of process is considered to be liquefaction of crystalline materials, for example,cellulose, because penetration of their structure by solvents is very limited. Generally, all the biomass components are decomposed into glucose particles and their low molecular weight derivatives, which can subsequently re-polymerize and combine with solvent particles leading to glycoside compounds [74, 75]. D'Souza and Yan [76] analyzed the influence of the liquefaction temperature on the chemical composition of resulting polyol using spectroscopic and chromatographic techniques (NMR, FTIR, GPC). They found out that lower reaction temperature (in that case 90°C) resulted in low yield, hardly exceeding 20%, which was related to only partial conversion of sugars and practically no conversion of lignin. Increasing of the temperature up to 160°C noticeably increased biomass conversion. Spectroscopic analysis revealed also that liquefaction process converted the phenolic structures present in biomass into aliphatic alcohols, which is very important for the stability of urethane linkages.

Because of these structural dependences and the necessity of using the excess of the solvent leading to its high content in final product, the selection of the proper solvent or mixture of solvents is crucial for the yield of the process. Moreover, solvent not only has to enable possibly fast and effective liquefaction of biomaterial, but also needs to show beneficial properties as a potential polyol. That is why so many compounds have been analyzed as potential solvents for biomass liquefaction. According to literature reports there have been used phenolic compounds [77, 78], supercritical ethanol [79], subcritical water [80], however by far, the two most popular compounds among the researchers are polyethylene glycol and glycerol, very often used in combination to provide the proper degree of branching of resulting polyol[81–83], which is essential for the manufacturing of polyurethane materials.

Jin et al. [84] analyzed the liquefaction of lignin isolated from the enzymatic hydrolysis residues of the biomass, with the mixture of polyethylene glycol (with molecular weight of 400, 600 and 1000 g/mol) and glycerol. Authors noticed that the increase of PEG molecular weight resulted in the reduction of hydroxyl number of liquefaction product, which can be very useful for preparation of polyols for flexible materials. Moreover, the influence of the catalyst content, reaction time and solvent: biomass ratio on the effectiveness of the process was analyzed. According to presented data, for high hydroxyl value of resulting polyol and satisfactory yield of liquefaction, content of catalyst and biomass in reaction mixture should be kept at lower levels, temperature should not exceed 140°C and reaction time should not exceed 1 hour. Increasing of reaction time resulted in noticeable lowering of hydroxyl numbers, which was also confirmed by other researchers [85].

Yana et al. [86] investigated cork liquefaction with glycerol and its mixtures with polyethylene glycol. Authors analyzed the influence of the type and concentration of catalyst (acidic or alkaline), reaction time and temperature, cork content and addition of PEG 400 on the yield of process. Higher yields were obtained in case of alkaline catalysis; authors suggest that it was strongly related to the chemical composition of cork. Replacement up to 50 wt. % of glycerol with polyethylene glycol noticeably increased yield under acidic conditions, simultaneously confirming results presented by Zhang et al. [87]. Further optimization of the liquefaction process and should result in increasing the attractiveness of the process for potential applications.

Xu et al. [88] prepared polyols with hydroxyl number exceeding 1000 mg KOH/g, by liquefaction of sawdust with glycerol and methanol as mix solvents. Next, they used these polyols to prepare rigid polyurethane foams which met the requirements of Chinese specification for rigid foams. Authors suggested that liquefaction of biomass results in higher reactivity of the polyols, comparing to other manufacturing methods, because of the increased content of primary hydroxyl groups in polyol structure, which confirms suggestions of D'Souza and Yan regarding transition of phenolic groups into aliphatic alcohols [89].

Chen and Lu [90] performed wheat straw liquefaction with combination of polyethylene glycol and glycerol. Prepared polyols were characterized and then used for preparation of rigid polyurethane foams. Replacement up to 30 wt. % of petroleum based polyol with wheat straw liquefaction product allowed the increase of compressive strength from 173 to 208 kPa and enhanced thermal stability of prepared material. In case of higher additions of biopolyol, in homogeneity of morphology was observed, resulting in deterioration of foam performance.

Liquefaction process can be also enhanced by application of microwave radiation, which generally accelerates the decomposition of biomaterials [91, 92].

Xue et al. [93] obtained bio-based polyols by liquefaction of lignin with mixture of glycerol and polyethylene glycol. Authors applied sulfuric acid as a catalyst and microwave radiation in order to reduce the reaction time. After liquefaction prepared polyol was neutralized using 40% solution of NaOH and then the influence of reaction time (5, 10, 20 and 30 minutes) on yield of process and properties of final product was analyzed. Properties of polyols are prepared in Table 33.1.

As it can be seen, incorporation of both, catalyst and microwave radiation allowed to obtain very high yields of the process at short reaction times. It can be seen that the viscosity of prepared polyols were increasing with reaction time, which can be related to the increase in molecular weight. However, the rate of viscosity increase was much lower, because of the high polydispersity of materials obtained with longer reaction times. Those

TABLE 33.1 Properties of Polyols Obtained Liquefaction of Lignin [93]

Polyol	P5	P10	P20	P30
Reaction time, min	5	10	20	30
Yield, %	97.5	98.3	98.8	97.2
Viscosity, mPas	1035	1116	1161	1266
Molecular weight, g/mol	525	673	725	1108
Polydispersity	1.13	1.44	1.51	2.43
Hydroxyl value, mg KOH/g	863	861	846	821

differences are related to the shift of the balance between degradation of cellulose, hemicellulose, lignin and other compounds present in biomaterial and re-polymerization processes occurring during lignocellulose biomass liquefaction [94, 95]. Degradation results in decomposition to simpler, low molecular weight compounds which are able to react with solvents, while the opposite effect, production of insoluble material, occurs through re-polymerization of various compounds. At the early stages of the process, the main mechanism of liquefaction is decomposition of biomass and reaction with solvent particles, resulting in polyol production. It occurs because at the beginning of the process solvent is present in the system in great excess, however in later stages balance begins to shift towards re-polymerization of low molecular weight compounds, which significantly affects the polydispersity of obtained polyol[96, 97].

Xie et al. [98] also synthesized bio-polyols through the microwave-assisted liquefaction. Bamboo residues were liquefied in the mixture of glycerol and methanol and using sulfuric acid as a catalyst. Prepared polyols were analyzed by FTIR and GC-MS techniques. Hydroxyl values of crude and fractionated polyol were respectively 564 and 824 mg KOH/g. Both types of material were used subsequently for the synthesis of rigid polyurethane foams.

Cinelli et al. [99] prepared polyols using microwaves to liquefy the Kraft lignin. Biomass was dried for 24 hours in 80°C and then, the specific amount of lignin and solvents (glycerol and polyethylene glycol) was placed in glass flasks and subjected to microwave radiation (180 W, 135°C, 3 min). Hydroxyl values of prepared polyols, depending on the used ratio of substrates, varied from 645 to 661 mg KOH/g. Authors used prepared polyols to synthesize different types of polyurethane foams. Depending on the used chain extender (polypropylene glycol triol or castor oil) foams showed glass transition temperature around −50°C or in the range of 10–40°C, which obviously have a great impact on the mechanical performance of prepared foams [100].

Other consequence of application of solvent in excess is the large use of petroleum-based compounds, which significantly increases the cost of resulting polyols and successfully disqualify the biomass liquefaction as a commercial method of polyols manufacturing. For this reason, researchers are constantly looking for alternative solvents, which will make this process economically feasible. Crude glycerol obtained from biodiesel production can be employed instead of pure glycerol, which will obviously decrease the cost of final product, because of the noticeable difference in price of these materials (see Table 33.2)[101].

Hu et al. [103] analyzed the influence of liquefaction parameters on the preparation and properties of biopolyols and rigid polyurethane foams. They found out that the most beneficial conditions for liquefaction do not have to be preferential for further preparation of foamed material. Although conversion of biomass was constantly increasing with the temperature of the process changing from 120 to 240°C, foam obtained from resulting polyol showed decreasing compressive strength above 200°C. Such phenomenon was related to the increase in polyol chemical structure and viscosity. Analysis performed by the authors suggested that the liquefaction temperature has the statistically significant effect on the density and mechanical performance of resulting foams. By investigating other parameters, preferential conditions for production of PU foams with suitable properties were determined – 240°C, 3h, and 3% of sulfuric acid as catalyst. Presented results suggest that materials obtained through lignocellulosic biomass liquefaction with crude glycerol can be promising alternative for petroleum-based polyols.

Wang et al. [104] prepared biopolyols by liquefaction of corn stover with an industrial biodiesel glycerol as a solvent and compared the results with ones obtained for pure glycerol. Although the impurities present in crude glycerol decreased the liquefaction yield (around 52% for different crude glycerol/corn stover ratio, around 83% for pure glycerol), and increased the insoluble residue content, resulting polyols showed comparable parameters to those of polyol based on crude glycerol. Differences in the yield of the process could be compensated by the differences between crude and pure glycerol cost (see Table 33.2).

Hu and Li [105] used Box-Behnken experimental design method to investigate the effect of crude glycerol organic impurities on the properties of liquefaction-derived polyols and polyurethane foams. Although free fatty acids and fatty acids methyl esters showed negative effect on biomass conversion, they were found to be very important in producing polyols with

TABLE 33.2 Annual Average Price of Glycerol from 2007 to 2012 [102]

Type of glycerol	Price ($ per ton)					
	2007	2012	2009	2010	2011	2012
Refined	1551	748	902	836	814	748
Crude	220	88	132	154	44	88

appropriate molecular weight and hydroxyl number for rigid foam applications. Presence of at least 20 wt.% of both, FFA and FAME in starting glycerol reduced apparent density of rigid foams from 71 to around 40 kg/m^3, simultaneously increasing their compressive strength from 127 to more than 140 kPa. On the other hand, increasing of content of glycerides resulted in deterioration of foams' properties. Changing the concentration of glycerides from 5 to 10 wt. % resulted in the increase of apparent density with simultaneous drop of compressive strength. Results suggest that crude glycerol containing impurities may be potentially applied in polyurethane industry, which is related to the reduction of the secondary hydroxyl group's content in prepared polyol. Such groups are less reactive than primary hydroxyls, which could lead to slow or even incomplete curing of resulting polyurethane foam.

As it can be seen, comparing to pure glycerol, the impurities present in crude glycerol show negative influence on the yield of the process and biomass conversion, which can be solved by application of proper catalytic system.

Hu and Li [106] presented two-step sequential liquefaction process resulting in preparation of biopolyols, using different types of crude glycerol as solvent. Applied solvents varied in the content of impurities – free fatty acids, fatty acids methyl esters and glycerides, which accounted for 26 to 40 wt.%. During the first stage of the process, concentrated sulfuric acid was applied as a catalyst, what resulted in high yield of liquefaction and esterification of fatty acids and glycerol or other hydroxyl compounds. In the second step, detailed amount of sodium hydroxide, based on the previously determined acid number, was added to the reaction mixture to reach the desired NaOH content. During this step, there have been observed mainly condensation reactions, such as transesterification or etherification, resulting in the increase of molecular weight of resulting polyol. Method proposed by authors successfully combined advantages of both types of catalysis, however obvious drawback of presented solution is the complexity of the process and necessity of application of two types of catalysts.

33.2.1   OLIGOMERIZATION AND POLYMERIZATION

Oligomerization and polymerization of glycerol, are responsible for the fact that glycerol resulting from transesterification of triglycerides is treated as a by-product. Because of its polymerization in high temperatures it cannot be directly added to fuels to avoid clogging of the engine [107]. Cosmetics and food industries are mainly interested in di- and triglycerols, which are used in production of toothpastes or deodorant sticks, but also as emulsifiers in chocolate products [108, 109]. On the other hand, plastics industry is rather interested in high molecular weight polyglycerols. These are very versatile and useful compounds with high degree of branching, high functionality and high reactivity of hydroxyl groups. They have been repeatedly reported as valuable compounds, which can be successfully applied in production of polyurethanes, but also polyesters [110, 111], polycarbonates [112], epoxy resins [113–115] and other types of polymeric materials [116, 117]. Thanks to their biocompatibility, assessed by in vitroand in vivoexperiments, they can be also used in various biomedical applications [118, 119]. Polyglycerols are known for a decades and number of them is already commercially available, although traditionally their production is related to the polymerization of glycidol, considered carcinogenic to humans [120, 121]. Other factor, which makes replacement of glycidol with crude glycerol very attractive is high price of glycidol-based polyglycerols, resulting from complexity of the process. Glycidol-based oligomerization is considered neither selective nor quantitative process, so resulting mixture has to be distilled in order to obtain satisfactory level of purity [122]. To avoid such treatment, glycidol can contain in its structure protective groups, which should be removed after the process, which obviously generates additional costs [123]. Lately, number of attempts of glycerol polymerization have been reported in the literature [124, 125]. Nowadays, the attention is focused on the development of process parameters and catalysts resulting in satisfactory selectivity and yield of high molecular weight polyglycerol in order to reduce post process complicated purification steps [126, 127]. Various type of catalysts have been investigated in order to provide the best reaction pathway for etherification, such as oxides, [128, 129], acids [130] metal hydroxides [131], carbonates [132], ion-exchange resins [133, 134]], zeolites [135] or mesoporous materials [136, 137]. Each type of catalyst has advantages, but also some drawbacks. Application of acidic catalyst results in high degree of polymerization, but at the same moment promotes formation of cyclic isomers, which is affecting the selectivity of desired product [138, 139]. Salehpuor and Dubé[140] used sulfuric acid in order to obtain polyglycerols with high molecular weight. Performed NMR spectroscopic studies revealed that branched structure was successfully formed and the catalyst concentration had only little influence on the degree of branching. Moreover, during the use of acidic catalysts at elevated temperatures color properties of the product are affected [141]. Incorporation of basic catalysts reduce the formation of cyclic isomers, so they are more commonly used in industry, especially carbonates, whose solubility in glycerol is higher in comparison to oxides and hydroxides, leading to higher catalytic activity [142]. In order to avoid the complex and often expensive separation steps, researchers are obviously looking for the possibility of application of heterogeneous catalysts, which are much easier to recover after the process [143, 144]. NikSiti et al. [145] performed microwave assisted crude glycerol polymerization without the use of catalysts. The highest yield of polyglycerol exceeded 94% for crude glycerol containing 12.5 wt. % of soap.

Klukowska-Majewska et al. [146] patented the method of light colored polyglycerol production with use of sodium hydroxide as catalyst at 200–260°C, under inert gas flow. Solid sodium hydroxide should be crumbled into particles with the diameter of 0.2–05 mm. Such low size of catalyst particles allows its use in low concentrations, without the decrease of polyglycerol yield. Moreover, low concentration of catalyst results in low condensation degree of obtained product. Resulting polyglycerol shows hydroxyl number of 1320–1620 mg KOH/g and acid number lower than 1.1 mg KOH/g.

Wirpsza and Banasiak [147] patented the method of crude glycerol-based polyols manufacturing. According to patent, polyols were obtained from biodiesel-based glycerol containing also impurities such as water, fatty

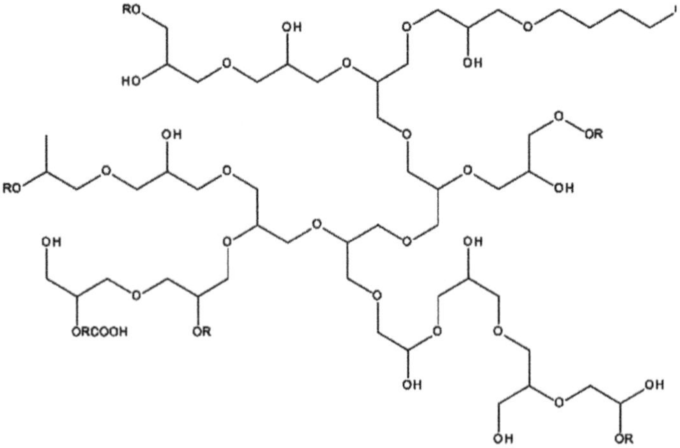

FIGURE 33.3   Potential structure of polyglycerol prepared by crude glycerol polymerization.

acids salts, natural oil, methanol, glycerides and catalyst of the biodiesel production process. Polymerization should be carried out at 260–300°C under atmospheric pressure and should be catalyzed by 1–6 wt.% of sodium or potassium hydroxide. Obtained product, after cooling, should be neutralized with sulfuric acid and purified from residual water and inorganic salts. Depending on applied conditions, resulting polyglycerols were characterized by hydroxyl number of 250–430 mg KOH/g and molecular weight of 1500–1700 g/mol.

Soi et al. [148] also patented the method of manufacturing of polyglycerol from unpurified crude glycerol. Fatty acid salts of alkaline metals were used as a catalyst so preferably crude glycerol should contain at least 10 wt.% of soaps formed during transesterification. Reaction parameters were similar to those proposed by Wirpsza and Banasiak[149], temperature in the range of 200–290°C and atmospheric pressure. After 3–5 hours of reaction, soaps should be removed from the polyglycerol by acidification with mineral acid and centrifugation in order to separate solid products and fatty acids. HPLC analysis revealed that there is very little or no evidence of cyclic diglycerol, so process can be claimed to be selective for linear polyglycerols.

Hu [150] investigated thermochemical polycondensation of crude glycerol. Author analyzed the influence of time and temperature of the process and soap content of glycerol on the composition and properties of final polyol. Under optimized conditions (120 min, 190°C, 6.6% of soap) obtained polyglycerol contained less than 1% of residual FFA and less than 2.5% of residual FAME, indicating high conversion of crude glycerol. Resulting polyol showed hydroxyl value of 378 mg KOH/g, acid value lower than 5 mg KOH/g, functionality of 4.7 and molecular weight around 700 g/mol, which is quite similar to commercially available products such as RokopolRF551(PCCRokita S. A.).

33.2.2 APPLICATIONS OF POLYGLYCEWROL IN POLYURETHANES

Polyglycerols, because of their branched structure can be considered very interesting substrates for synthesis of highly cross-linked polyurethane materials, such as adhesives or rigid polyurethane foams.

Gómez et al. [151] investigated biodegradability of polyurethane foams made from crude glycerol- and petroleum-based polyols during composting, anaerobic digestion and soil incubation. They found out that replacement on petrochemical polyol with biodiesel glycerol-based one, resulted in faster degradation of the material. Basing on spectroscopic analysis, the main sites of microbial attack are ester segments of the material. However, major degradation of bio-based foams was attributed to the decomposition of FFA and FAME present in the polyol part of the polymer.

Ionescu and Petrovic[152] prepared polyether polyols based on polyglycerol and used them to prepare rigid polyurethane foams. Polyglycerols, obtained by polycondensation of glycerol, were combined with propylene oxide or sucrose. Residual potassium alcoholate of polyglycerol was used as a catalyst for ring opening polymerization of propylene oxide. Process resulted in high functionality polyols(from 5 to 7), with hydroxyl numbers at the range 400–500 mg KOH/g. Polyols were successfully used to produce rigid polyurethane foams, which showed satisfactory physical and mechanical properties, especially dimensional stability and low friability.

Luo et al. [153] analyzed the influence of time (60–180 min), temperature (150–210°C) and catalyst concentration (0–4 wt.%) on the properties of crude glycerol-based biopolyols. Under preferential conditions (90 min, 200°C, 3 wt.% of sulfuric acid loading) authors obtained polyol, which showed hydroxyl number of 481 mg KOH/g, acid number of 5 mg KOH/g and viscosity 25 Pa·s. Subsequently, authors prepared rigid polyurethane foams, which showed apparent density of 43 kg/m^3 and compressive strength of 184.5 kPa, which is comparable to petroleum-based materials.

Li et al. [154] performed thermochemical conversion of crude glycerol. Authors obtained bio-polyols through non-catalytic condensation of biodiesel-based glycerol under vacuum conditions. Resulting polyols, prepared at 2.5 and 5 hours, showed hydroxyl numbers of 391 mg KOH/g and 346 mg KOH/g, respectively. Acid number of first polyol was 7 mg KOH/g, while in case of second one was below the detection limit. According to presented results, incorporation of prepared polyols into polyurethane foams resulted in the improvement of cell morphology and enhancement of thermal insulation properties. Authors attributed these changes to the increased content of monoglycerides in applied bio-polyols.

Piszczyk et al. [155] used two types of commercially available polyglycerols from Eco Innova, produced by thermocatalyticpolycondensation of crude glycerol, according to previously mentioned patent [149]. In Table 33.3, properties of rigid polyurethane foams obtained by replacement of 35 wt.% of commercial RokopolRF551 by polyglycerols Pole and PGK are presented. Such modification hardly affects the cellular structure of prepared foams, which show very similar cell size, closed cell content and thermal conductivity coefficient. Changes in apparent density of the material, are related to the differences in the molecular weight and density of applied polyols. Increase of apparent density simultaneously enhanced compressive strength of foams, since during compression the stiffness of the material arises from buckling of cell walls and higher density is obviously related to the more compact cellular structure. Moreover, incorporation of crude glycerol-based polyol into rigid foam resulted in enhanced thermal stability, with the increase of the onset temperature of degradation by 15°C.

Piszczyk et al. [156] used crude glycerol-based polyols to prepare polyurethane/ground tire rubber composite foams. Foams were characterized with satisfactory mechanical and thermal properties. Simultaneously, incorporation of ground tire rubber into bio-based polyurethane matrix could further enhance the ecological aspect of material.

33.3 CONCLUSIONS

Many studies show that materials from renewable resources can almost fully substitute their petrochemical analogs. Also polyurethane industry shows increasing interest in polyols derived from renewable resources, from which crude glycerol and biomass are the most economical, but further scientific and technological studies are needed aimed at obtaining of polyols with well defined structure and properties.

TABLE 33.3 Properties of Rigid Polyurethane Foams Prepared with Crude Glycerol-Based Polyglycerol[155]

Properties	Foam symbol		
	P_0	$P_{Pole\ 35}$	$P_{PGK\ 35}$
Apparent density, kg/m^3	21,7±1,2	30,3±0,9	25,5±1,2
Compressive strength at 20% deformation, kPa	140±3	170±4	164±3
Thermal conductivity coefficient, mW/m·K	26,1±0,8	25,8±0,7	26,3±0,8
Closed cell content, %	82	83	82
Average pore diameter, μm	110±9	111±8	107±8

KEYWORDS

- **biocomposites**
- **biopolyols**
- **crude glycerol**
- **polyurethanes**

REFERENCES

1. Directive 2009/28/EC of the European Parliament and of the Council of 23 April 2009 on the promotion of the use of energy from renewable sources and amending and subsequently repealing Directives 2001/77/EC and 2003/30/EC.
2. The 2020 climate and energy package. Available from: http://ec.europa.eu/clima/policies/package/index_en.htm.
3. Datta, J., Głowińska, E., (2014). Effect of hydroxylated soybean oil and bio-based propanediol on the structure and thermal properties of synthesized biopolyurethanes. *Ind. Crop. Prod., 61*, 84.
4. Prociak, A., (2007). Properties of polyurethane foams modified with natural oil-based polyols. *Cell. Polymer., 26*, 381–392.
5. Pawlik, H., Prociak, A., Pielichowski, J., (2009). CzasopismoTechniczne, 1-Ch., 111.
6. Zhang, L., (2008).Dissertation, Graduate School of The University of Minnesota, Minneapolis, Minnesota.
7. Yeganeh, H., Mehdizadeh, M. R., (2004). *Eur. Polym. J., 40*, 1233.
8. Zlatanic, A., Lava, C., Zhang, W., Petrovic, Z. S., (2004). *J.Polym. Sci B: Polym. Phys., 42*, 809.
9. Sharma, V., Kundu, P. P., (2008). *Prog. Polym. Sci., 33*, 1199.
10. Wilkes, G. L., Pechar, T. W., (2005). Structure-Property Studies of Polyurethane Networks that Incorporate Soy-Based Polyols; United Soybean Board, Technical Advisory Panel (TAP) Meeting: Detroit, April 27.
11. Ionescu, M., (2005). Chemistry and Technology of Polyols for Polyurethanes, Rapra Technology Limited, Shawbury, Shrewsbury, Shropshire.
12. Prociak, A., (2007). *RynekChemiczny, 7–8*, 17.
13. Głowińska, E., Datta, J., (2012). *Przem. Chem., 91*, 1234.
14. Ionescu, M., Petrovic, Z. S., Javini, I., Stojadinov, J., (2005). Polyurethane Technical Conference and Trade Fair, 17–19 October 2005, Houston.
15. Petrovic, Z. S., Zlatanic, A., Lava, C. C., Sinadinovic-Fiser, S., (2002). *Eur. J. Lipid Sci. Technol., 104*, 293.
16. Vlcek, T., Petrovic, Z. S., (2006). *J. Amer. Oil Chem. Soc., 83*, 247.
17. Petrovic, Z. S., Guo, A., Zhang, W. (2000). *J. Polym. Sci. Part A: Polym. Chem., 38*, 4062.
18. Guo, A., Demydov, D., Zhang, W., Petrovic, Z. S., (2002). *J. Polymers and Environment, 10*, 49.

19. Kandanarachchi, P., Guo, A., Petrovic, Z., (2002). *Journal of Molecular Catalysis, A., 184*, 65.
20. Kandanarachchi, P., Guo, A., Demydov, D., Petrovic, Z. J. (2002). *Amer. Oil Chem. Soc., 79*, 1221.
21. Petrović, Z., (2008). *Polym. Review., 48*, 109.
22. Petrovic, Z. S., Guo, A., Javni, I., Cvetkovic, I., Hong, D. P., (2008). *Polymer International, 57*, 275.
23. Tran, P., Graiver, D., Narayan, R. (2005). *J. Amer. Oil Chem. Soc., 82*, 653.
24. Graiver, D., Tran, P., Laura, P., Farminer, K. W., Narayan, R., (2005). *Degradable Polymers and Materials. Principles and Practice*, Oxford Univ. Press: New York, USA.
25. Graiver, D., Narayan, R., (2006). *Lipid Tech., 17*, 31.
26. Pat. JP2004083695. (2002).
27. Desai, S. D., Patel, J. V., Sinha, V. K., (2003). *Int. J. Adhesion and Adhesives, 23*, 393.
28. Pat. EP1260497. (2002).
29. Hoydonckx, H. E., Vos, D. E. D., Chavan, S. A., Jacobs, P. A., (2004). *Topics in Catalysis, 27*, 83.
30. Petrovic, Z. S., Lukic, M., Zhang, W., Shirley, W., (2005). Fatty Acid-Based Polyols And Polyurethanes, Academy of Science and Arts of Serbian Republic, Scientific Sessions vol VII, Section of natural, Mathematical and Technical Sciences: Banja Luka.
31. Petrovic, Z. S., Xu, Y., Zhang, W., (2007). *Polymer Preprints, 48*, 852.
32. European Commission. [updated 2015 Mar 9; cited 2015 Jun 25]. Energy from renewable sources. Available from: http://ec.europa.eu/eurostat/statisticsexplained/index.php/Energy_from_renewable_sources.
33. The International Council of Clean Transportation. [updated 2015 Jun 24; cited 2015 Jun 25]. CO_2 Standards. Available from: http://www.theicct.org/issues/co2-standards.
34. Puri, M., Abraham, R. E., Barrow, C. J. (2012). Biofuel production: prospects, challenges and feedstock in Australia. *Renew Sust. Energ. Rev., 16*, 6022–6031.
35. Energy Information Administration. [updated 2012 Oct 31; cited 2015 Jun 25]. Biofuels Issues and Trends. Available from: http://www.eia.gov/biofuels/issuestrends/pdf/bit.pdf.
36. Choi, W. J. (2008). Glycerol-based biorefinery for fuels and chemicals.*Recent Pat. Biotechnol. 2*, 173–180.
37. Hu, S., Luo, X., Wan, C., Li, Y., (2012). Characterization of crude glycerol from biodiesel plants.*J. Agric. Food Chem., 60*, 5915–5921.
38. Yazdani, S. S., Gonzalez, R., (2007). Anaerobic fermentation of glycerol: A path to economic viability for the biofuels industry. *Curr. Opin. Biotech. 18*, 213–219.
39. Siricharnsakunchai, P., Simasatitkul, L., Soottitantawat, A. A. (2012). Use of reactive distillation for triacetin production from crude glycerol: Simulation and performance analysis. In: Karimi, I. A., Srinivasan, R., Eds. 11th International Symposium on Process System Engineering.Proceedings of the 11th International Symposium on Process System Engineering; 2012 Jul 15–19; Singapore, Singapore. Amsterdam: Elsevier, 165–169.
40. Yang, F., Hanna, M., Sun, R., (2012). Value-added uses for crude glycerol—a by product of biodiesel production. *Biotechnol. Biofuels, 5*, 13–22.
41. Saifuddin, N., Refal, H., Kumaran, P., (2014). Rapid Purification of Glycerol byproduct from Biodiesel Production through Combined Process of Microwave Assisted Acidification and Adsorption via Chitosan Immobilized with Yeast, *Res. J. Appl. Sci. Eng. Technol., 7*, 593–602.

42. Tan, H. W., Abdul Aziz, A. R., Aroua, M. K., (2013). Glycerol production and its applications as a raw material: A review. *Renew. Sust. Energ. Rev., 27*, 118–127.

43. Petrov, K., Petrova, P., (2010). Enhanced production of 2, 3-butanediol from glycerol by forced pH fluctuations, *Appl. Microbiol. Biotechnol., 87*, 943–949.

44. Szymanowska-Powałowska, D., (2014). 1,3-Propanediol production from crude glycerol by Clostridium butyricumDSP1 in repeated batch. *Electronic Journal of Biotechnology, 17*, 322–328.

45. Zhu, S., Qiu, Y., Zhu, Y., Hao, S., Zheng, H., Li, Y., (2013). Hydrogenolysis of glycerol to 1, 3-propanediol over bifunctional catalysts containing Pt and heteropolyacids, *Catal.Today, 212*, 120–126.

46. Lee, C. S., Aroua, M. K., Daud, W. M. A. W., Cognet, P., Pérès-Lucchese, Y., Fabre, P. L., et al. (2015). A review: Conversion of bioglycerol into 1, 3-propanediol via biological and chemical method. *Renew. Sust. Energ. Rev., 42*, 963–972.

47. Casiello, M., Monopoli, A., Cotugno, P., Milella, A., Dell'Anna, M. M., Ciminale, F., et al. (2014). Copper(II) chloride-catalyzed oxidative carbonylation of glycerol to glycerol carbonate. *J. Mol. Catal. A-Chem; 381*, 99–106.

48. Ezhova, N. N., Korosteleva, I. G., Kolesnichenko, N. V., Kuzmin, A. E., Khadzhiev, S. N., Vasileva, M. A., et al., (2012). Glycerol carboxylation to glycerol carbonate in the presence of rhodium complexes with phosphine ligands, *Petrol. Chem., 52*, 91–96.

49. Algoufi, Y. T., Hameed, B. H., (2014). Synthesis of glycerol carbonate by trans esterification of glycerol with dimethyl carbonate over K-zeolite derived from coal fly ash, *Fuel. Process. Technol., 126*, 5–11.

50. Esteban, J., Fuente, E., Blanco, A., Ladero, M., Garcia-Ochoa, F., (2015). Phenomenological kinetic model of the synthesis of glycerol carbonate assisted by focused beam reflectance measurements. *Chem. Eng. J., 260*, 434–443.

51. Nguyen, N., Demirel, Y., (2013). Economic Analysis of Biodiesel and Glycerol Carbonate Production Plant by Glycerolysis.*Journal of Sustainable Bioenergy Systems, 3*, 209–216.

52. Jagadeeswaraiah, K., Kumar, C. R., Sai Prasad, P. S., Loridant, S., Lingaiah, N., (2014). Synthesis of glycerol carbonate from glycerol and urea overtin-tungsten mixed oxide catalysts.*Appl. Catal. A-Chem., 469*, 165–172.

53. Liang, L., Mao, Z., Li, Y., Wan, C., Wang, T., Zhang, L., et al., (2006). Liquefaction of crop residues for polyol production, *Bioresources, 1*, 1–9.

54. Yoshioka, M., Nishio, Y., Saito, D., Ohashi, H., Hasbhimoto, M., Shiraishi, N., (2013). Synthesis of Biopolyols by Mild Oxypropylation of Liquefied Starch and Its Application to Polyurethane Rigid Foams, *J. Appl. Polym. Sci., 130*, 622–630.

55. Hassan, E. M., Shukry, N., (2008). Polyhydric alcohol liquefaction of some lignocellulosic agricultural residues.*Ind Crop Prod 27*, 33–38.

56. Lee, W. J., Lin, M. S., (2008). Preparation and application of polyurethane adhesives made from polyhydric alcohol liquefied Taiwan acacia and China fir, *J. Appl. Polym. Sci., 109*.

57. Alma, M. H., Basturk, M. A., Digrak, M., (2003). New polyurethane-type rigid foams from liquified wood powders, *J. Mater. Sci. Lett., 22*, 1225–1228.

58. Zhang, H. R., Pang, H., Zhang, L., Chen, X., Liao, B., (2013). Biodegradability of Polyurethane Foam from Liquefied Wood Based Polyols, *J. Polym. Environ., 21*, 329–334.

59. Lee, S. H., Yoshioka, M., Shiraishi, N., (2000). Liquefaction of corn bran (CB) in the presence of alcohols and preparation of polyurethane foam from its liquefied polyol.*J. Appl. Polym. Sci., 78*, 319–325.

60. Zhang, H., Ding, F., Luo, C., Xiong, L., Chen, X., (2012). Liquefaction and characterization of acid hydrolysis residue of corncob in polyhydric alcohols, *Ind. Crop Prod., 39,* 47–51.

61. Wang, Y., Wu, J., Wan, Y., Lei, H., Yu, F., Chen, P., et al., (2009). Liquefaction of corn stover using industrial biodiesel glycerol.*Int. J. Agric. Biol. Eng., 2,* 32–40.

62. Hu, S., Li, Y., (2014). Two-step sequential liquefaction of lignocellulosic biomass by crude glycerol for the production of polyols and polyurethane foams, *Bioresource Technol., 161,* 410–415.

63. Wang, H., Chen, H. Z., (2007). A novel method of utilizing the biomass resource: Rapid liquefaction of wheat straw and preparation of biodegradable polyurethane foam (PUF). *J. Chin. Inst. Chem. Eng., 38,* 95–102.

64. Hu, S., Wan, C., Li, Y., (2012). Production and characterization of biopolyols and polyurethane foams from crude glycerol based liquefaction of soybean straw. *Bioresource Technol., 103,* 227–233.

65. Soares, B., Gama, N., Freire, C., Barros-Timmons, A., Brandão, I., Silva, R., et al., (2014). Ecopolyol Production from Industrial Cork Powder via Acid Liquefaction Using Polyhydric Alcohols.*ACS Sustainable Chem. Eng., 2,* 846–854.

66. Xie, J., Hse, C., Shupe, T., Qi, J., Pan, H., (2014). Liquefaction behaviors of bamboo residues in a glycerol-based solvent using microwave energy, *J. Appl. Polym. Sci., 131,* 402–407.

67. Xie, J., Qi, J., Hse, C., Shupe, T. F., (2015). Optimization for microwave-assisted direct liquefaction of bamboo residue in glycerol/methanol mixtures.*J. Forest Res., 26,* 261–265.

68. Xie, J., Qi, J., Hse, C. J., Shupe, T. S. (2014). Effect of lignin derivatives in the biopolyols from microwave liquefied bamboo on the properties of polyurethane foams, *Bioresources, 9,* 578–588.

69. Jo, Y. J., Ly, H. V., Kim, J., Kim, S. S., Lee, E. Y., (in press). Preparation of biopolyol by liquefaction of palm kernel cake using PEG 400 blended glycerol. *J. Ind. Eng. Chem,* doi: 10.1016/j.jiec.2015.04.010.

70. Briones, R., Serrano, L., Llano-Ponte, R., Labidi, J., (2011). Polyols obtained from solvolysis liquefaction of biodiesel production solid residues, *Chem. Eng. J., 175,* 169–175.

71. Bernardini, J., Cinelli, P., Anguillesi, I., Coltelli, M. B., Lazzeri, A., (2015). Flexible polyurethane foams green production employing lignin or oxypropylated lignin. *Eur. Polym. J., 64,* 147–156.

72. Cinelli, P., Anguillesi, I., Lazzeri, A., (2013). Green synthesis of flexible polyurethane foams from liquefied lignin, *Eur. Polym. J., 49,* 1174–1184.

73. Xue, B. L., Wen, J. L., Sun, R. C., (2015). Producing Lignin-Based Polyols through Microwave-Assisted Liquefaction for Rigid Polyurethane Foam Production, *Materials, 8,* 586–599.

74. Yamada, T., Ono, H., (2001). Characterization of the products resulting from ethylene glycol liquefaction of cellulose, *J. Wood Sci., 47,* 458–464.

75. Yamada, T., Aratani, M., Kubo, S., Ono, S., (2007). Chemical analysis of the product in acid-catalyzed solvolysis of cellulose using polyethylene glycol and ethylene carbonate. *J. Wood Sci. 53,* 487–493.

76. D'Souza, J., Yan, N., (2013). Producing Bark-based Polyols through Liquefaction: Effect of Liquefaction Temperature. *ACS Sustainable Chem. Eng., 1,* 534–540.

77. Ono, H., Yamada, T., Hatano, Y., Motohashi, K., (1996). Adhesives from Waste Paper by Means of Phenolation.*J. Adhes., 59,* 135–145.

78. Mun, S. P., Jang, J. P., (2009). Liquefaction of cellulose in the presence of phenol using *p*-toluene sulfonic acid as a catalyst.*J. Ind. Eng. Chem., 15*, 743–747.
79. Brand, S., Susanti, R. F., Kim, S. K., Lee, H., Kim, J., Sang, B., (2013). Supercritical ethanol as an enhanced medium for lignocellulosic biomass liquefaction: Influence of physical process parameters. *Energy, 59*, 173–1 82.
80. Harry, I., Ibrahim, H., Thring, R., Idem, R., (2014). Catalytic subcritical water liquefaction of flax straw for high yield of furfural.*Biomass Bioenerg., 71*, 381–393.
81. Krzan, A., Zagar, E., (2009). Microwave driven wood liquefaction with glycols.*Bioresource Technol., 100*, 3143–3146.
82. Demirbas, A., (2008). Liquefaction of Biomass Using Glycerol, Energ.*Source Part, A., 30*, 1120–1126.
83. D'Souza, J., Yan, N., (2013). Producing Bark-based Polyols through Liquefaction: Effect of Liquefaction Temperature. *ACS Sustainable Chem. Eng., 1*, 534–540.
84. Jin, Y., Ruan, X., Cheng, X., Lu, Q., (2011). Liquefaction of lignin by polyethyleneglycol and glycerol.*Bioresource Technol., 102*, 3581–3583.
85. Lee, S. H., Yoshioka, M., Shiraishi, N., (2000). Liquefaction of corn bran (CB) in the presence of alcohols and preparation of polyurethane foam from its liquefied polyol.*J. Appl. Polym. Sci., 78*, 319–325.
86. Yona, A. M. C., Budija, F., Kricej, B., Kutnara, A., Pavlic, M., Pori, P., et al. (2014). Production of biomaterials from cork: Liquefaction in polyhydric alcohols at moderate temperatures. *Ind. Crop Prod., 54*, 296–301.
87. Zhang, H., Ding, F., Luo, C., Xiong, L., Chen, X., (2012). Liquefaction and characterization of acid hydrolysis residue of corncob in polyhydric alcohols, *Ind. Crop Prod., 39*, 47–51.
88. Xu, J., Jiang, J., Hse, C., Shupe, T. (2014). Preparation of polyurethane foams using fractionated products in liquefied wood. *J. Appl. Polym. Sci., 131*, DOI: 10.1002/app.40096.
89. D'Souza, J., Yan, N., (2013). Producing Bark-based Polyols through Liquefaction: Effect of Liquefaction Temperature. *ACS Sustainable Chem. Eng., 1*, 534–540.
90. Chen, F. G., Lu, Z. M., (2009). Liquefaction of wheat straw and preparation of rigid polyurethane foam from the liquefaction products.*J. Appl. Polym Sci. 111*, 508–516. (2013).
91. Xie, J., Hse, C., Shupe, T., Qi, J., Pan, H., (2014). Liquefaction behaviors of bamboo residues in a glycerol-based solvent using microwave energy, *J. Appl. Polym. Sci. 131*, 402–407.
92. Xie, J., Qi, J., Hse, C., Shupe, T. F., (2015). Optimization for microwave-assisted direct liquefaction of bamboo residue in glycerol/methanol mixtures, *J. Forest Res., 26*, 261–265.
93. Xue, B. L., Wen, J. L., Sun, R. C., (2015). Producing Lignin-Based Polyols through Microwave-Assisted Liquefaction for Rigid Polyurethane Foam Production, *Materials, 8*, 586–599.
94. Yamada, T., Ono, H., (1999). "Rapid Liquefaction of Lignocellulosic Waste by Using Ethylene Carbonate, " *Bioresour.Technol., 70*, 61.
95. Yamada, T. Y., Hu, H., Ono, H., (2001). "Condensation Reaction of Degraded Lignocellulose during Wood Liquefaction in the Presence of Polyhydric Alcohols," *Nippon SetchakuGakkaishi, 37*(12), 471.
96. Hui Wang, Hong-Zhang Chen, (2007). A novel method of utilizing the biomass resource: Rapid liquefaction of wheat straw and preparation of biodegradable polyurethane foam (PUF), *Journal of the Chinese Institute of Chemical Engineers 38*, 95–102.

97. Chen, F., .G, Lu, Z. M., (2009). Liquefaction of wheat straw and preparation of rigid polyurethane foam from the liquefaction products.*J. Appl. Polym. Sci. 111*, 508–516.

98. Xie, J., Qi, J., Hse, C. J., Shupe, T. S., (2014). Effect of lignin derivatives in the biopolyols from microwave liquefied bamboo on the properties of polyurethane foams. *Bioresources, 9,* 578–588.

99. Cinelli, P., Anguillesi, I., Lazzeri, A., (2013). Green synthesis of flexible polyurethane foams from liquefied lignin, *Eur. Polym. J., 49*, 1174–1184.

100. Bernardini, J., Cinelli, P., Anguillesi, I., Coltelli, M. B., Lazzeri, A., (2015). Flexible polyurethane foams green production employing lignin or oxypropylated lignin, *Eur. Polym. J., 64*, 147–156.

101. Jo, Y. J., Ly, H. V., Kim, J., Kim, S. S., Lee, E. Y., (in press). Preparation of biopolyol by liquefaction of palm kernel cake using PEG 400 blended glycerol. J. Ind. Eng. Chem, DOI: 10.1016/j.jiec.2015.04.010.

102. Talebian-Kiakalaieh, A., Amin, N. A. S., Hezaveh, H., (2014). Glycerol for renewable acrolein production by catalytic dehydration.*Renew. Sust. Energ. Rev., 40*, 28–59.

103. Hu, S., Wan, C., Li, Y., (2012). Production and characterization of biopolyols and polyurethane foams from crude glycerol based liquefaction of soybean straw, *Bioresource Technol., 103*, 227–233.

104. Wang, Y., Wu, J., Wan, Y., Lei, H., Yu, F., Chen, P., et al., (2009). Liquefaction of corn stover using industrial biodiesel glycerol, *Int. J. Agric. Biol. Eng., 2,* 32–40.

105. Hu, S., Li, Y., (2014). Polyols and polyurethane foams from base-catalyzed liquefaction of lignocellulosic biomass by crude glycerol: Effects of crude glycerol impurities, *Ind. Crop Prod., 57*, 188–194.

106. Hu, S., Li, Y., (2014). Two-step sequential liquefaction of lignocellulosic biomass by crude glycerol for the production of polyols and polyurethane foams.*Bioresource Technol. 161,* 410–415.

107. Pagliaro, M., Ciriminna, R., Kimura, H., Rossi, M., Della Pina, C., (2007). From Glycerol to Value-Added Products.*Angew. Chem. Int. Ed., 46*, 4434–4440.

108. Martin, A., Richter, M., (2011). Oligomerization of glycerol – a critical review, *Eur. J. Lipid Sci. Technol., 113*, 100–117.

109. Sunder, Krämer, M., Hanselmann, R., Mülhaupt, R., Frey, H., (1999). Molecular Nanocapsules Based on AmphiphilicHyperbranchedPolyglycerols, *Angew. Chem, Int. Ed., 38*, 3552–3555.

110. Parzuchowski, P., Grabowska, M., Jaroch, M., Kusznerczuk, M., (2009). Synthesis and Characterization of Hyperbranched Polyesters from Glycerol-based AB2 Monomer, *J. Polym. Sci. Part A: Polym. Chem., 47,* 3860–3868.

111. Zhao, X., Liu, L., Dai, H., Ma, C., Tan, X., Yu, R., (2009). Synthesis and application of water-soluble hyperbranchedpoly(ester)s from maleic anhydride and glycerol., *J. Appl. Polym. Sci., 113*, 3376–3381.

112. Parzuchowski, P., Jaroch, M., Tryznowski, M., Rokicki, G., (2008). Synthesis of New Glycerol-Based Hyperbranched Polycarbonates.*Macromolecules 41*, 3859–3865.

113. Parzuchowski, P., Kiźlińska, M., Rokicki, G., (2007). New hyperbranched polyether containing cyclic carbonate groups as a toughening agent for epoxy resin. *Polymer, 48*, 1857–1865.

114. Shibata, M., Teramoto, N., Takada, Y., Yoshihara, S., (2010). Preparation and Properties of Biocomposites Composed of Glycerol-Based Epoxy Resins, Tannic Acid, and Wood Flour, *J. Appl. Polym. Sci., 118*, 2998–3004.

115. Auvergne, R., Caillol, S., David, G., Boutevin, B., Pascault, J. P., (2014). Biobased Thermosetting Epoxy: Present and Future, *Chem. Rev., 114*, 1082−1–115.

116. Mamiński, M. Ł., Witek, S., Szymona, K., Parzuchowski, P., (2013). Novel adhesive system based on 1, 3-dimethylol-4, 5-dihydroxyethyleneurea (DMDHEU) and hyper-branchedpolyglycerols, *Eur. J. Wood. Prod., 71*, 267–275.

117. Oudshoorn, M. H. M., Rissmann, R., Bouwstra, J. A., Hennink, W. E., (2006). Synthesis and characterization of hyperbranchedpolyglycerol hydrogels.*Biomaterials 27*, 5471–5479.

118. Son, S., Shin, E., Kim, B. S., (2015). Redox-Degradable Biocompatible Hyper-branchedPolyglycerols: Synthesis, Copolymerization Kinetics, Degradation, and Bio-compatibility, *Macromolecules, 48*, 600–609.

119. Kim, Y. J., Kim, B., Hyun, D. C., Kim, J. W., Shim, H. E., Kang, S. W., et al., (2015). Photo-cross-linkable Poly(ε-caprolactone)-b-HyperbranchedPolyglycerol (PCL-b-hbPG) with Improved Biocompatibility and Stability for Drug Delivery, *Macromol. Chem. Phys., 216,* 1161–1170.

120. Gholami, Z., Abdulla, A. Z., Lee, K. T., (2014). Dealing with the surplus of glycerol production from biodiesel industry through catalytic upgrading to polyglycerols and other value-added products, *Renew. Sust. Energ Rev., 39,* 327–341.

121. Sunder, A., Hanselmann, R., Frey, H., Mulhaupt, R., (1999). Controlled Synthesis of HyperbranchedPolyglycerols by Ring-Opening Multibranching Polymerization. *Macromolecules, 32*, 4240–4446.

122. Ciriminna, R., Katryniok, B., Paul, S., Dumeignil, F., Pagliaro, M., (in press,). Glycerol-Derived Renewable Polyglycerols: A Class of Versatile Chemicals of Wide Potential Application. Org. Process Res. Dev, DOI: 10.1021/op500313x.

123. Thomas, A., Müller, S. S., Frey, H. (2014). *Biomacromolecules, 15,* 1935–1954.

124. Barrault, J., Clacens, J. M., Pouilloux, Y., (2006). Methods for etherifying glycerol, and catalysts for implementing said methods, Pat. EP1292557.

125. Salehpour, S., Dubé, M. A., (2012). Reaction monitoring of glycerol step-growth polymerization using ATR-FTIR spectroscopy.*Macromol. React. Eng., 6*, 85–92.

126. Gandini, A., Lacerda, T. M., (in press,). From monomers to polymers from renewable resources: Recent advances, Prog. Polym. Sci., DOI: 10.1016/j.progpolymsci.2014.11.002.

127. Lemke, D. W., Nivens, S., (2008). Process for preparing polyglycerol and mixed ethers. Pat. US20080306211.

128. García-Sancho, C., Moreno-Tost, R., Mérida-Robles, J. M., Santamaría-González, J., Jiménez-López, A., Torres, P. M., (2011). Etherification of glycerol to polyglycerols over MgAl mixed oxides.*Catal. Today 167*, 84–90.

129. Ruppert, A. M., Meelkijk, J. D., Kuipers, B. W., Erné, B. H., Weckhuysen, B. M., (2008). Glycerol etherification over highly active CaO-based materials: new mechanistic aspects and related colloidal particle formation, *Chem. Eur. J., 14*, 2016–2024.

130. Uda, S., Takemoto, E., (1986). Production of polyglycerol. Pat. JP61140534.

131. Lemke, D. W., (2003). Processes for preparing linear polyglycerols and polyglycerol esters. Pat. US6620904.

132. Stuhler, H., (1985). Process for the preparation of polyglycerol. Pat. US4551561.

133. Behr, A., Obendorf, L., (2002). Development of a process for the acid-catalyzed etherification of glycerin and isobutene forming glycerin tertiary butyl ethers, *Eng. Life Sci., 2*, 185–189.

134. Klepáčová, K., Mravec, D., Bajus, M., (2005). Tert-Butylation of glycerol catalyzed by ion- exchange resins. *Appl. Catal. A-Chem., 294,* 141–147.
135. Eshuis, J. I., Laan, J. A., Potman, R. P., (1997). Polymerization of glycerol using a zeolite catalyst. Pat. US5635588.
136. Barrault, J., Clacens, J. M., Pouilloux, Y., (2001). Procedesd'etherification du glycerol, etcatalyzeurs pour la mise en oeuvre de cesprocedes. Pat. WO0198243.
137. Clacens, J. M., Pouilloux, Y., Barrault, J., (2000). Synthesis and modification of basic mesoporous materials for the selective etherification of glycerol, *Stud. Surf. Sci. Catal., 143,* 687–695.
138. Gholami, Z., Abdulla, A. Z., Lee, K. T., (2014). Dealing with the surplus of glycerol production from biodiesel industry through catalytic upgrading to polyglycerols and other value-added products, *Renew. Sust. Energ. Rev., 39,* 327–341.
139. Hasenhuettl, G. L., (2008). Synthesis and Commercial Preparation of Food Emulsifiers. In: Hasenhuettl, G. L., Hartel, R. W., editors. Food Emulsifiers and Their Applications, New York: Springer Science + Business Media, pp. 11–37.
140. Salehpour, S., Dubé, M. A., (2011). Towards the sustainable production of higher-molecular-weight polyglycerol, *Macromol. Chem. Phys. 212,* 1284–1293.
141. Uda, S., Takemoto, E., (2004). Preparation of polyglycerol with slight coloration. Pat. JP6143627.
142. Aslan, N., (2008). Application of response surface methodology and central composite rotatable design for modeling and optimization of a multi-gravity separator for chromite concentration, *Powder Technol., 185,* 80–86.
143. Mat, R., Samsudin, R. A., Mohamed, M., Johari, A., (2012). Solid catalysts and their application in biodiesel production, *Bull. Chem. React. Eng. Catal., 7,* 142–149.
144. Chopade, S. G., Kulkarni, K. S., Kulkarni, A. D., Topare, N. S., (2012). Solid heterogeneous catalysts for production of biodiesel from trans-esterification of triglycerides with methanol: a review. *ActaChim. Pharm. Indica, 2,* 8–14.
145. NikSiti, M. N. M. D., Idris, Z., Shoot, K. Y., Hassan, H. A. (2013). Preparation of polyglycerol from palm-biodiesel crude glycerin, *J. Oil Palm. Res., 25,* 289–297.
146. Klukowska-Majewska, Z., Mroczek, T., Osiejuk, E., Kulczycka, A., Chmielarz, B., (1986). Sposóbotrzymywaniapoliglicerolu o jasnejbarwie. Pat. PL176878.
147. Wirpsza, Z., Banasiak, S., (2012). Sposóbbwytwarzaniaoligoeteroli. Pat. PL210779.
148. Soi, H. S., Bakar, Z. A., Mat Din, N. S. M. N., Idris, Z., Kian, Y. S., Abu Hassan, H., et al. (2014).Process of producing polyglycerol from crude glycerol. US8816132,
149. Wirpsza, Z., Banasiak, S., (2012). Sposóbbwytwarzaniaoligoeteroli.PL210779B1.
150. Hu, S., (2013).Production and Characterization of Bio-based Polyols and Polyurethanes from Biodiesel-derived Crude Glycerol and Lignocellulosic Biomass [dissertation]. Columbus: The Ohio State University.
151. Gómez, E. F., Luo, X., Li, C., Michel Jr., F. C., Li, Y., (2014). Biodegradability of crude glycerol-based polyurethane foams during composting, anaerobic digestion and soil incubation, *Polym. Degrad. Stabil., 102,* 195–203.
152. Ionescu, M., Petrovic, Z. S., (2010). High Functionality Polyether Polyols Based on Polyglycerol. *J. Cell. Plast., 46,* 223–237.
153. Luo, X., Hu, S., Zhang, X., Li, Y., (2013). Thermo chemical conversion of crude glycerol to biopolyols for the production of polyurethane foams.*Bioresource Technol., 139,* 323–329.
154. Li, C., Luo, X., Li, T., Tong, X., Li, Y., (2014). Polyurethane Foams Based on Crude Glycerol-Derived Biopolyols: One-Pot Preparation of Biopolyols with Branched Fatty

Acid Ester Chains and Its Effects on Foam Formation and Properties, *Polymer, 55*, 6529–6538.

155. Piszczyk, Ł., Strankowski, M., Danowska, M., Hejna, A., Haponiuk, J. T. (2014). Rigid polyurethane foams from a polyglycerol-based polyol. *Eur. Polym. J., 57*, 143–150.

156. Piszczyk, Ł., Hejna, A., Danowska, M., Strankowski, M., Formela, K., (2015). Polyurethane/ground tire rubber composite foams based on polyglycerol: processing, mechanical and thermal properties. *J. Reinf. Plast. Comp., 34*, 708–717.

CHAPTER 34

MODIFICATION OF POLYETHYLENE FILMS VIA PHOTOCHEMICAL REACTION

V. SMOKAL, O. KRUPKA, A. KOLENDO, and V. SYROMYATNIKOV

Kyiv Taras Shevchenko National University, Volodymyrska Str. 60, 01033, Kyiv, Ukraine, E-mail: vitaliismokal@gmail.com

CONTENTS

ABSTRACT

New azidobenzenesulfonamides were synthesized and their photochemical properties were investigated. The surface modification of polyethylene (PE) films by using new azidobenzenesulfonamides was carried out. Antimicrobial activity of PE films with sulfonamide moiety was screened for vitro in the following bacterials: *Staphylococcus aurous, Escherichia coli*, and *Mycobacterium luteum*. Preliminary screening of PE films with sulfonamide moiety revealed their inhibitory action against some kinds of bacteria and fungi.

34.1 INTRODUCTION

Surface modification is a convenient and sometimes the only way of polymer materials acquiring necessary properties since it does not change the polymer structure in mass. The development of modern technologies enables us to create polymer materials with a number of properties which would be impossible to realize in a single polymer. A polymer can acquire special properties either in the course of inter-chain polymer alloying by adding certain monomer units to the basic monomer being polymerized [1] or by modifying the ready-made sample [2–4].

Designing the structure of a modified surface with various biologically active groups is one of the issues of current importance in polymer chemistry.

The chemistry of azides and nitrenes has attracted the attention of chemists since the discovery of phenyl azide by Griess [5] and the first proposal of nitrenes as reactions intermediates by Tiemann. Since then, many derivatives are the subject of extensive investigation because of their important biological and industrial applications as photoaffinity labeling agents [6, 7], cross-linking reagents in photo resists [8], the formation of conducting polymers [9], and the light induced activation of polymer surfaces [10, 11]. In the other hand, it is well known that sulfonamides are widely used for therapeutic and prophylatic purpose in humans and other animals and sometime as additives in animal feed [12]. Therefore, the aim of our work was syntheses new sulfonamides and modeling compounds based on them with biological active groups. The ability of azides to the photochemical decomposition with resulting active biradical can be used for modification of the polymer surface [13, 14–16]. In our work, PE films were used as a polymer surface. It is well known that PE is one of the widespread polymers used for industrial and biomedical applications due to their special properties that include low density, flexibility and high chemical resistance [17–19]

In this chapter, we turn attention to the investigation of the influence different moiety on the biocide property of polymer surface. In order to obtain pharmacologically active properties, the *azidobenzenesulfonamides* were synthesized and biologically active compounds introduced on the polymer surface.

34.2 EXPERIMENTAL PART

34.2.1 MATERIALS

34.2.1.1 N-(3-acetylphenyl)-4-azidobenzenesulfonamide(a1)

The mixture of concentrated hydrochloric acid (2 ml) and water (10 ml) was added to a solution of 0.95 g (0.00327 mol) of N-(3-acetylphenyl)-4-aminobenzenesulfonamide in 5 ml of ethanol. The reaction mixture was stirred for 15 min at 0°C and a solution of 0.225 g (0.00327 mol) of $NaNO_2$ in 5 ml of water was added drop wise to the solution, keeping the temperature of the reaction mixture −5–10°C. The reaction mixture was stirred for 30 min at −5°C. The cooled solution of 0.6376 g (0.00981 mol) of NaN_3 in 3 ml of water was added drop wise to the reaction mixture. After this, the reaction mixture was stirred for 1 h at room temperature. The product was recrystallized from ethanol, m.p. 135°C (white crystals), yield 53%.

^{1}HNMR (400 MHz, DMSO-d_6), % (ppm): 7.18 (d, Ar, 2H), 7.76 (d, Ar, 2H), 3.84 (s, CH_3, 3H), 10.44 (s, SO_2 NH, 1H). UV-vis (ethanol) λ_{max}: 265 nm. IR (KBr, cm^{-1}): 2100, 1630, 1580, 1500, 740.

34.2.1.2 N-(1-naphthalenyl)-4-azidobenzenesulfonamide(a2)

a2 was prepared by the route of a1 preparation. The light yellow crystals were collected by filtration and recrystallized from ethanol. Yield 48%, m.p. 138°C.

^{1}HNMR (400 MHz, DMSO-d_6), %(ppm): 7.15 (d, Ar, 2H), 7.40 (d, Ar, 2H), 10.08 (s, SO_2NH, 1H), 7.14–8.04 (m, 7H). UV- VIS (ethanol) λ_{max}: 264 nm. IR (KBr, cm^{-1}): 2080, 1580, 1270.

34.2.1.3 4-azidobenzenesulfonamide(a3)

This compound was obtained according to the method described for a1, m.p. 145°C (white crystals), yield 63%. The product was recrystallized from ethanol.

^{1}HNMR (400 MHz, DMSO-d_6), %(ppm): 7.18 (d, Ar, 2H), 7.83 (d, Ar, 2H), 7.21 (s, SO_2NH_2, 2H). UV-Vis (ethanol) λ_{max}: 260 nm. IR (KBr, cm^{-1}): 2100, 1600.

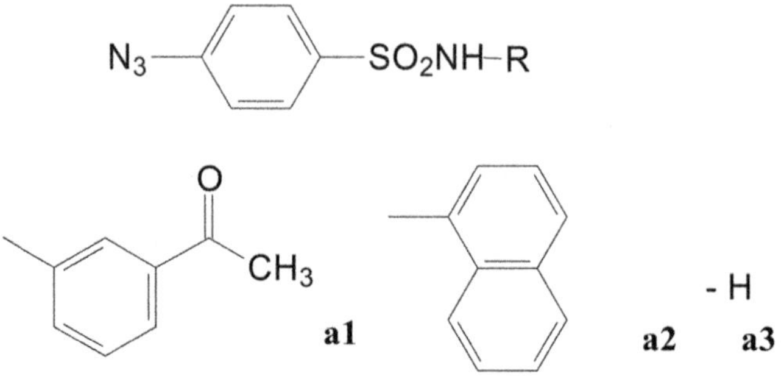

SCHEME 1 Chemical structures of azidobenzenesulfonamides.

The general structures of azidosulfonamides are shown in the Scheme 1.

34.2.2 CHARACTERIZATION

^{1}HNMR (400 MHz) spectra were recorded on a "Mercury-400" spectrometer using DMSO-d$_6$ as solvent. Chemical shifts are in ppm from the internal standard tetramethylsilane (TMS). UV-VIS measurements were performed at room temperature in ethanol in quartz curve (C = 10^{-5} mol/L) with a Specord UV-VIS spectrometer. The IR spectra were obtained on a UR-20 spectrometer in KBr. FT-IR-ATR spectroscopy was done over a range of 4000–800 cm^{-1} at room temperature.

34.2.3 OBJECTS OF STUDY

The solution of agar media 2% (beef-extract agar) was added to the sterile "Petri dishes". The media cooled and after the solidification the modified PE films were added. The surface of the films were treated by bacterial suspensions with 0.7% solutions of nutrient agar. The plates were incubated at 37°C for samples with bacteria. The time of incubation is 36 h. The results were analyzed by evaluation of the death rate of bacteria after the incubation with azidobenzenesulfonamides–modified PE films. The standard microbiological technique was used in order to estimate the concentration (colony counting) of micro-organism species, in the details was described by Meynell at all [20].

TABLE 34.1 The Evaluation of the Death Rate of Microorganisms in Relation to Non-modificated Film

The modified PE films	The evaluation of the death rate of microorganisms in relation to non-modificated film, %		
	Escherichia coli	*Staphylococcus aureus*	*Mycobacterium luteum*
PE-a1	25	13	30
PE-a2	30	34	90
PE-a3	50	49	90

The antibactericidal screening has been done for bacteria namely *Staphylococcus aureus, Escherichia coli, Mycobacteriumluteum*. The results of evaluation the death rate of microorganisms in comparison with non-modificated films are presented in the Table 34.1.

34.3 RESULTS AND DISCUSSION

34.3.1 PHOTOCHEMICAL PROPERTIES OF AZIDOBENZENESULFONAMIDES

Phenyl azides are known to undergo the photo-reaction under UV irradiation. The reaction of photolysis with formation active nitrenes (either singlet or triplet) has been observed [21]. Singlet phenylnitrene is primary reactive intermediate formed upon the photolysis of phenyl azide. Although triplet phenylnitrene, as well as didehydroazepine, has been directly observed by means of IR and UV spectroscopy in matrices and in solutions and the 3*H*-azepinederivative, respectively [22, 23]. Nitrenes possess high reaction ability and take part in photoreactions.

The photolysis process with strong spectral changes which took place during UV irradiation (313 nm) of the solution of **a1**, **a2**and **a3** in ethanol was shown in Figure 34.1. A strong decrease of the absorption in the region in 245–290 nm (**a1**), 250–300 (**a2**), 238–292 (**a3**) and their increase in the region 295–400 nm (**a1**), 220–235 nm (**a3**), 230–250 nm, 308–385 nm (**a2**) were observed.

In order to investigate photoactive abilities of new azidobenzenesulfonamides photolysis quantum yields were determined. High photochemical activity of azidobenzenesulfonamides was confirmed by the received results **a1**φ = 0.48 and for**a3**φ = 0.51. In the case of **a2** more low value of

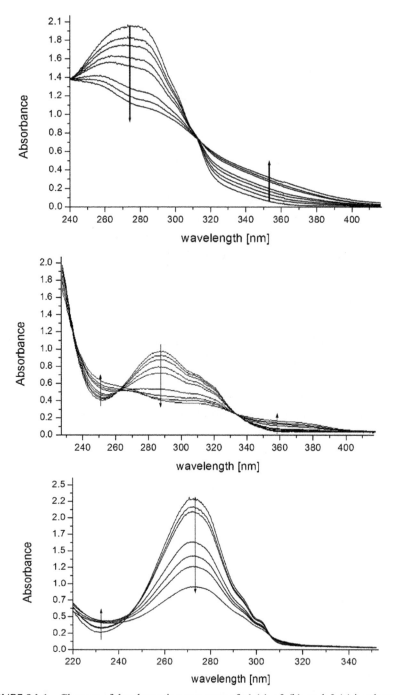

FIGURE 34.1 Changes of the absorption spectrum of a1 (a), a2 (b), anda3 (c) in ethanol (C = 10^{-5} mol/L) during 30 min of UV-irradiation (313 nm).

photolysis quantum yield ($\varphi = 0.14$) was explained by existence of effective singlet-triplet and triplet-triplet conversion of the excitation energy on the naphthalene π-electron system. This was described with details previously [24].

34.3.2 MODIFICATION OF PE FILMS

The possibilities of surface photo-modification of PE films LDPE (low density polyethylene, manufactured by Lyondell Basell Industries, product name – GGNX 18 D003) by new azides containing biological active sulfonamide group **a1**, **a2**, **a3** have been studied. PE films (size 4×7 cm) were prepared. Polymer films were sheeted by 1% azidobenzenesulfonamides using acetone as solvent and dry samples were irradiated by UV lamp DRT-1000 during 30 minutes at 20°C. The properties of modified polymer surface were established by IR-spectroscopy and direct contact angle measurement [13].

FT-IR-ATR (PE films after irradiation): The important bands in examining degradation mechanisms include the broad v (C = C) between 1585 and 1600 cm^{-1}, δ (C-H) at 1000 cm^{-1}, δ (SO$_2$N) at 1360 and 1280 cm^{-1}, v (C = O) at 1676 cm^{-1} δ (C-H) at 840 cm^{-1}.

34.3.3 BIOCIDE PROPERTIES OF PE FILMS

The syntheses have proceeded in two stages: preparation of the sulfonamides and azides based on them. New azidobenzenesulfonamides were synthesized by diazotization of corresponding amines and treatment by diazonium salt with aqueous solution of sodium azide [25]. The additional procedure of surface photo-modification of PE was used. For microbiological analysis of the modified films, bacteria *Staphylococcus aureus*, *Escherichia coli*, *Mycobacterium luteum* were chosen.

The films were screened for their inhibition activity against Gram-positive and Gram-negative bacteria. The film **PE-a1** exhibited practically the identical activity against *Escherichia coli* (Gram- negative) and *Mycobacterium luteum* (Gram-positive) bacteria. In the same time, **PE-a2** film showed high inhibition activity against Gram-positive (*Staphylococcus aureus* and *Mycobacterium luteum*) bacteria. **PE-a3** exhibited the highest biocide properties against *Mycobacterium luteum*. Finally, based on these preliminary results, it was found that modified films by azidobenzenesulfonamides are very promising material for practical application, but further consider investigation of toxic effect towards the mammalian cells is needed.

Statistical analysis of the results showed that the antimicrobial properties of modificated films were significantly more effective in comparison with non-modificated film. All the modificated films demonstrated a high activity against bacteria *Mycobacterium luteum* in comparison with non-modificated film. It is important to note that the activity of the modificated films was different for different type of bacteria. A modificated film **PE-a3**contains the amino group in the structure and more active than others samples against bacteria. The same principle of the structure–activity relationship for other models was observed in research [22].

34.4 CONCLUSIONS

The experimental procedures were standardized for obtaining optimum yield. The modificated PE films with anti-microbial activity have been prepared. All sulfonamide derivatives were characterized by NMR- and IR-spectroscopy. The investigation of photochemical properties has been shown ability new azidobenzenesulfonamides for polymer surface modification. New modificated PE films **(PE-a1, PE-a2, PE-a3)** have higher antimicrobial activity in vitrothan the initial non-modificated PE film. The PE surface modificated by **a3** is shown higher antimicrobial activity in comparison with PE surfaces modificated by **a1** and**a2** against *Escherichia coli* in vitro.

The achieved results confirmed, that modification of polymer surfaces by azides based on benzenesulfonamides enables to use them as perspective materials, for example, as biostabilizers for plastic masses.

ACKNOWLEDGMENTS

The authors would like to thank Prof. V. Novikov and Prof. A. Doroshenko for fruitful discussion.

KEYWORDS

- azides

- biocide properties
- photoreactions
- polyethylene modification

- sulfonamides
- UV irradiation

REFERENCES

1. Krupka, O., Kolendo, A., Yu, Kushnir, K., Blazejowski, J., (2005). Synthesis and investigation of a methacrylic monomer based on 4-[4-(phthalimido)-phenylsulphamido]-2, 6-dimethoxypyrimidine. *Mol. Cryst. Liq. Cryst, 427*, 233.
2. Sacristan, J., Reinecke, H., Mijangos, C. J. (2000). Surface modification of PVC films in solvent-non solvent mixtures. *Polymer 41*, 5577.
3. Spanring, J., Buchgrabcr, C., Ebel, M. F., Svagera, R., Kern, W., (2006). Trialkylsilanes as reagents for the UV-induced surface modification of polybutadiene. *J. Polymer, 47*, 156.
4. Jia, X., Herrera-Alonso, M., McCarthy, T., (2006). Nylon surface modification. Targeting the amide groups for selective introduction of reactive functionalities. *Polymer 47*, 4916.
5. Griess, J. P., (1864). On a new series of bodies in which nitrogen is substituted for hydrogen.*Philos. Trans. R. Soc. Lond.154*, 664.
6. Cai, S. X., Glenn, D. J., Gee, K. R., Yan, M. D., Cotter, R. E., Reddy, N. L., Weber, E., Keana, J., (1993). Chlorinated Phenyl Azides as Photolabeling Reagents.Synthesis of an o, o'-Dichlorinated Aryl Azido PCP Receptor Ligand Bioconjugate. *Chem. 4*, 545.
7. Bayley, H., Staros, J., (1984). In: Azides and Nitrenes – Reactivity and Utility. Academic Press Inc. New York.
8. Cai, S. X., Glenn, D. J., Kanskar, M., Wybourne, M. N., Keana, J. F. (1994). Development of Highly Efficient Deep-UV and Electron Beam Mediated Cross-linkers: Synthesis and Application of Bis(perfluorophenyl). *Azides. Chem. Mater.,6*, 1822.
9. Meijer, E. W., Nijhuis, S., Vroonhoven, F. C. (1988). Poly-1, 2-azepines by the photopolymerization of phenyl azides.Precursors for conducting polymer films. *J. Am. Chem. Soc. 110*, 7209.
10. Nahar, P., Wali, N., Gandhi, R., (2001). Photo immobilization of unmodified carbohydrates on activated surface. *Anal. Biochem. 294*, 148.
11. Cai, S. X., Glenn, D. J., Keana, J. F. (1992). Toward the Development of Radio labeled FluorophenylAzide-Based Photo labeling Reagents – Synthesis and Photolysis of Iodinated 4-Azidoperfluorobenzoates and 4-Azido-3, 5, 6-Trifluorobenzoates, *J. Org. Chem. 57*, 1299.
12. Long, A. R., Hsieh, L. C., Malbrough, M. S., Short, C. R., Barker, S. A. (1990). *J. Agric. Food Chem. 38*, 423.

13. Krupka, O., Smokal, V., Wilczek, M., Kostrzewa, M., Syromyatnikov, V., Kolendo, A., (2008). *Mol. Cryst. Lyq. Cryst. 497*, 323.
14. Kenawy, E., Abdel-Hay, F. I. (2006). *React. Funct. Polym. 66*, 419.
15. Khandare, J., Minko, T., (2006). *Prog. Polym. Sci. 31*, 359.
16. Donaruma, L. G., Dombroski, J. R. (1971). *J. Med. Chem.,14*, 460.
17. Fang, L., Leng, Y., Gao, P., (2005). *Biomaterials, 26*, 3471.
18. Wang, J., Huang, N., Yang, P., Leng, Y. X., Sun, H., Liu, Z. Y., Chu, P. K. (2004). *Biomaterials, 25*, 3163.
19. Lee, H., Hong, S., Yang, K., Choi, K., (2006). *Microelectron. Eng. 83*, 323.
20. Meynell, G., Meynell, E., (1965). Theory and Practice in Experimental Bacteriology, Cambridge University Press, London.
21. Tsao, M. L., Platz, M. S. (2003). *J. Am. Chem. Soc. 125,* 12014.
22. Hayes, J. C., Sheridan, R. S. (1990). *J. Am. Chem. Soc. 112*, 5879.
23. Li, Y. Z., Kirby, J. P., George, M. W., Poliakoff, M., Schuster, G. B. (1988). *J. Am. Chem. Soc. 110*, 8092.
24. Krupka, O., Kolendo, A., Doroshenko, A. (2005). *J. Vop. Khim. Tekh. Ukr. 2*, 103.
25. Smith, P. A., Hall, J. H. (1962). *J. Am. Chem. Soc. 84*, 480.

INDEX

Printed and bound by CPI Group (UK) Ltd, Croydon, CR0 4YY

23/10/2024

01777705-0012